LECTURE NOTES ON TURBULENCE

LECTURE NOTES ON TURBULENCE

Lecture Notes from the NCAR–GTP Summer School
June 1987

Edited by
Jackson R. Herring
James C. McWilliams

World Scientific
Singapore • New Jersey • London • Hong Kong

Published by

World Scientific Publishing Co. Pte. Ltd.,
P O Box 128, Farrer Road, Singapore 9128
USA office: 687 Hartwell Street, Teaneck, NJ 07666
UK office: 73 Lynton Mead, Totteridge, London N20 8DH

LECTURE NOTES ON TURBULENCE

ISBN 9971-50-805-2
9971-50-827-3 (pbk)

Printed in Singapore by Utopia Press.

PREFACE

Turbulence is an ubiquitous aspect of natural fluid flows, with scale ranges from astronomical to centimeters. The phenomena encompassed are similarly varied. These facts have, perhaps, contributed to a fragmentation of the field with workers in the disparate disciplines having little contact and opportunities for cross fertilization. At the same time, the subject is difficult with slow progress especially in its theoretical aspects. The summer school on turbulence research—held at NCAR in June, 1987, the basis for this volume—was motivated by the belief that presenting certain of the current challenging problems to a select group of graduate students and beginning post-doctoral researchers would be a rewarding investment. Thus, at the beginning of their careers they would have a chance to view the broad range of topics and techniques and see first hand the level of understanding and unsolved issues. This volume extends the experience to a wider audience.

The four lecturers are well known for their contributions to turbulence research, and for their ability to transmit with enthusiasm the more interesting, challenging, and rewarding problems. They are Hendrik Tennekes, engineering and certain aspects of meteorological turbulence; David Montgomery, plasma and magnetohydrodynamic turbulence; Douglas Lilly, meteorology; and Uriel Frisch, who covered the more theoretical and mathematical aspects. We hope the resulting notes produced here may convey some of the enthusiasm evident at the school, although they must be somewhat pale compared to the actual event.

Although these notes cover a wide range of topics, they are not intended to be comprehensive. Rather, they present certain of the more interesting research areas with sufficient background and references, so that the interested reader will know what the challenging problems are, and be able to pick up the thread of previous research from the references.

Of the many who have contributed to this volume, we must first credit the students, whose lecture notes were the *urtext*. They worked in pairs, checked, and corrected their product with the lecturers at later conferences.

These are: Rémi Abgrall, Raymond Arritt, Hamid Biglari, Xavier Carton, Héctor Castaneda, Tzihong Chiueh, Arnab Choudhuri, Gary Coleman, Evangelos Coustais, William Dannevik, Edward DeLuca, Stephen Derbyshire, Mark Hadfield, Hamid Johari, Leroy Pascal, Andy Leonard, Jens-Peter Lynov, Michael Montgomery, Noh Yign, Ivan Orsolini, Lorenzo Polvani, Zhongshan Qian, John Reynders, Steven Roy, Geoffrey Schladow, Jeanne Schneider, John Shebalin, William Smyth, Michael Theobald, Alain Vincent, Robert Walko, and Jeff Weiss.

Special thanks must be given to the several student editors who coordinated the collection of notes, and contributed to their preliminary editing. These were W. Dannevik, for the lectures of H. Tennekes and M. Montgomery; L. Polvani, for those of U. Frisch; S. Sheblein, for those of D. Montgomery; and M. Hatfield and R. Walko, for those of D. Lilly.

John Wyngaard offered continual support and good advice, and proof read sections of the final draft. Sheridan Garcia coordinated the events of the summer school, and the colloquium that followed.

The task of recording in TEX the entire manuscript, and checking, correcting, and formatting the final version fell to the able shoulders of Frances Huth. Without her organizational skills and hard work, these notes would doubtless not yet be finished. We also wish to thank Eileen Weppner for her editorial skills, and to Jim Adams for the final paste-up work.

Finally, we must express our gratitude to our institutional sponsors, not only for essential financial support, but also for their encouraging enthusiasm for this project. These are Dr. David Bohlin, Dr. John Theon, and Dr. Stanley Shawan, Atmospheric and Radiation Branch Earth Science and Application Division, National Aeronautical and Space Administration (NASA); Dr. George Kolstad, Division of Engineering and Geosciences, Office of Basic Energy Sciences, Department of Energy (DOE); Dr. Bernard McDonald, Special Projects Program, Division of Mathematical Sciences, National Science Foundation; Dr.

Doug Dwoyer, ICASE, NASA Langley; Parvis Moin of the Center for Turbulence Research of NASA Ames and Stanford; and Dr. John Firor of the Advanced Studies Program of NCAR. Special thanks are due to Dr. Jack Eddy of UCAR's Office for Interdisciplinary Earth Studies, who served as liaison to NASA and DOE.

Jackson R. Herring
James C. McWilliams
National Center for Atmospheric Research*

* The National Center for Atmospheric Research is sponsored by the National Science Foundation.

TABLE OF CONTENTS

Lectures on Turbulence and Lattice Gas Hydrodynamics by Uriel Frisch

PART I: TWO- AND THREE-DIMENSIONAL TURBULENCE

PREFACE

The students in the summer school have done a fine job of preparing notes from sometimes rather turbulent and chaotic lectures. They have wisely decided not to reproduce most of the written material that was distributed in preparation for the lectures. The principal printed sources were my paper on the comparative pathology of atmospheric turbulence in two and three dimensions, prepared for the 1983 Enrico Fermi summer school in Varenna, Italy (Tennekes, 1985), and John Wyngaard's chapter on boundary-layer modeling in the Nieuwstadt-van Dop book on atmospheric turbulence and air-pollution modeling (Wyngaard, 1982). Time limitations prevented me from pursuing the analogy between flux maintenance in two- and three-dimensional turbulence. Second-order modeling in boundary-layer meteorology shares several diagnostic methods with general-circulation research. I had planned to expound on this with the aid of my paper on the general circulation of two-dimensional turbulence on a beta plane (Tennekes, 1977). The manuscript version of that paper is included in these notes as Chapter 11; together with the notes prepared by the students, the present material becomes a reasonable approximation of my original intentions. Unfortunately, the written text cannot represent the lively and penetrating discussions that characterized all meetings of the National Center for Atmospheric Research (NCAR) summer school; the spiritual power of that many close encounters of inquiring minds has made an indelible impression on my mind.

H. Tennekes
Royal Netherlands Meteorological Institute
2 June 1988

Tennekes, H., 1977: *J. Atmos. Sci.*, **34**, 702.

Tennekes, H., 1985: Comparative pathology of atmospheric turbulence in two and three dimensions. In: *Turbulence and Predictability in Geophysical Fluid Dynamics and Climate Dynamics*. M. Ghil and R. Benzi, eds., 45–70. North Holland, Amsterdam and New York.

Wyngaard, J. C., 1982: Boundary-layer modeling. In: *Atmospheric Turbulence and Air Pollution Modelling*. F. T. M. Nieuwstadt and H. van Dop, eds., 69. Reidel Publishing Co., Dordrecht/Boston.

Chapter 1

INTRODUCTION

The purpose of this first chapter is to warn researchers of some of the major pitfalls on the road to a satisfactory description of turbulent flows. We will start by examining the equation of motion for a Newtonian fluid

$$\frac{\partial u_i}{\partial t} + u_j \frac{\partial u_i}{\partial x_j} = -\rho^{-1} \frac{\partial p}{\partial x_i} + \nu \frac{\partial^2 u_i}{\partial x_j \partial x_j} , \tag{1.1}$$

in which u_i is the velocity in the x_i direction, ρ the density and ν the kinematic viscosity. We will perform routine scale analysis and represent a characteristic velocity by V_o and length scale by L. So, a time scale for advection is given by L/V_o. The terms in (1.1) can be estimated as follows:

$$V_o \left(\frac{L}{V_o} \right)^{-1} , \frac{V_o^2}{L}, \frac{V_o^2}{L}, \nu \frac{V_o}{L^2} . \tag{1.2}$$

The relative effect of friction is estimated by dividing the viscous friction term by the advection term:

$$\frac{\nu V_o}{L^2} / \frac{V_o^2}{L} = \frac{\nu}{V_o L} . \tag{1.3}$$

Now, let us look at an everyday example. For a car having a width of 2 m and which travels at a speed of 30 m s^{-1}, the relative effect of friction is

$$Re^{-1} \equiv \frac{\nu}{V_o L} = \frac{1.5 \times 10^{-5}}{30 \times 2} = 2.5 \times 10^{-7} , \tag{1.4}$$

which appears to mean that we can neglect friction for practical purposes. This is clearly nonsense since we know that for even the sleekest of designs, cars of today are observed to have drag coefficients $\left[C_d = D/(1/2 \rho V^2 A) \right]$ of about 0.3.

Prandtl was the first to recognize the problem with this approach, and introduced various scales into fluid mechanics. Returning to the equation of motion (1.1), we may introduce a different length scale δ into the viscous term. The viscous term is balanced by the other terms in the equation of motion:

$$\frac{V_o^2}{L} \sim \nu \frac{V_o}{\delta^2} \Rightarrow \frac{\delta}{L} \sim Re^{-1/2} . \tag{1.5}$$

This tells us that viscosity plays an important role in only a small fraction of the field, and the Reynolds number (Re) can be viewed as a factor determining the extent of the "viscous layer." Later, this type of analysis developed into the classical boundary-layer theory.

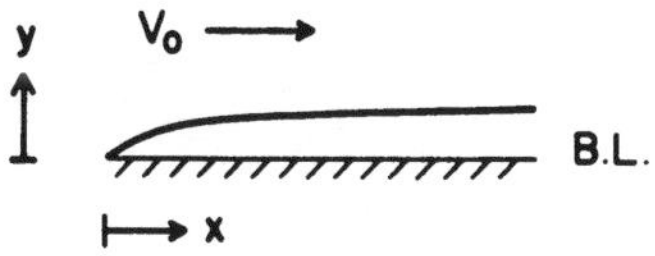

Figure 1.1: The development of a boundary layer over a flat plate.

Another example is appropriate here. Let us consider a laminar boundary layer over a flat plate (Fig. 1.1). The dissipation rate of kinetic energy (ϵ) can be estimated as

$$\epsilon = \nu \left(\frac{\partial u}{\partial y}\right)^2 \sim \nu \left(\frac{V_o}{\delta}\right)^2 . \tag{1.6}$$

Noting from (1.5) that $\delta \sim L Re^{-1/2}$, we find

$$\epsilon \sim \nu V_o^2 (L^{-2} Re) = \frac{V_o^3}{L} . \tag{1.7}$$

The interesting point of this result is that the dissipation rate is completely determined by the large-scale parameters of the flow, and not by the viscosity.

We can approach the problem in a different manner. The extent of a boundary layer on a flat plate is determined by the process of diffusion of low-momentum fluid close to the plate in the vertical direction. The diffusion equation for a quantity ϕ, with diffusion coefficient γ, is given by

$$\frac{\partial \phi}{\partial t} = \gamma \nabla^2 \phi . \tag{1.8}$$

Dimensional considerations reveal the length scale for diffusion as $L_d \sim (\gamma t_d)^{1/2}$, with t_d the diffusion time scale. Equating the diffusion and advection time scales provides us with the growth law for the laminar boundary layer

$$\left.\begin{array}{l} t_d \sim \delta^2 \nu^{-1} \\ t_a \sim x V_o^{-1} \end{array}\right\} \delta \sim Re^{-1/2} x, \text{ with } Re = \frac{V_o x}{\gamma} . \tag{1.9}$$

Clearly, interpretation of various parameters must be done with care. This is especially true for turbulent flows because of the possibility of a wide spectrum of relevant length scales.

Another very important note here is the general property

$$\lim_{\nu \to 0}(\phi) \neq \phi(\nu = 0) \ , \tag{1.10}$$

i.e., a flow with very small viscosity should never be confused with an inviscid flow! If inviscid calculations are performed, we must be certain that we do not proceed to large times: $t \gg L^2\nu^{-1}$, since $\delta \sim (\nu t)^{1/2}$. If, for example, no friction is taken into account in meteorological calculations of the general circulation, the prediction turns out completely unrealistic.

If we start with the equation of motion (1.1) and rearrange the advection term $u_j\partial u_i/\partial x_j$ as follows

$$\begin{aligned} u_j\frac{\partial u_i}{\partial x_j} &= u_j\left(\frac{\partial u_i}{\partial x_j} - \frac{\partial u_j}{\partial x_i}\right) + u_j\frac{\partial u_j}{\partial x_i} \\ &= -\boldsymbol{\omega} \times \boldsymbol{u} + u_j\frac{\partial u_j}{\partial x_i} \ , \end{aligned} \tag{1.11}$$

we arrive at (for the incompressible case)

$$\frac{\partial u_i}{\partial t} = -\boldsymbol{\omega} \times \boldsymbol{u} - \frac{\partial}{\partial x_i}\left(\frac{p}{\rho_o} + \frac{1}{2}u_j u_j\right) + \nu\frac{\partial^2 u_i}{\partial x_j \partial x_j} \ , \tag{1.12}$$

where $\boldsymbol{\omega}$ is the vorticity vector $\nabla \times \boldsymbol{u}$. This is a very useful form for incompressible flows. For inviscid, irrotational flows, the Bernoulli relation falls out immediately. In this form, it is easy to incorporate the Coriolis force $(-2\boldsymbol{\Omega} \times \boldsymbol{u})$ or Lorentz force $(\rho_o^{-1}\boldsymbol{J} \times \boldsymbol{B})$. Here, $\boldsymbol{\Omega}$ is the external rotation of the reference frame and $\boldsymbol{J},\boldsymbol{B}$ are conduction current and magnetic field vectors, respectively. As a result of this analogy the term "vortex force" is appropriate for $-\boldsymbol{\omega} \times \boldsymbol{u}$. Note that all of these forces act in a direction perpendicular to the field direction.

Let us consider two examples. A current-carrying wire in a magnetic field produces an induced $\boldsymbol{B}$-field which curves the magnetic field lines. The force on the wire is given by $\boldsymbol{F} = \boldsymbol{J} \times \boldsymbol{B}$, where $\boldsymbol{J} = \nabla \times \boldsymbol{B}$. The same is true for a wing in a uniform velocity field with speed U_o.

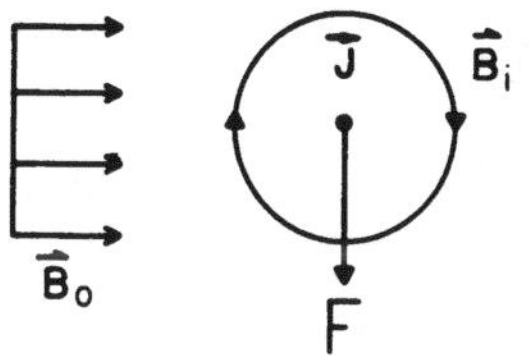

Figure 1.2: A current-carrying wire in a uniform magnetic field.

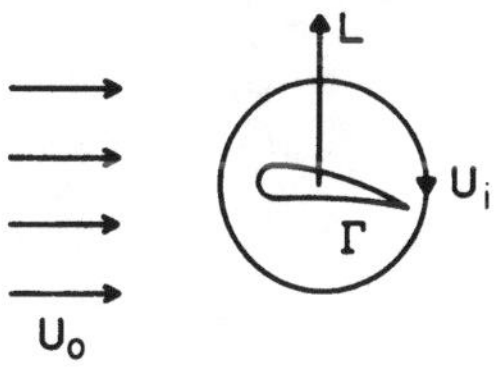

Figure 1.3: A wing in a uniform velocity field.

The wing creates an induced velocity field which bends the velocity-field lines (streamlines). The circulation Γ provides a lift force $L = \rho U_o \Gamma$, where

$$\Gamma \equiv \int_{wing} \boldsymbol{U} \cdot d\boldsymbol{l} = \oint_{surface} \boldsymbol{\omega} \cdot d\boldsymbol{s} \ . \tag{1.13}$$

The vorticity, like the electric current density, is the curl of a vector, and therefore has to be solenoidal: $\nabla \cdot \boldsymbol{\omega} = 0$ (nondivergent, divergence-free). As a result, we get tip vortices on a wing with finite span; they must continue and must finally close on themselves. Thus, the so-called start-up vortex is left behind at the airport. Vorticity is generated by viscosity at a boundary (on a wing) and then goes along for a ride along with the mean flow. Incidentally, since Γ is proportional to L/U_o, the most dangerous tip vortices occur behind a slow-flying Boeing 747. Other aircraft would get in real trouble if the air traffic controller forgot to maintain sufficient separation.

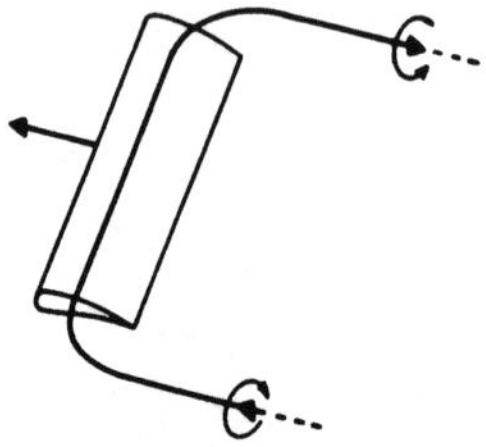

Figure 1.4: The vortex on a wing with finite span.

Chapter 2

VORTICITY DYNAMICS

The vorticity equation for an incompressible Newtonian fluid in an inertial frame of reference is

$$\frac{\partial \zeta_i}{\partial t} + u_j \frac{\partial \zeta_i}{\partial x_j} = \zeta_j e_{ij} + \nu \frac{\partial^2 \zeta_i}{\partial x_j \partial x_j} , \tag{2.1}$$

where e_{ij} is the deformation rate tensor, i.e., the symmetric part of the velocity gradient tensor,

$$e_{ij} = \frac{1}{2}\left(\frac{\partial u_i}{\partial x_j} + \frac{\partial u_j}{\partial x_i}\right) . \tag{2.2}$$

The first term on the right-hand side (r.h.s.) of the vorticity equation is the most interesting one; it describes the stretching and turning of the vortex lines. The figures below show the y component of vorticity being increased by (a) stretching and (b) turning.

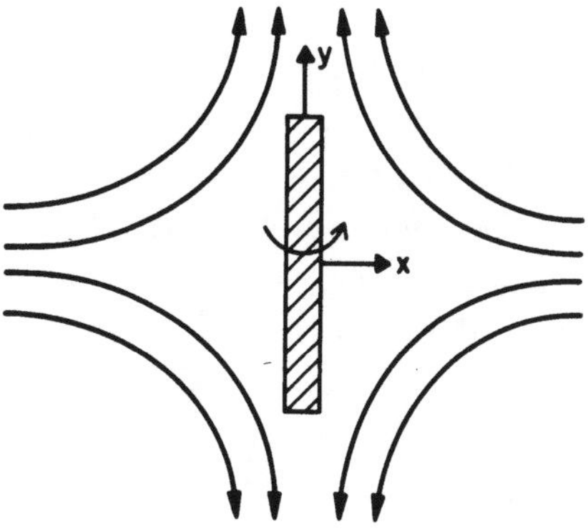

Figure 2.1a: The increase of ζ_y by stretching ($\zeta_x = \zeta_z = 0$).

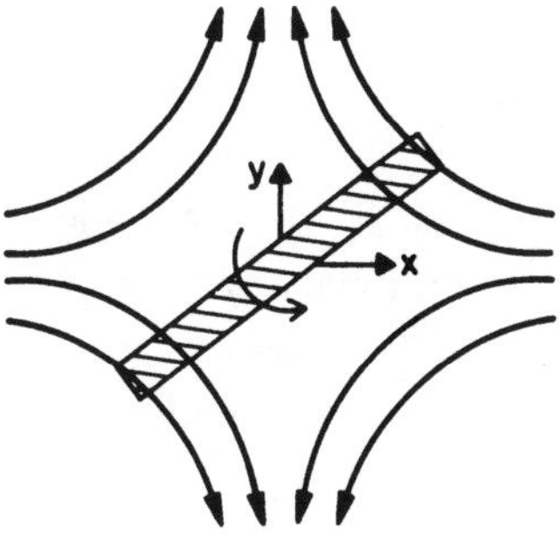

Figure 2.1b: The change of ζ_y by turning the vorticity vector.

The stretching/turning term can also be written as $\zeta_j(\partial u_i/\partial x_j)$. A similar term appears in the magnetohydrodynamic equation for magnetic field density in the kinetic regime

$$\frac{\partial B_i}{\partial t} + u_j \frac{\partial B_i}{\partial x_j} = B_j \frac{\partial u_i}{\partial x_j} + \cdots , \tag{2.3}$$

and in the hydrodynamic equation for a scalar gradient

$$\frac{\partial}{\partial t}\frac{\partial T}{\partial x_i} + u_j \frac{\partial}{\partial x_j}\frac{\partial T}{\partial x_i} = -\frac{\partial T}{\partial x_j}\frac{\partial u_j}{\partial x_i} + \cdots \,, \tag{2.4}$$

but not in the equation for the scalar itself. The negative sign on the stretching/turning term in the scalar gradient equation implies that scalar gradients are *weakened* by stretching whereas vorticity and magnetic field density are *intensified.*

Consider the equation for a single component of vorticity, say, the vertical component:

$$\frac{\partial \zeta_3}{\partial t} + u_j \frac{\partial \zeta_3}{\partial x_j} = \zeta_1 e_{13} + \zeta_2 e_{23} + \zeta_3 e_{33} + \nu \frac{\partial^2 \zeta_3}{\partial x_j \partial x_j}. \tag{2.5}$$

The first two terms on the r.h.s. describe turning and the third describes stretching. In two-dimensional flow ($u_3 = 0$) these three terms vanish and the vertical vorticity (just ζ now) behaves like a scalar:

$$\frac{\partial \zeta}{\partial t} + u_j \frac{\partial \zeta}{\partial x_j} = \nu \frac{\partial^2 \zeta}{\partial x_j \partial x_j}. \tag{2.6}$$

Clearly, ζ can be transported and diffused but there is no amplification mechanism. Vorticity *gradients* can still be amplified because the $\partial \zeta / \partial x_i$ vector is not perpendicular to the plane of flow. A slight relaxation of two-dimensionality ($2 + \epsilon$ turbulence) allows $\partial u_3 / \partial x_3 \neq 0$ but still has negligible ζ_1 and ζ_2. Then the stretching term is retained and the conserved property is not vorticity but some form of potential vorticity (for example, ζ / H, where H is the depth of a fluid layer).

Returning to the full three-dimensional vorticity equation, if this is multiplied by ζ_i it becomes an equation for the square of the magnitude of the vorticity:

$$\frac{\partial}{\partial t}\frac{1}{2}\zeta_i\zeta_i + u_j \frac{\partial}{\partial x_j}\frac{1}{2}\zeta_i\zeta_i = \zeta_i\zeta_j e_{ij} + \nu \frac{1}{2}\frac{\partial^2 \zeta_i\zeta_i}{\partial x_j \partial x_j} - \nu \frac{\partial \zeta_i}{\partial x_j}\frac{\partial \zeta_i}{\partial x_j} \,. \tag{2.7}$$

There is now a dissipative viscous term as well as a diffusive one.

The Reynolds-averaged equations for mean energy, $\overline{u_i' u_i'}/2$, and enstrophy, $\overline{\zeta_i' \zeta_i'}/2$, are given in Tennekes and Lumley (1972, equations 3.2.1 and 3.3.38). The major terms in the respective budgets are

$$-\overline{u_i' u_j'}\frac{\partial U_i}{\partial x_j} = \nu \overline{\zeta_i' \zeta_i'}, \tag{2.8}$$

and

$$\overline{\zeta_i'\zeta_j'e_{ij}'} = \nu\overline{\frac{\partial\zeta_i'}{\partial x_j}\frac{\partial\zeta_i'}{\partial x_j}}. \tag{2.9}$$

Vorticity fluctuations are involved in the dissipation of both energy and enstrophy. In a three-dimensional turbulent flow the vortex lines undergo a continual process of stretching and tangling, so the enstrophy tends to increase, even if external forcing is absent. The turbulence energy, on the other hand, must be extracted from the mean flow.

Tennekes, H., and J. L. Lumley, 1972: *A First Course in Turbulence.* MIT Press, Cambridge, Mass.

Chapter 3

THE PHENOMENOLOGY OF TURBULENCE

We will begin by considering a quotation from Gregory Bateman: "All creative systems are necessarily divergent." How does this relate to our approach to turbulence? First we will define the important terms in the above quotation. "Creative" will be understood to mean capable of generating new coherent structures. We use "divergent" to mean that any two initially close states will, on the average, tend to separate as they evolve. "Necessarily" implies that convergent systems will wind up on a limit cycle; they are predictable, periodic, and uncreative. If one considers the case of two creative scientists who are initially studying the same phenomenon it is clear that their careers will necessarily diverge.

If we are interested in being creative, then we must attempt to generate coherent structures. That is, a coherent picture of the physical process. Such pictures are very powerful because they are more easily remembered than the equations and approximations under which they apply. This has both a positive and negative side: on the positive side, one is able to obtain an intuitive understanding of an extremely complex system which will help indicate which processes are important and which can safely be ignored; on the negative side, one often, through laziness or stupidity, extends the pictures to physical conditions that are completely different from those under which the picture was originally made. The proper use of phenomenology is as a way of looking at science that increases the sophistication of our imagination. The images produced can provide both guidance and inspiration for the mathematics.

It is useful to look at some examples of unsophisticated thinking. In a previous chapter Frisch has correctly expressed the energy flux in an inertial range as

$$\mathrm{v\,v\,v}\,\nabla \sim \epsilon, \tag{3.1}$$

but C. M. Tchen has argued that the expression should read

$$\mathrm{v\,v}\,\frac{\partial U}{\partial z} \sim \epsilon. \tag{3.2}$$

This can be interpreted as the most distant encounters, not the closest (in k space), being the most important. Now we have

$$v^2 = \int_{\Delta k} E(k)dk \sim kE(k), \tag{3.3}$$

(all wavenumbers between k and $2k$ are included in the integral over an eddy). The result is an energy spectrum with a k^{-1} power law:

$$E(k) = \frac{\epsilon}{k}\left(\frac{\partial U}{\partial z}\right)^{-1}. \tag{3.4}$$

This result is generally wrong, but may apply to a flow with a very rapid distortion, at early times.

As a second example, consider the *Direct Interaction Approximation* of Kraichnan (1959). First we recall that $-\overline{uw}\partial U/\partial z = \epsilon$. The Reynolds stress $-\overline{uw}$ is assumed to arise from the shear in the unresolved motions. We therefore introduce an eddy viscosity, ν_t, such that $-\overline{uw} = \nu_t \partial U/\partial z$. We know that ν_t must be proportional to $v^2 t$, so the question arises as to how to choose the time scale in the turbulent viscosity ν_t. We choose $\tau \sim 1/(kq)$, the time scale during which eddies with length scale k^{-1} are advected by the most energetic eddies with velocity q. Thus, $\epsilon = (\partial U/\partial z)^2/(kq)^2$. Under these circumstances, we eliminate $\partial U/\partial z$ using (3.2), and find:

$$E(k) \sim \epsilon^{1/2} q^{1/2} k^{-3/2}\ . \tag{3.5}$$

This was a brilliant mistake. The correct choice for the time scale of the spectral transfer is the only time scale that can arise from k and ϵ alone: $t \sim \epsilon^{-1/3}k^{-2/3}$ (as argued by Onsager, 1949). This yields the Kolmogorov Law once more. (Please note the circular reasoning involved in these discussions. They are not intended to prove anything, only to show the implications of previously made assumptions.)

> "The localness of energy transfer assumed for dimensional analysis in the Kolmogorov $k^{-5/3}$ law might be wrong. Invariance of ϵ is not a fundamental physical property." (Frisch)
>
> "Actually, numerical simulations of two-dimensional turbulence show that ϵ is not local." (Montgomery)

In three-dimensional turbulence, maintained by shear in the mean motion (or by straining of the large-scale eddies), the kinetic energy budget for the eddy motion reads

$$\frac{\partial}{\partial t}\frac{1}{2}\,\overline{u_iu_i} + \overline{u_j}\,\frac{\partial}{\partial x_j}\,\overline{u_iu_i} = -\frac{\partial}{\partial x_j}\left(\frac{1}{\rho}\,\overline{u_jp} + \frac{1}{2}\,\overline{u_iu_iu_j}\right) + 2\nu\frac{\partial}{\partial x_j}\,\overline{u_ie_{ij}} - \overline{u_iu_j}\;\overline{e_{ij}} - 2\nu\;\overline{e_{ij}e_{ij}}\;. \tag{3.6}$$

In homogeneous turbulence, moreover, we have

$$2\nu\overline{e_{ij}e_{ij}} = \nu\overline{\zeta_i\zeta_i}\;. \tag{3.7}$$

Excluding flux-divergence terms, the overall energy level can be maintained only if production and dissipation of eddy kinetic energy are in balance:

$$-\overline{u_iu_j}\;\overline{e_{ij}} = \nu\overline{\zeta_i\zeta_i}\;. \tag{3.8}$$

It is not at all clear that the left-hand side (l.h.s.) term $-\overline{u_iu_j}\;\overline{e_{ij}} > 0$, as is required for mechanical production of three-dimensional turbulence. We will attempt to use phenomenological reasoning to understand the interaction of a boundary-layer flow with $\overline{e_{ij}} = \partial U/\partial x_2 > 0$ with some small-scale eddies as shown below.

The eddies will be distorted as shown in Fig. 3.1. This implies that the horizontal u and vertical w motions will be positively correlated: $u > 0$ and $w > 0$, or $u < 0$ and $w < 0$. The l.h.s. of (3.8) will then be negative!! What did we do wrong? First, note that within this frame of reference the velocity field has just vortex turning, and no vortex stretching (Chapter 2). The vorticity in the x_3-direction is not amplified ($u_3 = 0$, $\partial U_i/\partial x_3 = 0$), as discussed in the previous paragraph. The vorticity is most effectively increased if the vorticity is aligned at an angle of 45° to the x_1-direction as shown below.

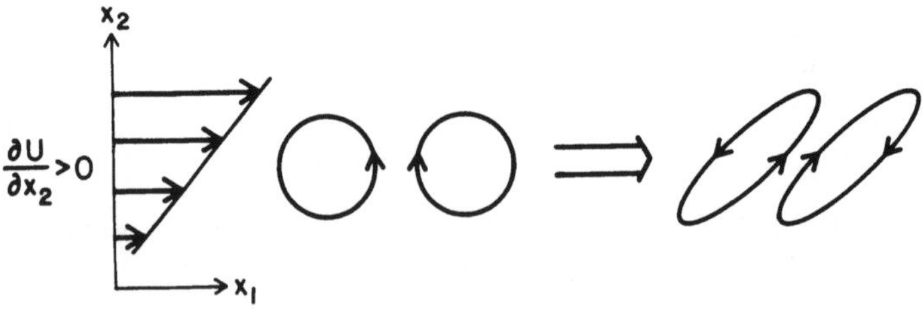

Figure 3.1: The deformation of an eddy pair in a uniform shear flow.

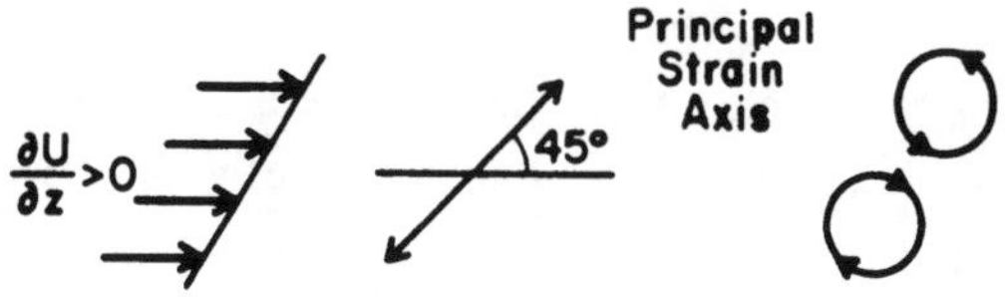

Figure 3.2: The amplification of the vorticity in a uniform shear flow has a maximum in the direction of the "principal axis."

This can be seen more clearly by considering the vorticity equation (2.1), which reads for this case

$$\frac{\partial \zeta_1}{\partial t} = \zeta_2 \frac{1}{2}\frac{\partial U_1}{\partial x_2} + \cdots , \quad \frac{\partial \zeta_2}{\partial t} = \zeta_1 \frac{1}{2}\frac{\partial U_1}{\partial x_2} + \cdots , \tag{3.9}$$

from which we can obtain

$$\frac{1}{2}\frac{\partial\left(\zeta_1^2 + \zeta_2^2\right)}{\partial t} = \zeta_1\zeta_2 \frac{\partial U_1}{\partial x_2} + \cdots . \tag{3.10}$$

This tells us that $\zeta_i\zeta_i$ increases faster for larger values of $\zeta_1\zeta_2$. The largest value for $\zeta_1\zeta_2$ is obtained when the vorticity vector is aligned at an angle of 45° to the x_1-direction.

Townsend (1951) has suggested the following schematic for an eddy life cycle. Eddies are formed with a random orientation, spend their mature life resisting the straining motion (this process intensifies the strength of the eddy vortex), and finally are aligned with the flow (and contribute to an inverse cascade?) before losing their identity.

The spectral energy flux is a constant over a wide range of scales if the source term is confined to the largest scales and viscous dissipation to the smallest scales; i.e., if there is scale separation. The cascade rate is a process whose rate is dictated by the instabilities in the larger eddies; viscosity merely mops up whatever winds up on the floor. This means that

$$\overline{u_i u_j}\ \overline{e_{ij}} = \epsilon = \text{constant}, \tag{3.11}$$

and ϵ becomes one of the principal parameters in three-dimensional turbulence.

As a final illustration, let us consider the derivation of the logarithmic wind profile in a neutral surface layer. The vertical momentum flux, $-\overline{uw}$, is independent of z. Therefore, we may express $-\overline{uw} = u_*^2$, where u_* is the friction velocity. Now the rate of energy conversion from the large scales to the small is given by $-\overline{uw}\partial U/\partial z$, which must be balanced by ϵ, as stated above. Frisch in his lectures has also made the argument that $\epsilon \sim$ v v v ∇. Since the energy cascade is initiated by instabilities of eddies with a velocity proportional to u_* we find that

$$u_*^2 \frac{\partial U}{\partial z} \sim \frac{u_*^3}{kz}. \tag{3.12}$$

(We have chosen $l = z$ because locally there are no other relevant length scales.) This leads to the logarithmic law.

Kraichnan, R., 1959: *J. Fluid Mech.*, **5**, 497.
Onsager, L., 1949: *Nuovo Cimento*, **6**, suppl. No. 2, 279.
Townsend, A. A., 1951: *Proc. Roy. Soc., London*, **A208**, #1095, 534.

Chapter 4

CONSEQUENCES OF THE KOLMOGOROV THEORY

The intuitive perception of the microscales is difficult. In this chapter, we will try to understand the vigor of the 3D turbulent microstructure, and explore some of the consequences of the Kolmogorov scaling. We already know that there is much less energy in 2D microstructure, because there is no vortex stretching. Therefore, we start with the equation for the energy describing the equilibrium between production and dissipation

$$\overline{-u_i u_j} \cdot \frac{\partial U_i}{\partial x_j} = \nu \overline{\zeta_i \zeta_i} = \epsilon \, . \tag{4.1}$$

We know that ϵ is not determined by the viscosity, but by the large scales. This equation relies on the transfer of energy between the large-scale mean flow and the small-scale fluctuations. The next equation — for the enstrophy in a steady regime — does not require a large-scale mean flow; it is self-sustaining (cf. Section 3-3 of Tennekes and Lumley, 1972):

$$\overline{\zeta_i \zeta_j \frac{\partial u_i}{\partial x_j}} = \nu \overline{\frac{\partial \zeta_i}{\partial x_j} \cdot \frac{\partial \zeta_i}{\partial x_j}} \, . \tag{4.2}$$

When we want to study small scales, we treat ϵ as an independent parameter. We can rewrite the r.h.s. of (4.1) as:

$$\overline{\zeta_i \zeta_i} = \frac{\epsilon}{\nu} \, . \tag{4.3}$$

Here, ϵ is a large-scale parameter, ν is a small-scale parameter. If ϵ increases, or ν decreases (or equivalently Re increases), the enstrophy increases. The difference of turbulence with low or high Re is the vigor at small scales and the resolution and detail of the microstructure. For low Re, we have a coarse microstructure; at high Re, there is a fine-grain microstructure.

It is conventional to derive the Kolmogorov microscale by dimensional analysis. But here we derive the Kolmogorov microscale η from (4.2):

$$\left(\frac{\epsilon}{\nu}\right)^{1/2} \cdot \left(\frac{\epsilon}{\nu}\right)^{1/2} \cdot \left(\frac{\epsilon}{\nu}\right)^{1/2} = \nu \frac{\left(\frac{\epsilon}{\nu}\right)^{1/2}}{\eta} \cdot \frac{\left(\frac{\epsilon}{\nu}\right)^{1/2}}{\eta} \, . \tag{4.4}$$

There are a number of assumptions involved in this seemingly offhand estimate. For one, we have assumed that the components of the strain-rate tensor have magnitudes comparable to those of the vorticity vector. That seems reasonable, because vorticity and strain rate are given by

$$\zeta_2 = \frac{\partial u_1}{\partial x_3} - \frac{\partial u_3}{\partial x_1} , \tag{4.5}$$

$$e_{ij} = \frac{\partial u_i}{\partial x_j} + \frac{\partial u_j}{\partial x_i} . \tag{4.6}$$

We have also assumed that the vorticity components in (4.2) are well correlated, and that the vorticity vector lines up fairly well with the principal axis of the strain rate (see Chapter 3). In other words, we assume that vorticity amplification (aligning and stretching) is a fairly efficient mechanism.

From (4.4), we get the value

$$\eta = \left(\frac{\nu^3}{\epsilon}\right)^{1/4} \rightarrow \frac{\eta}{l} = Re^{-3/4} , \tag{4.7}$$

with $Re = l^{4/3}\epsilon^{1/3}/\nu$. We can see that η decreases when the kinematic viscosity ν decreases or the rate of energy transfer ϵ increases. By introducing a finite η, we avoid the presence of singularities. The small-scale time $\tau \sim (\nu/\epsilon)^{1/2}$ can be defined the same way, as well as the small-scale acceleration

$$a_\eta = \left(\frac{\epsilon^3}{\nu}\right)^{1/4} \tag{4.8}$$

Note here the problem of singularities; as $\nu \rightarrow 0$, this implies that $a_\eta \rightarrow \infty$. We now compare a_η with the acceleration at large scales a_l

$$\frac{a_\eta}{a_l} = \frac{\epsilon^{3/4}}{\nu^{1/4}} \cdot \frac{l}{u^2} = \frac{u^{9/4}}{l^{3/4}} \cdot \frac{l}{u^2 \nu^{1/4}} = Re^{1/4} . \tag{4.9}$$

So, in turbulence with, for example, $Re = 10^8$, a_η is a factor of 10^2 larger than a_l. We can also compare energies at different scales,

$$\frac{v_\eta^2}{u^2} = Re^{-1/2}, \tag{4.10}$$

which tells us that small scales of motion have small velocities, but large rotation rates and, in consequence, high accelerations (for high Reynolds number). In the same way, we compare characteristic times

$$\frac{\tau}{l/u} = Re^{-1/2} \ . \tag{4.11}$$

The characteristic periods of small-scale eddies are smaller than those of large eddies, for high Re. In numerical 3D experiments, at high Re, the acceleration is mostly high-frequency noise. This has been verified by laboratory experiments. This result suggests that we must be careful when we study the Navier-Stokes equation numerically. In numerical calculations, the time step must be smaller than the Kolmogorov time scale; the same goes for the space scale.

Let us now compute the self-frequency of the smallest eddies,

$$\omega_s \sim \left(\frac{\epsilon}{\nu}\right)^{1/2} , \tag{4.12}$$

which is the intrinsic (diffusive) frequency at small scales. The advective frequency at small scales is

$$\omega_a \sim \frac{u}{\eta} , \tag{4.13}$$

so that their ratio is

$$\frac{\omega_s}{\omega_a} \sim \frac{u^{\frac{3}{2}}\,\nu^{\frac{3}{4}}\,l^{\frac{1}{4}}}{l^{\frac{1}{2}}\,\nu^{\frac{1}{2}}\,u^{\frac{3}{4}}u} = Re^{-\frac{1}{4}} \ . \tag{4.14}$$

This result is valid even with intermittency.

We can then compare the two energy spectra. In the Lagrangian frame the advection is removed because we are moving with the flow. This spectrum is experienced by a particle moving with the flow at all times; it is an ordinary Kolmogorov-type spectrum, a function of ϵ and ω. The slope $\epsilon\omega^{-2}$ comes from the assumption that the spectrum is independent of ν in the inertial range. In the Eulerian frame without mean motion, however, we have to account for advection by the large eddies. Therefore, the spectrum also depends upon q, the characteristic velocity of the larger eddies, defined by

$$\frac{1}{2}\,q^2 = \frac{1}{2}\,\overline{u_i u_i} \ . \tag{4.15}$$

Here we see that ϵ and k are not enough to determine the form of the spectrum. The two spectra are shown schematically in Figs. 4.1 and 4.2. Note from the figures that the viscous cut-off does not occur at the same ω.

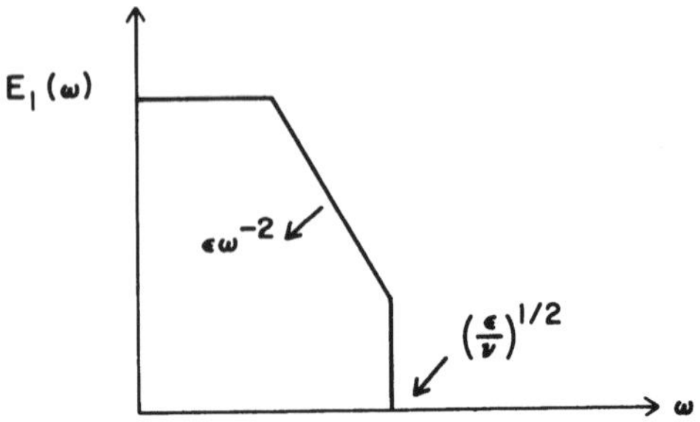

Figure 4.1: Energy spectrum in the Lagrangian frame of reference.

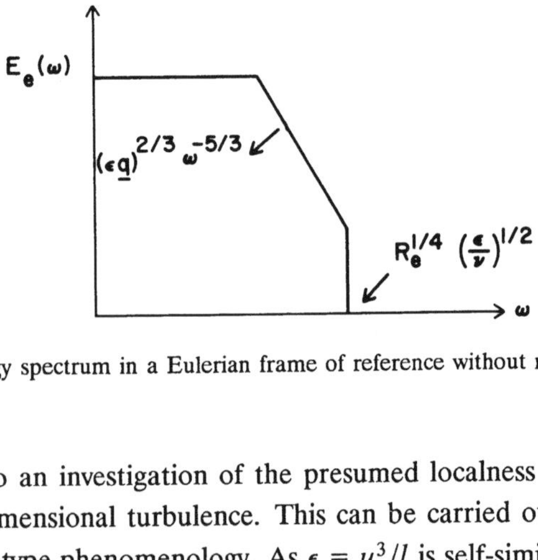

Figure 4.2: Energy spectrum in a Eulerian frame of reference without mean motion.

We now turn to an investigation of the presumed localness of the spectral cascade in three-dimensional turbulence. This can be carried out with conventional Kolmogorov-type phenomenology. As $\epsilon = u^3/l$ is self-similar for any part of the spectrum, we can derive the Kolmogorov spectrum from

$$\epsilon = k\mathrm{v}^3(k) \sim \mathrm{v} \cdot \mathrm{v} \cdot \mathrm{v} \cdot \nabla \ , \tag{4.16}$$

so that

$$\mathrm{v}^3(k) \sim \epsilon k^{-1} \ , \tag{4.17}$$

or

$$\mathrm{v}^2(k) \sim \epsilon^{2/3} k^{-2/3} \ , \tag{4.18}$$

and

$$kE(k) \sim \epsilon^{2/3} k^{-2/3} \rightarrow E(k) = \epsilon^{2/3} k^{-5/3} \ , \tag{4.19}$$

which is Kolmogorov's 3D energy spectrum.

Now we know that the typical v at wavenumber k goes as

$$\mathrm{v}(k) \sim \epsilon^{1/3} k^{-1/3} , \tag{4.20}$$

so that the eddy viscosity (ν_T) at k can be computed as

$$k^{-1}\mathrm{v}(k) \sim \epsilon^{1/3} k^{-4/3} , \tag{4.21}$$

the strain rate at k as

$$k\mathrm{v}(k) \sim \epsilon^{1/3} k^{2/3} , \tag{4.22}$$

and the energy flux across the wavenumber shell at k

$$\overline{u'w'}\,\frac{\partial U}{\partial z} = \nu_T \left(\frac{\partial U}{\partial z}\right)^2 , \tag{4.23}$$

where $\partial U/\partial z$ is the strain rate at the large scales of motion, or

$$\epsilon^{1/3}k^{-4/3} \cdot \epsilon^{1/3}k^{2/3} \cdot \epsilon^{1/3}k^{2/3} = \epsilon , \tag{4.24}$$

which is the obvious result. The energy transfer in wavenumber space is localized because the interaction between the different scales is localized on the wavenumber shell. We can now schematically understand the physical process at wavenumber k (Fig. 4.3). The large-scale shear ($k' < k$) interacts with the small-scale Reynolds stress ($k' > k$); both are most effective at the cut-off wavenumber k.

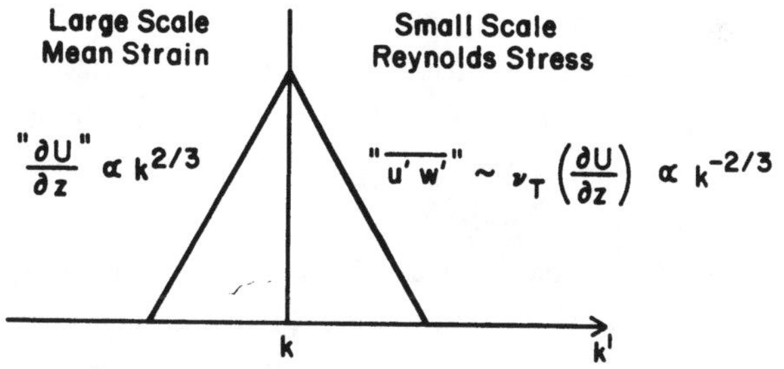

Figure 4.3: The large-scale mean strain and small-scale Reynolds stress as a function of k'.

Tennekes, H., and J. L. Lumley, 1972: *A First Course in Turbulence*. MIT Press, Cambridge, Mass.

Chapter 5

PREDICTABILITY AND OTHER PROBLEMS

In this chapter we will tackle the problem of backward error propagation in the k-spectrum and use the general method we have developed for Kolmogorov's spectrum to find the spectrum of a passive contaminant and of vorticity (in two-dimensional turbulence).

First, let us consider the example of two particles initially very close and drifting apart. The distance r which separates two trajectories obeys the following law

$$\frac{dr}{dt} = rs, \tag{5.1}$$

in a self-similar framework where s is a Lyapounov exponent. In a Kolmogorov spectrum, s is assumed to be the strain rate at scale k^{-1}. Then its dimension is $[velocity] \times [length]^{-1}$ and can be estimated as :

$$s \sim \epsilon^{1/3} k^{2/3} \ . \tag{5.2}$$

So we can conclude that the small eddies are the most unpredictable.

5.1 Calculation of the Predictability Horizon

In most numerical experiments, the smallest simulated scales ($\sim k_0^{-1}$) are larger than the Kolmogorov dissipation scale ($\sim k_K^{-1}$). So deterministically speaking, these estimates are wrong. Thus, the error automatically introduced in the microstructures (because the Lyapounov exponent is higher at small scales) progressively contaminates the whole spectrum. That means that the error cascade is always backward.

That is why Métais and Lesieur (1986) propose the following relation to govern the decrease of k_e which is, at time t, the minimum wavenumber at which the spectrum has been severely contaminated:

$$\frac{dk_e}{dt} = -\frac{k_e}{\tau} \ . \tag{5.3}$$

Here self-similarity is assumed and $\tau^{-1} \sim \epsilon^{1/3} k^{2/3}$ (Lyapounov exponent). This is valid in a statistical sense but not applicable to particular cases, such as coherent structures. So the equation to solve is

$$\frac{dk}{dt} = -k\epsilon^{1/3}k^{2/3} = -\epsilon^{1/3}k^{5/3}, \tag{5.4}$$

which leads to

$$k_e^{-2/3} - k_0^{-2/3} = c\epsilon^{1/3}t \ , \tag{5.5}$$

and, assuming $k_0^{-2/3}$ is small enough,

$$k_e^{-2/3} = c\epsilon^{1/3}t \ . \tag{5.6}$$

In particular, the largest scales $\mathcal{L}$ will be contaminated in a finite time T such that

$$\left(\frac{\mathcal{L}}{2\pi}\right)^{2/3} = c\epsilon^{1/3}T \ . \tag{5.7}$$

Since $\epsilon \sim u^3/\mathcal{L}$, this amounts to $T \sim \mathcal{L}/u$; i.e., all predictability is lost within one or two large-eddy turnover times.

5.2 Diffusion of a Scalar

The well-known equation for a passive scalar contaminant similar to temperature, θ is

$$\frac{\partial\theta}{\partial t} + u_i\frac{\partial\theta}{\partial x_i} = \kappa\frac{\partial^2\theta}{\partial x_i\partial x_i} \ . \tag{5.8}$$

When u_i and θ are split into an average and a fluctuation ($\theta = \theta' + \Theta$), one gets the following balance equation for production and dissipation:

$$-\overline{u_i\theta}\frac{\partial\Theta}{\partial x_i} = \kappa\overline{\frac{\partial\theta}{\partial x_i}\frac{\partial\theta}{\partial x_i}} \ . \tag{5.9}$$

We want to derive the temperature spectrum when $\kappa \ll \nu$. This means that the smallest active scales of temperature are smaller than the smallest active scales of velocity. This is illustrated by the following spectrum, where $E_\theta(k)$ is the temperature spectrum related to the temperature fluctuations by

$$\int_0^\infty E_\theta(k)dk = \overline{\theta'^2} \ . \tag{5.10}$$

In the octave surrounding k, these fluctuations are estimated by $kE_\theta(k) \sim \overline{\theta'^2}(k)$. In the range ($a$), the flux of temperature ϵ_θ is constant because the thermal dissipation occurs mainly at the spectral cut-off. It is estimated as $\epsilon_\theta(k) \sim \overline{\theta'^2}(k)/\tau \sim kE_\theta(k)\tau^{-1}$ where τ is the appropriate characteristic time $\tau \sim (\nu/\epsilon)^{1/2}$. This is the characteristic time associated with the smallest eddies in the

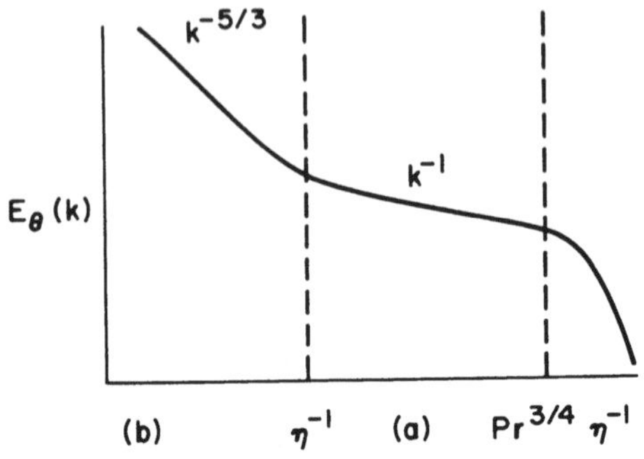

Figure 5.1: Spectrum of temperature variance for $Pr = \nu/k \gg 1$.

Kolmogorov inertial range. This leads to $E_\theta(k) = \epsilon_\theta\, (\nu/\epsilon)^{1/2}\, k^{-1}$ (Batchelor, 1962) spectrum, in the k-range (a). The cut-off takes place at $\eta_\theta = \left(k^3/\epsilon\right)^{1/4} = Pr^{-3/4}\eta$.

5.3. Convection in Two-Dimensional Turbulence (2DT)

Temperature θ and vorticity ζ obey equations which look similar :

$$\frac{\partial\theta}{\partial t} + u_i\frac{\partial\theta}{\partial x_i} = \kappa\frac{\partial^2\theta}{\partial x_i\partial x_i}\ ; \tag{5.11}$$

$$\frac{\partial\zeta}{\partial t} + u_i\frac{\partial\zeta}{\partial x_i} = \nu\frac{\partial^2\zeta}{\partial x_i\partial x_i}. \tag{5.12}$$

Equation (5.12) shows that there is no vorticity amplification mechanism even if vorticity gradients can be amplified. Let us follow Batchelor's arguments. We define the vorticity spectrum $E_\zeta(k)$ related to the vorticity fluctuation by

$$\int_0^\infty E_\zeta(k)dk = \overline{\zeta'^2}\ . \tag{5.13}$$

An estimate in an octave band leads to

$$kE_\zeta(k) \sim \overline{\zeta'^2}(k)\ , \tag{5.14}$$

assuming a constant vorticity flux rate ϵ_ζ :

$$\epsilon_\zeta \sim \frac{\overline{\zeta^2}(k)}{\tau} \sim \frac{kE_\zeta(k)}{\tau}\ . \tag{5.15}$$

τ is a characteristic time related to the enstrophy flux rate ϵ_ζ. Since ϵ_ζ has the dimension of $(\sec)^{-3}$, the appropriate τ is $\epsilon_\zeta^{-1/3}$. This is independent of k. Therefore, $E_\zeta(k) \sim k^{-1}$. Note that this spectrum has the same slope as the Batchelor spectrum in three-dimensional turbulence. This similarity is due to the scale independence of the controlling cascading time scale.

Remark : Since $\zeta'(k)$ is independent of k, the Lyapounov exponent [which is related to $\zeta'(k)$] does not depend upon the scale k; therefore the error-increasing rate will be the same at each scale. This means that there is no separation of time scales in the enstrophy-cascading range of 2DT. In the absence of scale separation, *non*-local interactions in k-space are potentially just as effective as local interactions.

We can conclude that 2DT phenomenology is much more complicated than 3DT phenomenology. For another example, see Chapter 11.

Batchelor, G. K., 1962: *J. Fluid Mech.*, **5**, 113.

Métais, O., and M. Lesieur, 1986: *J. Atmos. Sci.*, **43**, 857.

Chapter 6

PREDICTABILITY OF A SIMPLE SYSTEM

The topic of second-order modeling may also be phrased as "flux maintenance dynamics." Some of the questions which can be addressed by second-order modeling are: Why can eddy viscosities (or diffusivities) become negative? How are momentum fluxes (e.g., $\overline{u'w'}$) maintained? How are the fluxes generated or dissipated, and on what time scales?

Second-order modeling is confronted with the closure problem. Averaging of the mean equations results in unknown terms in second moments. If we attempt to solve for the second moments, unknown terms then arise involving the third moments. The problem cannot be solved by writing equations for still higher orders; to the contrary, extending the treatment to higher moments results in a larger number of equations to be solved. In a sense, the problem is divergent. The quote from Bateson: "All creative systems are necessarily divergent," may have a corollary: "Creativity is needed to deal with divergent systems." At some level, one needs creativity to close the system. The problem resembles the divergence of paths on a strange attractor. Therefore, second-order modeling of the behavior of a strange attractor will be used as an introduction.

The problem is that of forecasting forecast errors, or "meta-predictability." Even with large, fast computers and modern techniques, weather forecasts are often vague and inaccurate — the forecast is typically expressed in terms such as "probably cloudy today." Scientists should work on predicting the errors in forecast models as well as developing more sophisticated forecast models.

The problem of predicting errors in prediction was seen first by Thompson (1957) and Epstein (1969) and studied recently by Thompson (1984). The measurement of the initial state of a system has some error associated with it and a probability distribution for the error. It is necessary not only to predict the evolution of the system but also to predict the evolution of the error probability distribution. An important question is how long will it be before a prediction becomes useless.

Consider the system of Lorenz equations and a distribution of initial states about some value. The trajectory of the center of gravity may be computed by ensemble averaging over many trajectories. The system is given by

$$\frac{dx}{dt} = -\sigma x + \sigma y \ , \tag{6.1}$$

$$\frac{dy}{dt} = -xz + rx - y \ , \tag{6.2}$$

$$\frac{dz}{dt} = xy - bz \ , \tag{6.3}$$

where $\sigma = 10$, $r = 28$, and $b = 8/3$. Decompose the variables into means and fluctuations to obtain

$$x = X + x' \ , \tag{6.4}$$

$$y = Y + y' \ , \tag{6.5}$$

$$z = Z + z' \ . \tag{6.6}$$

Reynolds averaging produces

$$\frac{dX}{dt} = \sigma Y - \sigma X \ , \tag{6.7}$$

$$\frac{dY}{dt} = -XZ - \overline{x'z'} + rX - Y \ , \tag{6.8}$$

$$\frac{dZ}{dt} = XY + \overline{x'y'} - bZ \ . \tag{6.9}$$

The trajectory of the center of gravity differs from the individual trajectories because of the effect of the covariances. The equations for the variances are

$$\frac{d}{dt}(\frac{1}{2}\overline{x'x'}) = -\sigma\overline{x'x'} + \sigma\overline{x'y'} \ , \tag{6.10}$$

$$\frac{d}{dt}(\frac{1}{2}\overline{y'y'}) = \overline{x'y'}(r - Z) - X\overline{y'z'} - \overline{y'y'} - \overline{x'y'z'} \ , \tag{6.11}$$

$$\frac{d}{dt}(\frac{1}{2}\overline{z'z'}) = X\overline{y'z'} + Y\overline{x'z'} - b\overline{z'z'} + \overline{x'y'z'} \ . \tag{6.12}$$

If the errors are small (e.g., $x' \ll X$), the triple correlations may be neglected. This will commonly be valid for short times. The equations for the variances contain unknown covariances. Thus, the system will be closed upon obtaining the covariances. The dissipation of the variance is the same as the rate of contraction of volume in phase space, i.e., $-(\sigma + b + 1)$. If $\overline{x'x'}$ is to grow, $\overline{x'y'}$ must be positive more often than not. Thus, the covariances are necessary for error production. The equations for the covariances are

$$\frac{d}{dt}\overline{x'y'} = -(\sigma + 1)\overline{x'y'} + (r - Z)\overline{x'x'} + \sigma\overline{y'y'}$$
$$-X\overline{x'z'} + Y\overline{x'x'} - \overline{x'x'z'} , \tag{6.13}$$

$$\frac{d}{dt}\overline{x'z'} = -(\sigma + b)\overline{x'z'} + \sigma\overline{y'z'} + X\overline{x'y'} + Y\overline{x'x'} + \overline{x'x'y'} , \tag{6.14}$$

$$\frac{d}{dt}\overline{y'z'} = -X\overline{z'z'} + X\overline{y'y'} + Y\overline{x'y'} + (r - Z)\overline{x'z'}$$
$$-(b + 1)\overline{y'z'} + \overline{x'y'y'} - \overline{x'z'z'} . \tag{6.15}$$

The covariance equations all contain terms for dissipation. For example, the equation for $\overline{x'y'}$ shows that nonzero initial $\overline{y'y'}$ will produce $\overline{x'y'}$. This in turn will increase $\overline{x'x'}$.

The challenge is to see if these correlations can lead to mixing-length behavior or if some phenomenology can be developed. This is a link between metapredictability and second-order modeling. The objective of studying the early error evolution in a problem with just a few degrees of freedom is not to obtain detailed information on the evolution of errors in systems with many degrees of freedom (such as the earth's atmosphere), but to make a start on the almost completely unexplored issue of developing a consistent methodology (and/or phenomenology) for problems of this kind.

Epstein, E. S., 1969: *J. Appl. Meteor.*, **8**, 190.

Thompson, P. D., 1957: *Tellus*, **9**, 275.

Thompson, P. D., 1984: "A Review of Predictability Problems," in *Predictability of Fluid Motion*, American Institute of Physics, A.I.P. Conference Proceedings #106, Greg Holloway and Bruce West, eds., p. 1–10.

Chapter 7

SELF-CONSISTENT SCALING IN THE $K - \varepsilon$ MODEL

In this chapter we introduce a second-order model of turbulence and apply it to a plane shear flow. We shall see how effectively we can model the decay of homogeneous turbulence. When we speak of "second-order model," we consider a model which contains variance and covariance terms. Our particular model will contain just variance equations.

Consider a wind tunnel with a grating near the inlet. We introduce a shear flow and wish to model the flow far from the grating. We note that by imposing a fixed shear flow, we impose a time scale

$$\frac{\partial U}{\partial z} = \frac{1}{\tau_i} \,. \tag{7.1}$$

Unfortunately, it is not as easy to deduce a length scale from the flow. The grating imposes an initial length scale (the grid spacing) which is not a natural scale of the wind tunnel geometry. The turbulence then evolves and starts to hunt for a length scale consistent with the geometry of the flow. Combining the natural variables of the system into terms such as $U/(\partial U/\partial z)$ will not give us appropriate length scales since they will not be Galilean invariant. Since there are no self-defining length scales, let us try to obtain one from a set of equations modeling the decay of wind-tunnel turbulence with and without plane shear.

The following equations are commonly referred to as the $K - \epsilon$ model. K refers to the kinetic energy (here called E); ϵ is the dissipation rate.

$$U\frac{\partial E}{\partial x} = -\overline{u'w'}\frac{\partial U}{\partial z} - \varepsilon + \frac{\partial}{\partial z}K_E\frac{\partial E}{\partial z} \,, \tag{7.2}$$

$$U\frac{\partial \varepsilon}{\partial x} = \left(-c_1\overline{u'w'}\frac{\partial U}{\partial z} - c_2\varepsilon\right)\frac{\varepsilon}{E} + \frac{\partial}{\partial z}K_\epsilon\frac{\partial \varepsilon}{\partial z} \,. \tag{7.3}$$

In the above equations (written for steady flow), we have:

U = mean flow in the x-direction;
u' = fluctuation in flow, x-direction;
w' = fluctuation in flow, z-direction;
E = turbulent kinetic energy of flow per unit mass;

ε = dissipation of kinetic energy per unit mass;
c_1 = constant;
c_2 = constant;
K_E = turbulent exchange coefficients for kinetic energy;
K_ε = turbulent exchange coefficient for ε.

In (7.2), which represents the change in E as it follows along a streamline of the mean flow, the first term on the right is shear generation of turbulent kinetic energy (which is the product of the turbulent momentum flux ($\overline{u'w'}$) and the imposed mean shear ($\partial U/\partial z$), the second term is energy dissipation, and the third term is the z-component of the flux convergence (triple correlations, pressure terms, etc., put in parameterized form).

In (7.3), which represents the change in ε along a streamline of the mean flow, the first term on the right is the difference between the energy production and dissipation terms, divided by E/ε. The latter is the self-relaxation time of the turbulent flow and can be thought of as a time scale representing the rate at which energy cascades down the spectrum. The last term represents the mean flux divergence or convergence of ε.

In our pursuit of length scales, we shall first simplify (7.2) and (7.3) and adopt a slightly different notation. We can initialize the flow with a homogeneous turbulent energy field, such that $\partial E/\partial z = 0$ and $\partial\varepsilon/\partial z = 0$ at $t = 0$. The flux divergence terms in (7.2) and (7.3) will then remain negligible for some distance. Therefore, we are left with

$$\frac{\partial E}{\partial t_m} = -\overline{u'w'}\frac{\partial U}{\partial z} - \varepsilon \tag{7.4}$$

$$\frac{\partial \varepsilon}{\partial t_m} = \left(-c_1\overline{u'w'}\frac{\partial U}{\partial z} - c_2\varepsilon\right)\frac{\varepsilon}{E}\,, \tag{7.5}$$

where we have introduced the notation $U(\partial E/\partial x) = \partial E/\partial t_m$, where t_m represents the time it takes for a particle to travel the distance along a streamline of the mean flow.

Let us now see how reasonable these equations are. In (7.4) we see that if we construct a shear flow in which shear production balances the energy dissipation we have $\partial E/\partial t_m = 0$, which implies a constant turbulent kinetic energy in the flow — a reasonable result. The second equation, however, is a little more complicated. We basically have the same terms as in (7.3) (less the third), each multiplied by a constant and both multiplied by ε/E, which essentially divides the terms by a time scale representing the rate at which energy cascades down the spectrum. The constants represent "two metal plates patching your ship" (McWilliams, private communication) with which you can fit the model to experimental results.

Nevertheless, assuming that this second-order model is reasonable, our next step in the diagnosis of this flow is to set the imposed shear ($\partial U/\partial z$) equal to zero. This reduces (7.4) and (7.5) to

$$\frac{\partial E}{\partial t_m} = -\varepsilon \ , \tag{7.6}$$

$$\frac{\partial \varepsilon}{\partial t_m} = -c_2 \frac{\varepsilon^2}{E} \ . \tag{7.7}$$

We can combine (7.6) and (7.7) to form an equation for the self-relaxation of the turbulence

$$\frac{\partial}{\partial t_m}\left(\frac{E}{\varepsilon}\right) = \frac{\partial}{\partial t_m}(\tau) = (c_2 - 1) \ , \tag{7.8}$$

which has as its solution

$$\tau = t_m (c_2 - 1) + \tau_o \ . \tag{7.9}$$

This equation makes sense. By removing the shear, we have deprived the flow of externally imposed characteristic time scales. Therefore, the only parameter left upon which the time scale can depend is t_m. This is similar to the argument explaining logarithmic profiles for flows with a constant shear (self-similarity, or, formally, intermediate asymptotics). Note: we expect $c_2 > 1$ since the time scales ought to increase in proportion to the running time $t_m = x/U$, not decrease with time.

Let us rewrite (7.9) in terms of E and ε:

$$\frac{E}{\varepsilon} = \frac{E_o}{\varepsilon_o} + (c_2 - 1)t_m \ . \tag{7.10}$$

Now, let us assume that the kinetic energy of the flow behaves like

$$E = E_o(1 + at_m)^{-n} \ . \tag{7.11}$$

In order for (7.10) and (7.11) to be satisfied, we require:

$$\frac{\varepsilon}{\varepsilon_o} = (1 + at_m)^{-n-1} \ , \tag{7.12}$$

$$a = \frac{(c_2 - 1)\,\varepsilon_o}{E_o} = \frac{(c_2 - 1)}{\tau} . \tag{7.13}$$

Now, we differentiate (7.11) to obtain

$$\frac{\partial E}{\partial t_m} = -E_o a n (1 + a t_m)^{-n-1} , \tag{7.14}$$

but, $\partial E/\partial t_m = -\varepsilon$; therefore, by (7.12)

$$-E_o a n (1 + a t_m)^{-n-1} = -\varepsilon_o (1 + a t_m)^{-n-1} , \tag{7.15}$$

which implies

$$n = \frac{\varepsilon_o}{a E_o} . \tag{7.16}$$

Using (7.13) we obtain our final result

$$n = (c_2 - 1)^{-1} . \tag{7.17}$$

This final equation cannot be solved. Instead, we have to refer to the results obtained from experimental data. The general consensus indicates that we should expect the kinetic energy of the flow to decay with $n \simeq 6/5$, which implies $c_2 \simeq 11/6$. Now that we have determined n, we may calculate the length scale from (7.11) and (7.12)

$$l = \frac{E^{3/2}}{\varepsilon} = \frac{E_o{}^{3/2}}{\varepsilon_o} \left(1 + \frac{t_m}{n \tau_o}\right)^{-n/2+1} , \tag{7.18}$$

which grows with an exponent of $2/5$ if $n = 6/5$.

The scaling of the Reynolds number may be obtained from (7.11) and (7.18)

$$Re = \frac{ul}{\nu} = \frac{E^{1/2} l}{\nu} = \frac{E_o{}^2}{\varepsilon_o \nu} \left(1 + \frac{t_m}{n \tau_o}\right)^{-n+1} , \tag{7.19}$$

which decays as $t_m^{-1/5}$ if $n = 6/5$. [Note that the invariant here is the Saffman invariant El^3 since E decays as $t_m^{-6/5}$ and l^3 increases as $t_m^{6/5}$. This differs from the Loitsianskii (1939) version which has El^5 as an invariant]. The various invariants are related to the spectral shape at very small wavenumbers, and thereby to the so-called "permanence of the large eddies" (Batchelor and Townsend, 1947). This is how lack of uniqueness in experimental results may be generated.

Thus, we see that by appealing to a second-order model of the flow we obtain a self-consistent set of scales defining the flow which could not be directly obtained from the equations of motion. Unfortunately, most people attending this lecture were very skeptical of the validity of the $K - \varepsilon$ model. Thus, much time was spent debating the assumptions made in the $K - \varepsilon$ model and the lack of replication of experimental results with slightly varying initial conditions. Nevertheless, it was agreed by all that an insight into our "turbulent roots" was an important goal of the summer school.

Batchelor, G. K. and A. A. Townsend, 1947: *Proc. Roy. Soc., London*, **194A**, 527.
Loitsianskii, L. G., 1939: *Trudy Tsentr. Aero.-Gidrodin. Inst.*, **440**, 3.

Chapter 8

THE DECAY OF TURBULENCE IN PLANE HOMOGENEOUS SHEAR FLOW

In the previous chapter we derived the equations for the evolution of energy and dissipation rate of the fluctuating part of the velocity field for a planar mean flow U with an imposed shear $\partial U/\partial z$:

$$U\frac{\partial E}{\partial x} = \frac{\partial E}{\partial t_m} = -\overline{u'w'}\frac{\partial U}{\partial z} - \epsilon + \frac{\partial}{\partial z}\left(K_E\frac{\partial E}{\partial z}\right) , \tag{8.1a}$$

$$U\frac{\partial \epsilon}{\partial x} = \frac{\partial \epsilon}{\partial t_m} = -\left(-c_1\overline{u'w'}\frac{\partial U}{\partial z} - c_2\epsilon\right)\frac{\epsilon}{E} + \frac{\partial}{\partial E}\left(K_\epsilon\frac{\partial E}{\partial z}\right) . \tag{8.1b}$$

Here, $E \equiv 1/2\overline{u_i'u_i'}$ and $\epsilon = \nu\overline{\zeta_i'\zeta_i'}$ are the kinetic energy and dissipation rate of the velocity fluctuations, respectively. Additionally, E/ϵ gives the relaxation time and $t_m \equiv x/U$ represents the advective time scale following the mean flow. It was found in the previous chapter that, in the absence of shear ($\partial U/\partial z = 0$), turbulence decayed as expected. In fact, the above system gives a linear evolution of the relaxation time and is consistent with experimental data (Champagne et al., 1970). Moreover, experimental data give a value of 11/6 for the constant c_2. It was noted that c_2 might not be "universal," especially as $K_E \to 0$, when initial conditions and the details of the turbulence generation may persist for a long time.

We now consider the effect of adding homogeneous shear to our flow. For our discussion to be relevant, the characteristic length ℓ of the turbulence (created by the screen that generates it at the entry point) must remain small compared to the width of the flow region h:

$$\ell << h . \tag{8.2}$$

This implies that there is "scale separation." Also, since the imposed shear is homogeneous ($\partial U/\partial z =$ const.), there is no *other* external length scale in this problem. Under these conditions it is reasonable to attempt a closure by expressing the Reynolds stress as the product of an eddy viscosity K and the shear

$$-\overline{u'w'} = K\frac{\partial U}{\partial z} , \tag{8.3a}$$

where K is proportional to a typical eddy velocity v and the length scale of the turbulence ℓ,

$$K \sim \mathrm{v}\ell\,. \tag{8.3b}$$

Furthermore, the expression for K can be rewritten in terms of the characteristic time τ as

$$K \sim \mathrm{v}\ell \sim \mathrm{v}^2\tau \sim E\tau\,. \tag{8.3c}$$

Finally, the characteristic time τ of the turbulence will be proportional to the kinetic energy divided by the dissipation rate,

$$\tau \sim \frac{E}{\epsilon}\,. \tag{8.3d}$$

Combining the four parts of (8.3) gives the following *closure* relation for the Reynolds stress:

$$-\overline{u'w'} = c_m E \cdot \frac{E}{\epsilon}\frac{\partial U}{\partial z} = c_m \frac{E^2}{\epsilon}\frac{\partial U}{\partial z}\,. \tag{8.4}$$

Inserting this into (8.1) for the evolution of the energy and dissipation rate yields the desired equations

$$\frac{\partial E}{\partial t_m} = c_m \frac{E^2}{\epsilon}\left(\frac{\partial U}{\partial z}\right)^2 - \epsilon + \frac{\partial}{\partial z}\left(K\frac{\partial E}{\partial z}\right), \tag{8.5a}$$

$$\frac{\partial \epsilon}{\partial t_m} = \left[c_1 c_m \frac{E^2}{\epsilon}\left(\frac{\partial U}{\partial z}\right)^2 - c_2\epsilon\right]\frac{\epsilon}{E} + \frac{\partial}{\partial z}\left(K_\epsilon \frac{\partial \epsilon}{\partial z}\right)\,. \tag{8.5b}$$

Before considering the effects of both shear and vertical diffusion simultaneously, it is of physical interest to neglect the diffusive aspects and focus our attention on the production and dissipation terms. Following the ideas of the previous chapter, the evolution of the turbulent time scale is given by

$$\begin{aligned}
\frac{\partial \tau}{\partial t_*} &= \frac{\partial}{\partial t_*}\left(\frac{E}{\epsilon}\right) = \frac{1}{\epsilon}\frac{\partial E}{\partial t_*} - \frac{E}{\epsilon^2}\frac{\partial E}{\partial t_*} \\
&= c_m\left(\frac{E}{\epsilon}\right)^2\left(\frac{\partial U}{\partial z}\right)^2 - 1 - \left[c_1 c_m \frac{E^2}{\epsilon}\left(\frac{\partial U}{\partial z}\right)^2 - c_2\right] \\
&= c_2 - 1 + c_m\left(\frac{E}{\epsilon}\right)^2\left(\frac{\partial U}{\partial z}\right)^2\{1 - C_1\}\,,
\end{aligned} \tag{8.5c}$$

and therefore

$$\frac{\partial \tau}{\partial t_*} = c_2 - 1 + c_m \left(\frac{E}{\epsilon}\right)^2 \left(\frac{\partial U}{\partial z}\right)^2 \{1 - C_1\} \, . \tag{8.6a}$$

Now, in order to have strong interaction between the mean flow and the turbulence, the time scale of the turbulence must match the time scale associated with the shear at all times. To insure this, we require that $\partial T/\partial t_m = 0$, for otherwise the turbulent time scale would run away from the imposed time scale. Consequently, (8.6a) implies

$$c_2 - 1 + c_m \tau^2 \left(\frac{\partial U}{\partial z}\right)^2 (1 - c_1) \simeq 0 \, . \tag{8.7}$$

Experiments on homogeneous shear turbulence give values of $c_1 \doteq 1.46$ and $c_m = 0.044$. The value of c_m is typical for an eddy viscosity and its smallness suggests that turbulence is not very effective in generating stresses.

Similar manipulations of (8.5) give the evolution of the kinetic energy, dissipation rate and the length scale of the turbulence. The results are:

$$\frac{E}{E_0} = \frac{\epsilon}{\epsilon_0} = \exp\left(\frac{c_2 - c_1}{c_2 - 1}\frac{t}{\tau}\right) , \tag{8.8a}$$

$$\ell = \frac{E^{3/2}}{\epsilon} = \frac{E_0^{3/2}}{\epsilon_0} \exp\left(\frac{c_2 - c_1}{2(c_1 - 1)}\frac{t}{\tau}\right) , \tag{8.8b}$$

where the length scale ℓ is obtained from dimensional analysis. Using the above values for the constants reveals that the exponents in (8.8a) are positive. The consequences of this are:

1. The kinetic energy and the dissipation rate increase exponentially with increasing distance from the generator grid. We expect behavior of this sort since the time scale $\tau \sim \ell/u$ is kept constant as we move downstream, so that the natural tendency for the turbulence to increase its ℓ (unless boundary conditions make that impossible) has to be compensated for by a corresponding increase in kinetic energy. In other words, though the time scales are matched, the length scale can escape, and with it the energy level. Even with perfectly matched time scales, the turbulence cannot be made stationary in a frame moving with the mean flow.

2. The length scale of the turbulence increases exponentially downstream. Of course, the exponential trends revealed in (8.8) cannot be used at large distances from the generation grid since the ultimate effect of the interaction between the eddies and the mean stream is to alter the original velocity profile *significantly*. No term in the equations captures the fact that the energy in the mean flow is depleted by that extraction, so that the turbulence energy should in fact saturate.

As an attractive alternative for finding the "universal" constant c_1 we note that the Lagrangian evolution for ℓ is given by

$$\frac{\partial \ell}{\partial t_m} = \frac{\partial}{\partial t_m}\left(\frac{E^{3/2}}{\epsilon}\right) = E^{1/2}\left\{\left(c_2 - \frac{3}{2}\right) - \left(c_1 - \frac{3}{2}\right) c_m \tau^2 \left(\frac{\partial U}{\partial z}\right)^2\right\} . \tag{8.9}$$

On dimensional grounds, ℓ cannot depend upon the shear because the shear is homogeneous and cannot impose a *length* scale. This requires that $c_1 = 3/2$, which is very close to the experimental values of $c_1 = 1.46$. From (8.9) it is clear that, irrespective of any dependence of $d\ell/dt_m$ on $\partial U/\partial z$, the principal dependence of ℓ on t_m is

$$\frac{\partial \ell}{\partial t_m} = \left(c_2 - \frac{3}{2}\right) E^{1/2} \sim u . \tag{8.10}$$

In other words, the turbulence, not being constrained by an imposed length scale, increases its own internal length scale at a rate proportional to its own r.m.s. intensity. This effect is unavoidable; the corresponding exponential growth (8.8a,b) of u and ℓ therefore is also unavoidable in this flow, even though τ is kept constant.

Champagne, F. H., V. G. Harris, and S. Corrsin, 1970: *J. Fluid Mech.*, **41**, 81.

Chapter 9

SECOND-ORDER MODELING: SOME EXAMPLES

This chapter examines the maintenance of variances and covariances in a convective surface layer. In particular, we consider the case of an infinitely large heated prairie over which the convective turbulence is horizontally homogeneous and stationary, and is governed by the Boussinesq approximation (i.e., the convection is relatively shallow). A mean shear $\partial U/\partial z$ is present. We focus on the constant flux layer, or the lowest 5 to 10% of the atmospheric boundary layer where the total change of vertical heat flux is very small.

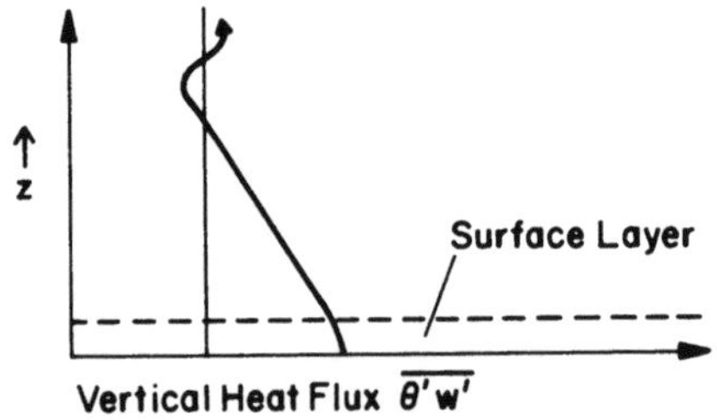

Figure 9.1: The vertical heat flux $\overline{\theta' w'}$ as a function of height.

The equation for the mean potential temperature is

$$\frac{\partial \theta}{\partial t} + U\frac{\partial \theta}{\partial x} + \text{ other advection terms } = -\frac{\partial \overline{\theta' u'}}{\partial x} - \frac{\partial \overline{\theta' v'}}{\partial y} - \frac{\partial \overline{\theta' w'}}{\partial z}$$

$$+ \text{radiation } + \text{ molecular diffusion} . \qquad (9.1)$$

The time derivative vanishes through the stationary assumption, and the advective terms vanish because θ is nearly constant within the convective layer. Likewise, the horizontal flux terms vanish due to horizontal homogeneity. Under these assumptions, we are left with

$$\frac{\partial}{\partial z}\overline{\theta' w'} \sim 0 \ ; \frac{\partial}{\partial z}\overline{u' w'} \sim 0 \ . \qquad (9.2)$$

We thus adopt the usual surface layer normalization

$$-\overline{u' w'} = u_*^2 \ ; \overline{\theta' w'} = \theta_* u_* \ , \qquad (9.3)$$

where u_* and θ_* are considered constants for the layer; u_* is the so-called friction velocity. The fluxes through the layer therefore equal the surface fluxes. We now write the equations for the three velocity variances

$$\frac{\partial}{\partial t}\left(\frac{1}{2}\overline{u'u'}\right) = 0 = -\overline{u'w'}\frac{\partial U}{\partial z} + \frac{1}{\rho}\overline{p'\frac{\partial u'}{\partial x}} - \frac{\partial}{\partial z}\left(\frac{1}{2}\overline{u'u'w'}\right) - \frac{1}{3}\epsilon \, , \tag{9.4}$$

$$\frac{\partial}{\partial t}\left(\frac{1}{2}\overline{v'v'}\right) = 0 = 0 + \frac{1}{\rho}\overline{p'\frac{\partial v'}{\partial y}} - \frac{\partial}{\partial z}\left(\frac{1}{2}\overline{v'v'w'}\right) - \frac{1}{3}\epsilon \, , \tag{9.5}$$

$$\frac{\partial}{\partial t}\left(\frac{1}{2}\overline{w'w'}\right) = 0 = \frac{g}{\theta}\overline{\theta'w'} + \frac{1}{\rho}\frac{\overline{p'\partial w'}}{\partial z} - \frac{\partial}{\partial z}\left(\frac{1}{2}\overline{w'w'w'} + \frac{\overline{p'w'}}{\rho}\right) - \frac{1}{3}\epsilon \, . \tag{9.6}$$

All terms of the l.h.s., i.e., the time derivatives and the advective terms (not shown) are zero from the assumptions of stationarity and horizontal homogeneity. The terms containing $\partial U/\partial z$ and g/θ represent production of variance by the mean wind shear and by buoyancy, respectively. All triple correlation terms other than those shown can be neglected due to horizontal homogeneity. The sum of these three equations would net a total dissipation rate ϵ. Because dissipation occurs mainly in the small scales where the turbulence is essentially isotropic, an equal partioning of dissipation into the three components of variance can be assumed. The terms containing p are responsible for generating isotropy and uniformly distributing energy and dissipation among the three components. The energy is injected as $\overline{u'w'}\,\partial U/\partial z$, and then redistributed into the components. The covariances $\overline{p'\partial u'}/\partial x$ are maintained principally by the larger scales where the turbulence is anisotropic.

To further examine the so-called return to isotropy problem (which implies the analysis of the mechanisms for redistribution of energy), we follow the work of Rotta (1951) and write the pressure-velocity correlation terms in compact tensor form as

$$\frac{1}{2\rho}\overline{p'\left(\frac{\partial u_i'}{\partial x_j} + \frac{\partial u_j'}{\partial x_i}\right)} = \frac{C}{\tau}\left(\overline{u_i'u_j'} - \frac{2}{3}\delta_{ij}E\right) \, , \tag{9.7}$$

where

$$E = \frac{1}{2}\overline{u_i'u_i'} \, , \tag{9.8}$$

τ is a characteristic time scale, and C is some constant. The factor inside the brackets of the r.h.s. of (9.7) represents a measure for the anisotropy of the turbulence. This equation uses the scaling argument that the intercomponent transfer rate of variances and covariances is proportional to the anisotropy of the tensor. This would not be valid near a solid surface or in cases of extreme buoyancy where pressure fluctuations arise directly from these effects.

The individual variance components of these equations are written as

$$\frac{1}{\rho}\overline{p'\frac{\partial u'}{\partial x}} = -\frac{C}{\tau}\left(\overline{u'u'} - \frac{1}{3}\overline{(u'^2 + v'^2 + w'^2)}\right) , \tag{9.9}$$

$$\frac{1}{\rho}\overline{p'\frac{\partial v'}{\partial y}} = -\frac{C}{\tau}\left(\overline{v'v'} - \frac{1}{3}\overline{(u'^2 + v'^2 + w'^2)}\right) , \tag{9.10}$$

$$\frac{1}{\rho}\overline{p'\frac{\partial w'}{\partial z}} = -\frac{C}{\tau}\left(\overline{w'w'} - \frac{1}{3}\overline{(u'^2 + v'^2 + w'^2)}\right) . \tag{9.11}$$

Of the two terms of the r.h.s. of each equation, the first expresses the variance in the individual velocity component, and the second expresses the variance in isotropic flow with the same energy. The equations illustrate the tendency of the pressure-velocity terms to distribute energy equally among the three variances. If any one variance has more than its 1/3 share of the total energy, it will tend to give its excess to the other components. If $\overline{u'u'}$ is much larger than its isotropic value, that implies that the energy is given back faster; for larger (small) anisotropy, we have a fast (slow) redistribution.

A similar development can be done for the covariance equations. For example,

$$\begin{aligned}\frac{\partial}{\partial t}(\overline{u'w'}) = 0 - \overline{w'w'}\frac{\partial U}{\partial z} + \frac{g}{\Theta}\overline{\theta'u'} - \frac{\partial}{\partial z}\ (\text{several terms})\\ + \frac{1}{\rho}\left(\overline{-p\left(\frac{\partial u'}{\partial z} + \frac{\partial w'}{\partial x}\right)}\right) ,\end{aligned} \tag{9.12}$$

and from the pressure-strain rate covariance (9.7), we have

$$\frac{1}{2\rho}\overline{p'\left(\frac{\partial u'}{\partial z} + \frac{\partial w'}{\partial x}\right)} = -\frac{C}{\tau}\overline{u'w'} . \tag{9.13}$$

Ignoring the time derivative and vertical gradient term as before, and considering the neutral buoyancy case yields to

$$\frac{\partial}{\partial t}(\overline{u'w'}) = 0 = -\overline{w'w'}\frac{\partial U}{\partial z} - \frac{2C}{\tau}(\overline{u'w'}) \ . \tag{9.14}$$

The first term of the r.h.s. denotes production of $-\overline{u'w'}$ by the mean shear, and this is balanced by the second term, representing a self-relaxation of $-\overline{u'w'}$ as the flow attempts to become isotropic. An interesting interpretation of eddy viscosity arises from rewriting this equation as

$$-\overline{u'w'} = \frac{\overline{w'w'}\tau}{2C}\frac{\partial U}{\partial z} \ . \tag{9.15}$$

The l.h.s. represents the Reynolds stress tensor, and the r.h.s. contains the mean shear $\partial U/\partial z$ multiplied by the factor

$$\frac{\overline{w'w'}}{2C}\tau \ , \tag{9.16}$$

consisting of a variance multiplied a time, which can be interpreted as an eddy viscosity coefficient. This reinterpretation of mixing-length theory is based on the production and relaxation rates of the covariance.

Remarks: The diagnostic methods of second-order modeling, of which examples are given in Chapters 8, 9, and 10, correspond fairly closely to the diagnostic tools used in general-circulation research. In both cases, flux-maintenance equations are used to help us understand why some fluxes go down the gradient, while others go the other way. Eddy diffusivities are not necessarily positive; in fact, the atmosphere's general circulation is maintained by *negative* eddy-viscosity phenomena and a *reverse* energy cascade. These processes are explored in some detail in Chapter 11.

Rotta, J. C., 1951: *Z. Phys.*, **129**, 547.

Chapter 10

PART I: SECOND-ORDER MODELING: THE SURFACE LAYER

The steady-state, horizontally homogeneous, diabatic surface layer is governed by a system of variance and covariance equations. With some fairly drastic simplifications, this system of equations becomes:

$$\overline{u'w'} \; : \; 0 = -\sigma_w^2 \frac{\partial U}{\partial z} + \frac{g}{\Theta}\overline{\theta' u'} - 2\frac{C_m}{\tau}\overline{u'w'} \,, \tag{10.1}$$

$$\overline{\theta' w'} \; : \; 0 = -\sigma_w^2 \frac{\partial \Theta}{\partial z} + \frac{g}{\Theta}\sigma_\theta^2 - \frac{C_\theta}{\tau}\overline{\theta' w'} \,, \tag{10.2}$$

$$\overline{\theta' u'} \; : \; 0 = -\overline{\theta' w'}\frac{\partial U}{\partial z} - \overline{u'w'}\frac{\partial \Theta}{\partial z} - \frac{C_\theta}{\tau}\overline{\theta' u'} \,, \tag{10.3}$$

$$\frac{1}{2}\sigma_u^2 \; : \; 0 = -\overline{u'w'}\frac{\partial U}{\partial z} - \frac{C_m}{\tau}\left(\sigma_u^2 - \frac{2}{3}E\right) - \frac{1}{3}\epsilon \,, \tag{10.4}$$

$$\frac{1}{2}\sigma_v^2 \; : \; 0 = -\frac{C_m}{\tau}\left(\sigma_v^2 - \frac{2}{3}E\right) - \frac{1}{3}\epsilon \,, \tag{10.5}$$

$$\frac{1}{2}\sigma_w^2 \; : \; 0 = \frac{g}{\Theta}\overline{\theta' w'} - \frac{C_m}{\tau}\left(\sigma_w^2 - \frac{2}{3}E\right) - \frac{1}{3}\epsilon \,, \tag{10.6}$$

$$\frac{1}{2}\sigma_\theta^2 \; : \; 0 = -\overline{\theta' w'}\frac{\partial \Theta}{\partial z} - \epsilon_\theta \,. \tag{10.7}$$

The two epsilons are parameterized as

$$\epsilon = 2\frac{C_\epsilon}{\tau}E = \frac{C_\epsilon}{\tau}\left(\sigma_u^2 + \sigma_v^2 + \sigma_w^2\right) \,,$$

$$\epsilon_\theta = \frac{C_{\epsilon\theta}}{\tau}\sigma_\theta^2 \,. \tag{10.8}$$

Let us suppose now that $\partial U/\partial z$ is positive and $\partial\Theta/\partial z$ is negative. The last case corresponds to a situation where the bottom of the domain is hotter than its top.

Since σ_θ^2 is always positive, (10.2) leads to $\overline{\theta' w'} \geq 0$. Because $\overline{u'w'} < 0$ if $\partial U/\partial z > 0$, and $\overline{\theta' w'} > 0$ if $\partial\Theta/\partial z < 0$, the first two terms in (10.3) are

$$-\overline{\theta' w'}\frac{\partial U}{\partial z} < 0 \tag{10.9}$$

and

$$-\overline{u'w'}\frac{\partial\Theta}{\partial z} < 0 \, . \tag{10.10}$$

Clearly, both terms participate in the maintenance of the horizontal component of the heat flux.

Mean Heat Flux Terms

a) $\overline{\theta' w'}$

From (10.2), we get:

$$\overline{\theta' w'} = -\frac{\tau}{C_\theta}\sigma_w^2\frac{\partial\Theta}{\partial z} + \frac{g}{\Theta}\sigma_\theta^2\frac{\tau}{C_\theta} \, . \tag{10.11}$$

We define $K_H = \sigma_\theta^2\tau/C_\theta$ which has the dimension of a viscosity. Thus,

$$\overline{\theta' w'} = -K_H\left[\frac{\partial\Theta}{\partial z} - \frac{g}{\Theta}\frac{\sigma_\theta^2}{\sigma_w^2}\right] \, . \tag{10.12}$$

The gravity term in the last equation is of the order:

$$\frac{g}{\Theta}\frac{\sigma_\theta^2}{\sigma_w^2} \sim \frac{10}{300}\frac{10^{-1}}{1} \sim 3\times 10^{-3} = 3°\mathrm{C/km} \, , \tag{10.13}$$

which is not negligible at all. So it is necessary to use this corrected gradient to judge convection.

b) $\overline{\theta' u'}$

In the nearly neutral case (the buoyancy terms are set to 0), we get from (10.1) and (10.2)

$$\overline{\theta' w'} = -\frac{\sigma_w^2\tau}{C_\theta}\frac{\partial\Theta}{\partial z} \, , \tag{10.14}$$

$$\overline{u'w'} = -\frac{\tau}{2C_m}\sigma_w^2\frac{\partial U}{\partial z} \, . \tag{10.15}$$

So we get

$$\overline{\theta' u'} = (K_H + K_m)\frac{\tau}{C_\theta}\frac{\partial U}{\partial z}\frac{\partial\Theta}{\partial z} \, , \tag{10.16}$$

where we have set

$$K_m = \sigma_w^2\frac{\tau}{2C_m} \, . \tag{10.17}$$

$K_m + K_H$ is a mixed diffusivity and $\tau / C_\theta \, \partial U / \partial z$ is a time-scale ratio.

We may conclude that, from a quite simple set of equations, we get expressions for the heat fluxes (and general circulation is always a matter of fluxes) which reflect quite well the physics of convection phenomena.

Let us define the surface-layer time scale τ as

$$\tau = kz/\sigma_w \ , \tag{10.18}$$

and the Monin-Obukhov length L as

$$L = \left(k \, \frac{g}{\theta} \, \frac{\theta_*}{u_*^2}\right)^{-1} . \tag{10.19}$$

In (10.2) and (10.3) we have shown that in the surface layer $\overline{w'\theta'}$ and $\overline{u'w'}$, and, therefore, also u_* and θ_* can be considered as constant. With (10.18) and (10.19), we then can obtain from (10.1) to (10.8) the dimensionless wind shear ($\phi_m = kz/u_* \, \partial U/\partial z$), temperature gradient ($\phi_h = kz/\theta_* \, \partial \theta / \partial z$), vertical velocity variance (σ_w / u_*), and temperature variance (σ_θ / θ_*) as functions of the independent variable $\zeta = z/L$.

Eliminating $\overline{\theta' u'}$ from (10.1) and (10.3), we obtain

$$\frac{\zeta}{C_\theta}(\phi_m + \phi_H) - \left(\frac{\sigma_w}{u_*}\right)^3 \phi_m + 2C_m \left(\frac{\sigma_w}{u_*}\right)^2 = 0 \, . \tag{10.20}$$

Equations (10.2) and (10.7) can be combined to give

$$\frac{\zeta}{C_{\epsilon\theta}} \phi_H - \left(\frac{\sigma_w}{u_*}\right)^3 \phi_H + C_\theta \left(\frac{\sigma_w}{u_*}\right)^2 = 0 \, . \tag{10.21}$$

The three equations for the variances (10.5) to (10.7) can be reduced to

$$2\left(\frac{\sigma_w}{u_*}\right)^3 C_m + \left(\frac{2C_\epsilon - 2C_m}{3C_\epsilon}\right) \phi_m + \left(\frac{4C_\epsilon + 2C_m}{3C_\epsilon}\right) \zeta = 0 \, . \tag{10.22}$$

Equations (10.7) and (10.8) give

$$\left(\frac{\sigma_\theta}{\theta_*}\right)^2 = \frac{1}{C_{\epsilon\theta}} \left(\frac{u_*}{\partial_w}\right) \phi_H \, . \tag{10.23}$$

We thus have four equations (10.20) to (10.23) with four unknowns (ϕ_m, ϕ_H, σ_w/u_* and σ/θ_*). The unknowns can be determined as a function of ζ when the constants, C_θ, c_m, $c_{\epsilon\theta}$, c_ϵ are given.

Under neutral conditions ($\zeta = 0$, $\phi_m = \phi_H = 1$), (10.20) to (10.23) reduce to

$$2c_m - c_\theta = \frac{\sigma_w}{u_*}, \; C_\epsilon = \frac{c_m}{24c_m^4 + 1}, \; c_{\epsilon\theta} = \left(\frac{\sigma_w}{u_*}\right)^{-1} \left(\frac{\sigma_\theta}{\theta_*}\right)^{-2} . \tag{10.24}$$

From observations we find that under neutral conditions $\sigma_w/u_* = 1.2$, and $\sigma_\theta/\theta_* = 2.5$. As a result we find

$$c_m = 0.6, \; c_\theta = 1.2, \; c_\epsilon = 0.146, \; c_{\epsilon\theta} = 0.133 . \tag{10.25}$$

In the free convection limit u_* has to drop out; therefore we have

$$\frac{\sigma_w}{u_*} = c_w\left(-\frac{z}{L}\right)^{1/3}, \; \phi_H = c_H\left(-\frac{z}{L}\right)^{-1/3}, \; \frac{\sigma_\theta}{\theta_*} = c_T\left(-\frac{z}{L}\right)^{-1/3} . \tag{10.26}$$

For ϕ_m, no relation can be obtained in this limit. In the equations for the variances, the shear has to be negligible and therefore ϕ_m can be neglected in (10.22). As a result we get from (10.21), (10.22), and (10.23)

$$\begin{aligned} c_w &= \left(\frac{2c_\epsilon + c_m}{3c_\epsilon c_m}\right)^{1/3} = 1.5028 , \\ c_H &= \frac{+c_\theta c_w^2}{\left(\frac{1}{c_{\epsilon\theta}} + c_w^3\right)} = 0.249 , \\ c_T &= \left(\frac{c_H}{c_{\epsilon\theta} c_w}\right)^{1/2} = 1.11 . \end{aligned} \tag{10.27}$$

These values are in reasonable agreement with observational data.

Chapter 11

PART II:
FLUX MAINTENANCE IN TWO-DIMENSIONAL TURBULENCE

1. Introduction

The study of two-dimensional turbulence of an incompressible fluid is motivated in part by problems that arise in the analysis of the statistical dynamics of nondivergent, barotropic models of atmospheric motion. If the fluid involved is inviscid, such motion conserves both its total kinetic energy and its total enstrophy (the enstrophy is defined as one-half of the vorticity variance). However, enstrophy is not a conservative quantity if the viscosity of the fluid is not identically equal to zero (Batchelor, 1969). This problem arises because random, two-dimensional flow fields tend to intensify vorticity gradients, thus forcing enstrophy toward smaller scales. If such an "enstrophy cascade" is permitted to proceed for a long enough time, the smallest scale containing significant enstrophy will become small enough to be affected by viscosity. The resulting irreversible viscous loss of enstrophy will gradually decrease the total enstrophy of the flow field, no matter how small the viscosity.

Due to the "spectral blocking" tendency of two-dimensional flow (Fjørtoft, 1953), it is difficult — if not impossible — to cascade kinetic energy toward ever smaller scales. In the limit as the viscosity approaches zero, the viscous dissipation rate of kinetic energy vanishes (Batchelor, 1969). For all practical purposes the total kinetic energy of two-dimensional turbulence at very high Reynolds numbers is a conservative quantity.

Barotropic, nondivergent flow of an almost-inviscid fluid can be maintained in a statistically steady state only if there is a source of vorticity fluctuations that balances the enstrophy cascade rate. In studies of the spectral dynamics of two-dimensional turbulence, one often employs random stirring at a given wavenumber. Such stirring, however, also acts as a source of eddy kinetic energy. This implies that energy has to be removed by some other process; a convenient way to do this is by introducing some kind of parameterized surface friction, which makes up for the absence of an energy cascade. In this way, the spectral dynamics of statistically steady states can be studied.

Most of the practical problems of turbulence dynamics, however, arise from the interactions between the mean flow and the velocity fluctuations. For example, what happens when two-dimensional turbulence is exposed to a mean wind shear in the plane of motion? There are reliable indications (Kraichnan, 1967) that one must expect a reverse energy cascade in two-dimensional flow: kinetic energy tends to migrate toward large scales of motion and eventually feeds into the mean flow. The midlatitude westerlies, for example, are maintained by converting eddy kinetic energy into mean-flow kinetic energy. In other words, two-dimensional flows tend to be barotropically stable. Because of this tendency, it is unlikely that a statistically steady general circulation with a non-trivial mean velocity distribution can be maintained without providing some source for eddy kinetic energy and a corresponding sink for mean-flow kinetic energy. If surface friction is a suitable sink, what would be a realistic source of eddy energy and eddy enstrophy if we are not interested in the consequences of artificial, random two-dimensional stirring?

In three-dimensional turbulence, there is no such dilemma. Though three-dimensional turbulence suffers from much greater geometrical intricacy than its two-dimensional counterpart, the maintenance of its kinetic energy can often be approximated by a simple balance between the major source term and the energy dissipation rate. The energy cascade mechanism in three dimensions is quite vigorous; eddy kinetic energy is dissipated quickly in the absence of energy sources (Tennekes and Lumley, 1972). The gross energetics of three-dimensional turbulence is fairly simple, because it is not linked closely to the issue of maintenance of a general circulation. By comparison, the energy cycles of two-dimensional turbulence are rather complicated.

Let us look at this problem posed here from a different perspective. Imagine that we want to study problems in climate dynamics by computing the general circulation of nondivergent, barotropic flow of an inviscid fluid. Would the results make any sense? Clearly not: a barotropic model merely advects vorticity around; it cannot begin to represent the energetics of the atmosphere's general circulation because it does not include dynamical interactions between the temperature field and the flow field. If isotherms and isobars are parallel everywhere, there is no temperature advection and no vorticity amplification in developing cyclones. The flow in a nondivergent, barotropic model cannot be baroclinically unstable and cannot feed on the mean meridional temperature gradient. We conclude that two-dimensional turbulence will have trivial energy cycles unless we provide a mechanism by which the flow field interacts in a sensible way with the temperature field.

If a nondivergent, one-level model of atmospheric motion is to have nontrivial energy cycles, it must somehow imitate baroclinic effects. It must allow vorticity amplification in developing disturbances and it must lead to conversion of eddy potential energy into eddy kinetic energy. In addition, it should be able to maintain midlatitude westerlies by a reverse energy cascade and it should relate all of the above features to the poleward eddy flux of heat and the mean temperature contrast between the equator and the poles.

In this chapter, we will attempt to show that nondivergent, two-dimensional flow of an almost incompressible, nearly inviscid gas with buoyant forcing is capable of exhibiting behavior that resembles that of the general circulation in the earth's atmosphere. Obviously, two-dimensional turbulence in which baroclinic forcing is replaced by buoyant forcing cannot be expected to represent accurately the subtle interactions between vertical and horizontal fields of motion that make atmospheric dynamics so interesting. Nevertheless, it turns out that the similarities are surprisingly close in many respects. We will, therefore, employ the terminology of climate dynamics when we discuss the statistical characteristics of the flow and temperature fields. Also, we will not hesitate to borrow adjectives from the arsenal of dynamic meteorology when the occasion warrants such use. In particular, we will use the adjective "baroclinic" frequently without quotation marks, though "buoyant" would be more appropriate from a formal point of view.

This is an exploratory chapter. In order to keep the analysis relatively simple, we shall employ a Cartesian beta plane and assume that the flow is of infinite extent in the zonal direction, but confined between rigid walls at two latitudes. The only source of energy is the temperature difference between the two boundaries; it is maintained by diabatic heating and cooling. The mean heating rate and the mean temperature field are taken to be independent of time and of longitude; whenever averages need to be taken, we shall take them both over time and over latitude "circles". Surface topography, continents, and oceans are absent; surface friction will be represented by linear Rayleigh terms. We study the zonally and temporally averaged circulation generated by the model equations; storm tracks and climate changes are beyond the scope of the lecture (though not beyond the scope of the equations).

2. Governing Equations

We shall investigate the dynamical and statistical characteristics of the following system of equations:

$$\frac{\partial u}{\partial x} + \frac{\partial \mathrm{v}}{\partial y} = 0 \,, \tag{11.1}$$

$$\frac{\partial u}{\partial t} + u\frac{\partial u}{\partial x} + \mathrm{v}\frac{\partial u}{\partial y} = -\frac{1}{\rho}\frac{\partial p}{\partial x} + f\mathrm{v} - \frac{K_s}{h^2}u \,, \tag{11.2}$$

$$\frac{\partial \mathrm{v}}{\partial t} + u\frac{\partial \mathrm{v}}{\partial x} + \mathrm{v}\frac{\partial \mathrm{v}}{\partial y} = -\frac{1}{\rho}\frac{\partial p}{\partial y} - fu - \frac{K_s}{h^2}\mathrm{v} + \frac{g^*}{T}\theta \,, \tag{11.3}$$

$$\frac{\partial \theta}{\partial t} + u\frac{\partial \theta}{\partial x} + \mathrm{v}\frac{\partial \theta}{\partial y} = R \,. \tag{11.4}$$

In the Boussinesq approximation employed here, the first law of thermodynamics is replaced by the temperature equation (11.4); for the time being we will take the thermal diffusivity to be negligible. The source term in R in (11.4) represents diabatic heating and cooling; it will have to be positive near the lower wall of the zonal channel and negative near the upper wall. The equations of motion include linear surface-friction terms; K_s is the surface exchange coefficient and h represents the scale height of the atmosphere. The acceleration of gravity, g^*, is directed toward the equator; this makes sure that warm air is forced toward the pole, while cool air "sinks" toward the equator. The magnitude of g^* will be scaled shortly in such a way that the amplification rate of unstable buoyant waves corresponds to that of unstable baroclinic waves in the earth's atmosphere. The fluid is nearly inviscid; we ignore viscous diffusion terms in the equations of motion, but we shall make provisions for the viscous dissipation of enstrophy caused by the presumed existence of an enstrophy cascade.

Terms similar to the buoyancy force in (11.3) occur in primitive equations written for sloping coordinate systems. The analogy with Fleagle's (1955) equations in stream-surface coordinates is particularly close; the effective acceleration of gravity (g^*) corresponds to the true acceleration (g), multiplied by the slope of stream surfaces. That slope is typically about one in one thousand; this suggests that g^* should be about 10^{-2} m $\mathrm{s}^{-1}\mathrm{ec}^2$. This estimate will be verified in the next chapter.

The temperature equation, (11.4), plays a role which resembles that of thickness advection in multilevel models of atmospheric motion. The height of an air column depends upon its average temperature; as the air is advected around, this leads to the kind of acceleration terms which the buoyancy term in (11.3) attempts to represent.

The analogy should not be carried too far, however. Energy conversions in the atmosphere depend upon the difference in slope between isentropic surfaces and stream surfaces (Holton, 1972); that difference depends upon horizontal scale and vertical hydrostatic stability. By comparison, g^* must be taken constant; it does not take into account that isentropic surfaces may have different slopes. Indeed, the buoyancy term in (11.3) represents an atmosphere in which all stream surfaces are horizontal, while isentropic surfaces always have the same slope. In addition, the buoyancy term occurs only in the equation for the meridional velocity component; zonal temperature differences do not create accelerations in the model equations. The shortcomings of the model should be kept in mind.

Let us investigate the vorticity dynamics associated with the equations given above. The relative vorticity ζ is defined by

$$\zeta = \frac{\partial \mathrm{v}}{\partial x} - \frac{\partial u}{\partial y} \, . \tag{11.5}$$

The evolution of ζ is governed by the following equation (we use $f = f_0 + \beta y$):

$$\frac{\partial \zeta}{\partial t} + u\frac{\partial \zeta}{\partial x} + \mathrm{v}\frac{\partial \zeta}{\partial y} = -\beta \mathrm{v} + \frac{g^*}{T}\frac{\partial \theta}{\partial x} - \frac{K_s}{h^2}\zeta \, . \tag{11.6}$$

Besides a surface friction term and the beta effect (advection of planetary vorticity), this equation incorporates a "baroclinic" source term. It is easy to see that this term would be absent if we had decided to make the buoyant forcing terms in (11.2) and (11.3) proportional to $\partial\theta/\partial x$ and $\partial\theta/\partial y$, respectively. The somewhat awkward buoyancy term in (11.3) is necessary if we want to incorporate vorticity amplification in our model. We note that the presence of $\partial\theta/\partial x$ in (11.6) corresponds to a similar term in the vorticity equation used by Fleagle (1955). Figure 11.1 shows how the zonal temperature gradient in an unstable baroclinic wave is related to the intensification of the disturbance.

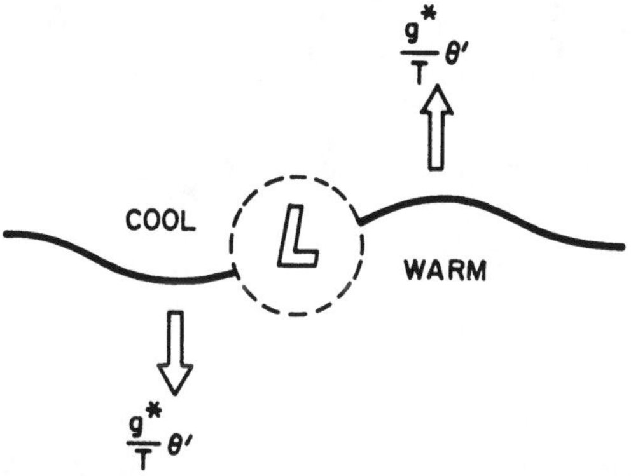

Figure 11.1: As warm air begins to move poleward and cool air moves toward the equator, the zonal temperature gradient $\partial\theta/\partial x$ becomes positive. This intensifies the vorticity in the low-pressure center, because the buoyancy forces exert a torque.

It is convenient to introduce a wind field which corresponds to a balance among the three principal forces involved. We define the "thermal wind" $\boldsymbol{v}_T = (u_T\ ,\ v_T)$ by

$$0 = -\frac{1}{\rho}\frac{\partial p}{\partial x} + f v_T\ , \tag{11.7}$$

$$0 = -\frac{1}{\rho}\frac{\partial p}{\partial y} = f u_T + \frac{g^*}{T}\theta\ . \tag{11.8}$$

Strictly speaking, $\boldsymbol{v}_T$ is the geostrophic-buoyant balance wind. However, as we shall see shortly, $\boldsymbol{v}_T$ in a certain sense imitates the behavior of the upper-level wind field in the earth's atmosphere; the choice of adjective thus is not altogether unjustified.

From (11.7) and (11.8) we obtain by differentiation:

$$\frac{g^*}{T}\frac{\partial\theta}{\partial y} = \beta v_T + f\nabla\cdot\boldsymbol{v}_T\ , \tag{11.9}$$

$$\frac{g^*}{T}\frac{\partial\theta}{\partial y} = \beta u_T - f\zeta_T + \frac{1}{\rho}\nabla^2 p\ . \tag{11.10}$$

Here ζ_T is the thermal vorticity, defined by

$$\zeta_T = \frac{\partial v_T}{\partial x} - \frac{\partial u_T}{\partial y}\ . \tag{11.11}$$

Substituting (11.9) into (11.6), we obtain

$$\frac{d\zeta}{dt} = -\beta\,(\mathrm{v} - \mathrm{v}_T) + f\nabla \cdot \boldsymbol{v}_T - \frac{K_s}{h^2}\zeta \ . \tag{11.12}$$

If we now define the unbalanced wind, $\boldsymbol{v}_A$, by

$$\boldsymbol{v} = \boldsymbol{v}_T + \boldsymbol{v}_A \ , \tag{11.13}$$

then, because $\boldsymbol{v}$ is nondivergent,

$$\nabla \cdot \boldsymbol{v}_A = -\nabla \cdot \boldsymbol{v}_T \ . \tag{11.14}$$

The vorticity equation thus can be written also in the following form:

$$\frac{d\zeta}{dt} = -\beta \mathrm{v}_A - f\nabla \cdot \boldsymbol{v}_A - \frac{K_s}{h^2}\zeta \ . \tag{11.15}$$

This states that the vorticity of a fluid parcel is changed by three processes. The first of these is the advection of planetary vorticity on the beta plane by the meridional component of the unbalanced wind, the second is the amplification of planetary vorticity caused by the convergence of the unbalanced wind, and the third is vorticity decay by surface friction. By contrast, (11.12) states that the vorticity is intensified by the divergence of the thermal wind. The unbalanced wind thus corresponds to low-level winds in a growing baroclinic disturbance, while the thermal wind imitates the divergence of the upper-level winds in a developing storm. Since the thermal wind is defined on the basis of a force balance, it is fair to state that the intensification of a storm is generated by the divergence of the thermal wind.

Figure 11.2 illustrates the geometry of the force vector diagram for a developing wave; Figure 11.3 shows how this corresponds to divergence of the thermal wind and convergence of the unbalanced wind.

Figure 11.2: The balance of forces east and west of a developing cyclone in the northern hemisphere. The pressure gradient P is directed toward the low-pressure center, the buoyancy force B is directed toward the pole in the warm sector and toward the equator in the cold sector, and the Coriolis force C_T of the "thermal" wind balances the other two.

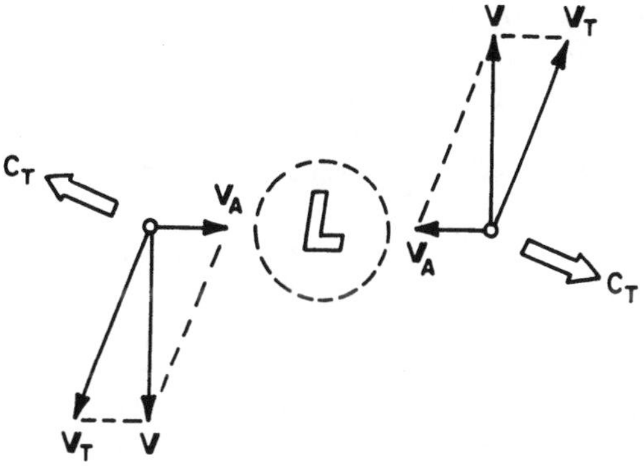

Figure 11.3: Divergence of the thermal wind and convergence of the unbalanced wind around a developing storm, seen from a frame of reference moving with the mean flow. For simplicity, it is assumed that the meridional component of $\boldsymbol{v}_A$ is zero. The direction of $\boldsymbol{v}_T$ is determined by the force balance sketched in Figure 11.2.

3. Stability Analysis

It is useful to verify the qualitative interpretation developed above by performing a linear stability analysis. Clearly, there is no need to study the isentropic stability characteristics of the model, since they are identical to those of a nondivergent barotropic model. For our baroclinic stability analysis, we assume that there is a uniform mean zonal flow $\overline{u}$ and a uniform mean temperature gradient $\partial\overline{\theta}/\partial y$. We use a perturbation stream function ψ' and a temperature perturbation θ'; they are defined by

$$\psi' = a_\psi \exp\,\{ik(x - ct)\}\ , \tag{11.16}$$

$$\theta' = a_\theta \exp\,\{ik(x - ct) + i\phi\}\ . \tag{11.17}$$

Here, ϕ is the phase difference between the temperature wave and the streamfunction wave, k is the zonal wavenumber, and c is the complex phase speed.

Substituting these expressions into the linearized vorticity and temperature equations and neglecting surface friction terms, we obtain the following results. The wavenumber of neutral stability is given by

$$-\frac{g^*}{T}\frac{\partial\overline{\theta}}{\partial y} = \left(\frac{\beta}{2k_0}\right)^2\ , \tag{11.18}$$

all waves with $k \geq k_0$ propagate at a speed

$$c_r = \overline{u} - \frac{\beta}{2k^2} , \tag{11.19}$$

and the amplification rate of those waves is given by

$$\alpha^2 = -\frac{g^*}{T}\frac{\partial \overline{\theta}}{\partial y} - \left(\frac{\beta}{2k}\right)^2 . \tag{11.20}$$

Here, α is the imaginary part of ck.

The air gets colder toward the pole, and the mean meridional temperature gradient destabilizes the flow. Equation (11.18) states that the wave number of neutral stability is reached when the imaginary Brunt-Väisälä frequency of the flow equals the "Rossby frequency" $\beta/2k$. Equation (11.19) indicates that unstable waves move westward relative to the mean zonal flow, and that their relative propagation speed is one-half of that of Rossby waves on a barotropic beta plane. Equation (11.20) shows that the amplification rate of baroclinic waves increases as the wavenumber increases, and that the amplification rate of very short waves is approximately equal to

$$\alpha_\infty = \left(-\frac{g^*}{T}\frac{\partial \overline{\theta}}{\partial y}\right)^{1/2} . \tag{11.21}$$

Unfortunately, the system has no high-wavenumber cut-off corresponding to the Rossby radius of deformation in atmospheric motion. In view of our earlier discussion, this shortcoming is not surprising. It is hard to speculate just how serious this defect is, but we may take some comfort from the fact that the general circulation model of the Geophysical Fluid Dynamics Laboratory (GFDL) of the Environmental Research Laboratories, National Oceanic and Atmospheric Administration, has similar problems (Gall, 1976). Also, the linear stability characteristics of the system do not necessarily correspond on a one-to-one basis with its nonlinear stability characteristics; randomizing tendencies at small wavelengths may very well create an effective nonlinear cut-off at high wavenumbers. This would have to be tested by computer simulations.

In principle it would be possible, at least in a spectral solution method of the model equations, to postulate an appropriate spectral cut-off for g^*. The point is that g^* controls the rate of conversion of potential energy into kinetic energy; if g^* were dependent upon wavenumber it would imitate the stabilization of short waves by the vertical stability of the earth's atmosphere. At the time of writing, we have not yet discovered a simple and internally consistent solution to this problem. We note that an explicit dependence of g^* upon wavenumber would do harm to the formulation of the primitive equations, because they are nonlinear.

Equations (11.18) and (11.21) permit a rough estimate of the value of g^* needed to get values of α and k_0 that correspond to those in the earth's atmosphere. We put $\partial\overline{\theta}/\partial y = -$ 3K/1000 km, T = 300K, and $\alpha_\infty = 10^{-5}$ sec $^{-1}$ (roughly 1 day^{-1}). We thus need to use $g^* = 10^{-2}$ m s^{-1}ec^2. The corresponding critical wavelength is about 12,500 km if $\beta \simeq 10^{-11}$ (m sec)$^{-1}$. Roughly speaking, waves above zonal wavenumber 2 (midlatitude) are unstable. The effective acceleration of gravity g^* is about one-tenth of one percent of g; this corresponds to horizontal motion across isentropic surfaces whose slope is about one in one thousand. These estimates appear to correspond fairly well with the numbers used in dynamical meteorology (see Table 11.1 for a complete list).

We can verify the validity of the numbers given above by scaling the source term in the vorticity equation. The divergence of the thermal wind should be of order 10^{-6} sec^{-1} in order to be realistic. Therefore, $(g^*/T)\partial\theta/\partial x$ should be of order 10^{-10} sec^{-1} in unstable baroclinic waves. If we estimate $\partial\theta/\partial x \simeq$ 3K/1000 km, and take T = 300 K as before, we also obtain $g^* \simeq 10^{-2}$ m s^{-1}ec^2. The corresponding buoyant acceleration in the equations of motion is about $3 \cdot 10^{-5}$ m s^{-1}ec^2 if the temperature differences are small (1K) and about $3 \cdot 10^{-4}$ m s^{-1}ec^2 if they are very large (10K). The buoyancy forces thus tend to be fairly small (but not *very* small!) compared to the Coriolis and pressure gradient forces. The flow is quasi-geostrophic all the time, and the buoyancy forces are sufficiently strong to overcome the weak retarding effects of surface friction (see also next chapter). Parenthetically, we note that the introduction of geostrophic winds in the primitive equations does not appear to be particularly useful, because the flow is strictly nondivergent.

4. Mean Flow and Eddy Kinetic Energy

We now turn to an analysis of the general circulation of our model. The climate of the model is defined by the meridional profile of the mean zonal wind $\overline{u}(y)$, the mean temperature distribution $\overline{\theta}(y)$, and the various variances and covariances of the fluctuations associated with the eddy motion. All averages used here are temporal averages of zonal averages; therefore the $\partial/\partial t$ and $\partial/\partial x$ of all averaged variables are zero.

The flow is confined between rigid boundaries at $y = \pm 1/2L$. Since the velocity field is nondivergent, $\bar{v}$ has to be zero everywhere. Maintenance of the mean zonal wind requires that the convergence of the poleward eddy flux of zonal momentum be equal to the surface friction force:

$$\frac{\partial \bar{u}}{\partial t} = 0 = -\frac{\partial}{\partial y}\,\overline{u'v'} - \frac{K_s}{h^2}\,\bar{u}\ . \tag{11.22}$$

Here, $u' = u - \bar{u}$, $v' = v$ because $\bar{v} = 0$, and averages are indicated by overbars. The mean momentum balance would be trivial in the absence of surface friction because the flow is nondivergent and does not allow circulations in the meridional-vertical plane. Since the convergence of the eddy momentum flux is equal to the vorticity flux $\overline{v'\zeta'}$, (11.22) can also be written as

$$0 = \overline{v'\zeta'} - \frac{K_s}{h^2}\bar{u}\ . \tag{11.23}$$

This equation suggests that midlatitude cyclones ($\zeta' > 0$ in the northern hemisphere) migrate toward the pole; because the mean zonal wind at middle latitudes blows from the west, storm tracks must be oriented toward the northeast.

The value of K_s can be estimated by taking the spin-down time constant of the mean zonal flow to be of order 10^6 sec (ten days) in the absence of an eddy vorticity flux. This means that K_s/h^2 has to be about 10^{-6} sec^{-1}. If we take the scale height h to be 10 km, we conclude that K_s must be about 10^2 m^2/sec. Alternatively, if the kinematic surface stress is of order 0.1 (m s^{-1}ec)^2, then the frictional deceleration is of order 10^{-5} m s^{-1}ec^2 for a flow whose depth is 10 km. If $\bar{u}$ is of order 10 m s^{-1}ec, this again implies that $k_s/h^2 \simeq 10^{-6}$ sec. These estimates also determine the order of magnitude of $\overline{v'\zeta'}$. We find

$$\overline{v'\zeta'} \simeq 10^{-5}\text{m/sec}^2\ . \tag{11.24}$$

Since $\sigma_v \simeq 6$ m s^{-1}ec and $\sigma_\zeta \simeq 10^{-5}$ sec^{-1} (Blackmon et al., 1977), the correlation between v' and ζ' is fairly poor (throughout this chapter the symbol σ is used to represent the standard deviation of fluctuations). If the meridional scale of variation of $\overline{u'v'}$ is about 1000 km, the order of magnitude of $\overline{u'v'}$ must be

$$\overline{u'v'} \simeq 10 \text{ m}^2/\text{sec}^2 \ . \tag{11.25}$$

Since $\sigma_u \simeq 10$ m s^{-1}ec and $\sigma_v \simeq 6$ m s^{-1}ec, the correlation between u' and v$'$ also is rather small (Blackmon et al., 1977).

It is useful to determine the equations governing the mean thermal wind and the mean unbalanced wind. We note that $\bar{v}_T = 0$ because $\partial\bar{p}/\partial x = 0$; since $\bar{v} = 0$, we also have $\bar{v}_A = 0$. Since $\bar{u}_T$ is a function of y only, the mean thermal wind is nondivergent: $\nabla \cdot \bar{v}_T = 0$. The zonal component of the mean thermal wind is given by

$$0 = -\frac{1}{\rho}\frac{\partial\bar{p}}{\partial y} - f\bar{u}_T + \frac{g^*}{T}\bar{\theta} \ ; \tag{11.26}$$

the value of $\bar{u}_A$ is determined by

$$f(\bar{u} - \bar{u}_T) = f\bar{u}_A = -\frac{\partial}{\partial y}\sigma_v^2 \ . \tag{11.27}$$

The r.h.s. of (11.27) is less than 10^{-4} m s^{-1}ec^2 if the meridional scale of variation of σ_v^2 is about 1000 km. This means that $\bar{u}_A \simeq 0.1$ m s^{-1}ec, which is negligible compared to $\bar{u}_T$. The mean zonal wind thus is in almost perfect thermal balance. Moreover, since the buoyancy term in (11.26) is one order of magnitude smaller than the other two terms, the mean zonal wind must be in approximate geostrophic balance. The pressure has extrema at latitudes close to those at which $\bar{u}_T$ changes sign; this creates a subtropical high-pressure belt south of the midlatitude westerlies in the northern hemisphere and a low-pressure belt north of the jet stream.

In our model the meridional profile of $\overline{v'\zeta'}$ is exactly the same as that of $\bar{u}$. This condition imposes severe constraints upon the profiles of $\bar{u}$, $\bar{\zeta}$, $\overline{u'v'}$, and $\overline{v'\zeta'}$ across the zonal channel. If the mean wind profile is symmetric about the centerline of the channel ($y = 0$), the various meridional profiles must look like the ones sketched in Figure 11.4.

For future reference, we note that there are two regions (shaded) near the boundaries of the channel where the sign of $\overline{u'v'}$ opposes that of $\partial\bar{u}/\partial y$, while there are also two narrow belts (solid black) where the sign of $\overline{v'\zeta'}$ opposes that of $\partial\bar{\zeta}/\partial y$.

Since $K_s/h^2 \simeq 10^{-6}$ sec^{-1}, the surface friction terms in the primitive equations are of order 10^{-5} m s^{-1}ec^2. This is two orders of magnitude smaller than the leading terms; it would thus be justified to ignore surface friction for the purpose of short-term forecasts. However, surface friction cannot be neglected

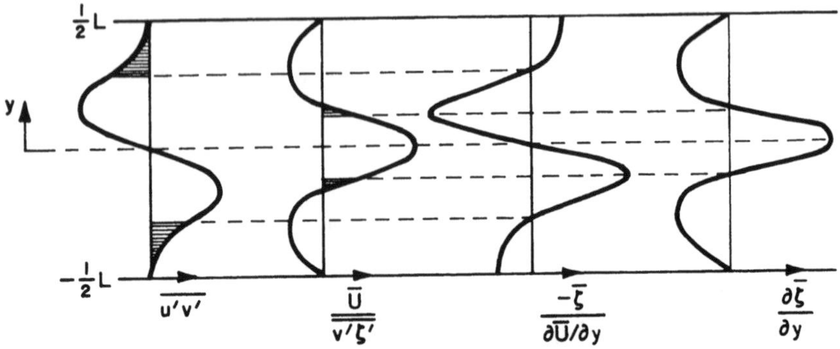

Figure 11.4: Schematic distributions of mean zonal wind, mean relative vorticity, mean vorticity gradient, eddy momentum flux, and eddy vorticity flux associated with a symmetric zonal-wind profile in the channel between $y = -1/2L$ and $y = 1/2L$. Since the boundaries are rigid, all eddy fluxes must vanish there. The relevance of the shaded regions is explained in the text.

in studies of the climate of the model, for the simple reason that (11.22) and (11.23) would become trivial if friction were absent.

The buoyancy term in the equations of motion is of order 10^{-4} m s^{-1}ec^{2} if $\theta/T \simeq 10^{-2}$. Buoyancy effects thus tend to be large compared to those of surface friction, and we must expect that significant baroclinic effects occupy only a relatively small fraction of the entire space-time domain. In other words, storms with large eddy fluxes must be relatively rare and fairly localized, and most of the flow field must be quasi-barotropic. If our model were equipped with mountain ridges and ocean-continent contrast, we expect that it would imitate the large eddy fluxes associated with the northern Atlantic and Pacific storm tracks (Blackmon et al., 1977).

We now are ready for a first look at the energetics of the general circulation. The equation for the maintenance of eddy kinetic energy is obtained by straightforward manipulation; it reads

$$\frac{\partial}{\partial t}\left(\frac{1}{2}\overline{q^2}\right) = 0 = -\overline{u'v'}\,\frac{\partial \bar{u}}{\partial y} + \frac{g^*}{T}\overline{\theta' v'} - \frac{K_s}{h^2}\,\overline{q^2} - \frac{\partial}{\partial y}\left(\frac{1}{2}\overline{q^2 v'} + \overline{p'v'}/\rho\right) . \tag{11.28}$$

Here, $1/2\ \overline{q^2} = 1/2\left(\sigma_u^2 + \sigma_v^2\right)$ is the mean kinetic energy per unit mass. With the aid of the definitions of the thermal and unbalanced winds, (11.28) can also be written as:

$$\frac{\partial}{\partial t}\left(\frac{1}{2}\overline{q^2}\right) = 0 = -\overline{u'v'}\,\frac{\partial \overline{u}}{\partial y} + f\left(\overline{u'v'_A} - \overline{v'u'_A}\right) - \frac{K_s}{h^2}\,\overline{q^2} - \frac{\partial}{\partial y}\left(\overline{\frac{1}{2}\,q^2 v'}\right) . \tag{11.29}$$

The kinetic energy budget (11.28) demonstrates the energy conversion processes that are necessary to maintain the general circulation of the model. Figure 11.4 shows that the sign of $\overline{u'v'}$ is almost everywhere the same as that of $\partial\overline{u}/\partial y$, except near the two zonal walls, where both the flux and the gradient are small. The first term of (11.28) thus is a sink for eddy kinetic energy. This conversion of eddy kinetic energy into the kinetic energy of the mean zonal flow represents one aspect of the reverse energy cascade in two-dimensional flow; it is required to maintain the midlatitude jet against surface friction. The model faithfully imitates one of the principal energy conversions of the atmosphere's general circulation. We note that the conversion rate is largest in the belts where the mean zonal wind changes sign. In the subtropical high-pressure belt, for example, both the eddy momentum flux and the mean shear are large (Fig. 11.4); the conversion rate $\overline{u'v'}\ \partial\overline{u}/\partial y$ thus has a very pronounced peak there. This suggests that the eddy kinetic energy will be fairly small in the subtropics, and that the flux divergence term of (11.28) will have to make up for some of the loss. The flux of eddy kinetic energy is likely to be down the energy gradient.

The only source term in (11.28) is the conversion rate $(g^*/T)\,\overline{\theta'v'}$ of eddy potential energy into eddy kinetic energy. The mean heat flux is from the equator toward the pole, so that $\overline{\theta'v'} > 0$ everywhere, with a broad maximum near the center of the channel if the distribution of the diabatic heating is symmetric with respect to $y = 0$. Equation (11.29) sheds more light on the dynamical processes involved. As is evident from Figs. 11.2 and 11.3, positive values of v' in growing baroclinic eddies tend to be associated with negative values of u'_A, and vice versa. The term $-f\overline{v'u'_A}$ thus tends to be positive in active storms. In the same way, the eastward acceleration of the flow in the warm sector of a developing cyclone must be associated with positive values of v'_A, because $du/dt = f\,v'_A - \left(K_s/h^2\right)u$. The deceleration in the cold sector, on the other hand, is associated with negative values of v'_A. The term $f\overline{u'v'_A}$ thus is also likely to be positive in the development stage of midlatitude storms. We note, however, that it is difficult to speculate on the sign of $f\left(\overline{u'v'_A} - \overline{v'u'_A}\right)$ outside active baroclinic disturbances.

It is not difficult to see that the principal source and sink terms in (11.29) will roughly balance each other. Since v'_A is typically about 10% of v', we must estimate that $\overline{u'v'} \simeq 10\overline{u'v'_A}$. The mean relative vorticity $\overline{\zeta}$, on the other hand, is of order 10^{-5} sec^{-1}, while $f \simeq 10^{-4}$ sec. The term $\overline{u'v'}\ \overline{\zeta}$ balances the term $f\overline{u'v'_A}$, because the two factors of ten involved cancel each other. It appears possible to maintain an approximate balance.

Eventually, all kinetic energy is dissipated by surface friction. If we include the frictional energy losses of the mean zonal flow, an overall kinetic energy balance [which ignores the flux-divergence terms of (11.28), because the fluxes vanish at the boundaries of the channel] may be approximated by

$$\frac{g^*}{T}\,\overline{\theta' v'} \simeq \frac{K_s}{h^2}\left(\overline{u}^2 + \sigma_u^2 + \sigma_v^2\right) . \tag{11.30}$$

If $K_s/h^2 = 10^{-6}$ sec^{-1}, $\overline{u} = 10$ m s^{-1}ec, $\sigma_u = 10$ m s^{-1}ec, $\sigma_v = 6$ m s^{-1}ec, $g^* = 10^{-2}$ m s^{-1}ec^2, and $T = 300$ K, we obtain

$$\overline{\theta' v'} \simeq 10 \text{ m K/sec} . \tag{11.31}$$

This estimate agrees with the data of Blackmon et al. (1977). Assuming that the correlation coefficient between θ' and v' is about 1/2 (a value borrowed from free convection in the atmospheric boundary layer), and that $\sigma_v \simeq 6$ m s^{-1}ec, we find that σ_θ must be about 3 K. This seems altogether reasonable.

5. Heat Flux and Temperature Variance

The poleward eddy flux of heat is determined by the diabatic heating term R in (11.4). Taking the average of (11.4), we find

$$\frac{\partial}{\partial y}\left(\overline{\theta' v'}\right) = \overline{R} . \tag{11.32}$$

Since the heat flux must vanish at the boundaries of the zonal channel, the meridional profiles of $\overline{\theta' v'}$ and $\overline{R}$ look like those sketched in Fig. 11.5. The shape of the heat-flux profile is fairly insensitive to the distribution of $\overline{R}$, because the former is an integral of the latter. The meridional scale of variation of $\overline{\theta' v'}$ near the boundaries presumably is about 1000 km. Since $\overline{\theta' v'} \simeq 10$ m K/sec by virtue of (11.31), the extrema of $\overline{R}$ are of order 10^{-5} K/sec. This corresponds to diabatic heating and cooling rates of about 1 K/day near the boundaries of the channel.

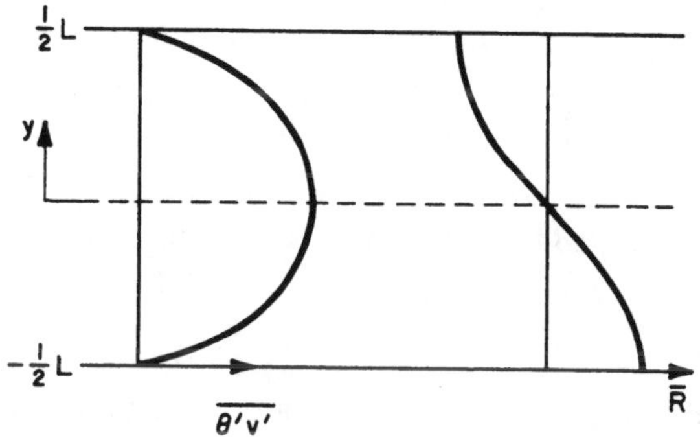

Fig. 11.5: The distribution of the poleward eddy heat flux can be specified at will by selecting an appropriate distribution of the mean diabatic heating rate $\overline{R}$. Energy conservation requires that the integral of $\overline{R}$ across the channel be zero.

The maintenance equation for temperature variance reads

$$\frac{\partial}{\partial t}\left(\frac{1}{2}\sigma_\theta^2\right) = 0 = -\frac{\partial}{\partial y}\left(\frac{1}{2}\overline{\theta'\theta' v'}\right) - \overline{\theta' v'}\,\frac{\partial \overline{\theta}}{\partial y} + \overline{R'\theta'} - \epsilon_\theta\ . \tag{11.33}$$

The first term on the r.h.s. of (11.33) is a flux-divergence term; it can only redistribute temperature variance inside the zonal channel because the flux must vanish at the boundaries. The second term of (11.33) is a source for temperature variance, because $\overline{\theta' v'} > 0$ and $\partial\overline{\theta}/\partial y < 0$ everywhere if the lower wall is heated and the upper wall is cooled. It is clear that we need sink terms, too. One way of providing a sink is to arrange the heating fluctuations R' in such a way that $\overline{R'\theta'} < 0$. This would mean that warmer air is cooled while colder air is heated. However, it is not clear that this would be a significant effect if some radiative model of heating and cooling were used, because the temperature fluctuations are quite small ($\simeq 3K$). This issue deserves further investigation.

The other sink term in (11.33) is ϵ_θ, which stands for the destruction of temperature variance caused by the spectral cascade of temperature fluctuations in an almost nonconducting gas. In two-dimensional turbulence, the cascade mechanism for scalar contaminants is similar to the enstrophy cascade (Kraichnan, 1976) because stochastic flow fields in two dimensions tend to intensify scalar fluctuation gradients in much the same way as vorticity gradients. The cascade rate presumably is determined by the large-scale dynamics of the flow field, not by the exact value of the thermal diffusivity if the latter is sufficiently small. This means that the numerical diffusivity needed for the solution of the primitive equations on a grid with finite mesh size will not significantly harm the dynamics if the spatial resolution is adequate.

Equation (11.31) states that $\overline{\theta' v'}$ must be of order 10 m K/sec. If the mean temperature gradient $\partial\overline{\theta}/\partial y$ is of order 3 K per 1000 km, the source term in (11.33) must be of order $3 \cdot 10^{-5}$ K^2/sec. In view of the analogy with the enstrophy budget (which will be discussed later), it appears reasonable to estimate ϵ_θ as

$$\epsilon_\theta \simeq \frac{1}{3}\sigma_\theta^2\sigma_\zeta \ . \tag{11.34}$$

With $\sigma_\theta \simeq 3$ K and $\sigma_\zeta \simeq 10^{-5}$ sec, the dissipation term must be of order $3 \cdot 10^{-5}$ K^2/sec. This agrees with our estimate for the source term. We conclude that the temperature variance cascade is vigorous enough to maintain a balance in (11.33).

The presence of ϵ_θ represents the likelihood that the two-dimensional eddies in this almost inviscid, almost nonconducting gas will create temperature and vorticity "fronts." While it is easy enough to suppress this tendency in numerical solutions of the system of equations, the conceptual implications will have to be explored in more detail. We will do so when we discuss the enstrophy budget.

The key feature of the temperature variance equation is that the potential energy available in the mean meridional temperature gradient is converted into eddy potential energy. In this model, eddy potential energy is not proportional to the temperature variance, but to the standard deviation of temperature. The conversion rate from eddy potential energy to eddy kinetic energy is $(g^*/T)\overline{\theta' v'}$; this implies that the eddy potential energy is proportional to $(g^*/T)\,\theta' y$ (apart from an additive constant related to the width L of the channel.) There is no useful analogy with the concept of available potential energy in atmospheric dynamics. It is here in particular that we pay a high price for the simplicity of our Boussinesq model.

6. Flux Maintenance Dynamics

So far, our discussion has skirted around the closure problem of turbulence dynamics. It is easy enough to write budget equations for $\overline{u' v'}$ and $\overline{\theta' v'}$. If we appeal to some gradient-transfer hypothesis (turbulent "mixing" or "diffusion"), it seems reasonable to expect that $\overline{\theta' v'}$ must be positive when $\partial\overline{\theta}/\partial y$ is negative. However, the eddy momentum flux, $\overline{u'v'}$, is not directed *down* the gradient $\partial\overline{u}/\partial y$, but *up* the gradient almost everywhere. How is a countergradient momentum flux maintained? It seems inevitable to seek wisdom from the maintenance equations for $\overline{\theta' v'}$ and $\overline{u'v'}$. However, those equations will introduce yet other covariances, and it is not at all clear that sensible closure assumptions can be made at that level of sophistication. Nevertheless, we are convinced that climate models will have to deal with the physics of flux maintenance, and that it will be necessary to face squarely the parameterization problems that arise in this context.

The maintenance equation for $\overline{\theta' v'}$ reads

$$\begin{aligned}\frac{\partial}{\partial t}\left(\overline{\theta' v'}\right) = 0 = &-\sigma_v^2\frac{\partial\overline{\theta}}{\partial y} + \frac{g^*}{T}\sigma_\theta^2 - \frac{K_s}{h}\overline{\theta' v'} \\ &-\frac{\partial}{\partial y}\left(\overline{\theta' v' v'}\right) - \frac{\overline{\theta'}}{\rho}\frac{\overline{\partial p'}}{\partial y} - f\overline{\theta' u'}\ .\end{aligned} \tag{11.35}$$

Fortunately, the structure of this equation is more transparent if the thermal balance equations, (11.7) and (11.8), and the definition (11.13) of the unbalanced wind are used. We obtain:

$$\begin{aligned}\frac{\partial}{\partial t}\left(\overline{\theta' v'}\right) = 0 = &-\sigma_v^2\frac{\partial\overline{\theta}}{\partial y} - f\overline{\theta' u'_A} - \frac{K_s}{h^2}\overline{\theta' v'} \\ &-\frac{\partial}{\partial y}\left(\overline{\theta' v' v'}\right)\ .\end{aligned} \tag{11.36}$$

Two of the terms in (11.35) and (11.36) can be estimated directly with the numbers given in Table 1. We obtain

$$-\sigma_v^2\frac{\partial\overline{\theta}}{\partial y} \simeq 10^{-4}\ \text{m K/sec}^2\ , \tag{11.37}$$

$$\frac{K_s}{h^2}\overline{\theta' v'} = 10^{-5}\ \text{m K/sec}^2\ . \tag{11.38}$$

The surface friction term thus can be ignored compared to the leading source term. The latter represents turbulent "mixing" of the mean temperature field: the mere presence of σ_v^2 tends to create a poleward heat flux if $\partial\overline{\theta}/\partial y$ is negative.

The flux divergence term of (11.35) and (11.36) represents redistribution of $\overline{\theta' v'}$ inside the channel (the flux vanishes at the boundaries). Since surface friction is an insufficient sink, the Coriolis term in (11.36) must be the major sink term. This poses a difficult problem. In growing baroclinic disturbances warm air ($\theta' > 0$) tends to be associated with negative values of u'_A, while cool air ($\theta' < 0$) tends to be correlated with positive values of u'_A (see Figs. 11.2 and 11.3). In developing storms the term $-f\overline{\theta' u'_A}$ thus is positive, adding to the effects of the mixing term and creating very large local values of $\overline{\theta' v'}$ because there is no appropriate sink. This implies that a significant fraction of $\overline{\theta' v'}$ must be associated with storm tracks, and that the remainder of the field transports relatively little heat. This qualitative conclusion agrees with the data of Blackmon et al. (1977).

On the average, however, the Coriolis term in (11.36) must be negative. This suggests that it probably switches sign as a storm enters its decay stage, and that the flux-divergence term evens out the differences. We do not expect that $-f\overline{\theta' u'_A}$ will have significant values outside storms, because there is reason to believe that the remainder of the field will be quasi-barotropic. One — very tentative — approach to this problem would be to speculate that storms in the northern hemisphere of the model originate south of the jet axis and dissipate north of it, and that the sign of $-\overline{\theta' u'_A}$ depends upon that of the mean shear $\partial \overline{u}/\partial y$. As a cyclone reaches maturity, warm air travels around the storm center and arrives to the west of it. The resulting sign reversal of u'_A (see Figs. 11.2 and 11.3) reduces the poleward heat flux and dissipates the storm. The problem is complicated because it relates to the exact nature of the imbalance among the buoyancy term, the Coriolis term, and the pressure-gradient term in (11.35). Numerical experiments with the general circulation of the model may give the data necessary to decide what is happening; they may also give further hints for suitable closure approximations.

We note parenthetically that $\overline{\theta' u'}$ probably is positive south of the jet axis (because $\overline{u' v'} > 0$ and $\overline{\theta' v'} > 0$ there), while it will be negative north of the jet axis ($\overline{\theta' v'} > 0$, but $\overline{u' v'} < 0$). The Coriolis term in (11.35) thus is a sink south of the jet and a source to the north. Again, this information is of little use because the pressure-gradient term and the buoyancy term in (11.35) determine what kind of imbalance is achieved.

The order of magnitude of $f\overline{\theta' u'_A}$ can be estimated as follows. Since u'_A is typically about 10% of u', we estimate

$$\overline{f\theta' u'_A} \simeq \frac{1}{10}\overline{f\theta' u'} \simeq 10^{-4} \text{ m K/sec}^2 \ . \tag{11.39}$$

Here we have used $\sigma_u \simeq 10$ m s^{-1}, $\sigma_\theta \simeq 3$ K, and we have assumed that the correlation coefficient is about 1/3. Apart from the problems with the sign of this term, there is no reason to suspect that it is not capable of balancing the mixing term (11.37).

If the maintenance of $\overline{\theta' \text{v}'}$ leads to difficult questions, what about the maintenance of $\overline{u'\text{v}'}$? The budget equation for $\overline{u'\text{v}'}$ reads

$$\begin{aligned}\frac{\partial}{\partial t}\overline{u'\text{v}'} = 0 = &-\sigma_\text{v}^2\frac{\partial \overline{u}}{\partial y} + \frac{g^*}{T}\overline{\theta' u'} - 2\frac{K_s}{h^2}\overline{u'\text{v}'} \\ &+ f\left(\sigma_\text{v}^2 - \sigma_u^2\right) - \frac{1}{\rho}\left(\overline{\text{v}'\frac{\partial p'}{\partial x} + u'\frac{\partial p'}{\partial y}}\right) \\ &- \frac{\partial}{\partial y}\left(\overline{u'\text{v}'\text{v}'}\right) \ .\end{aligned} \tag{11.40}$$

If we simplify this by introducing the unbalanced wind, we obtain

$$\begin{aligned}\frac{\partial}{\partial t}\overline{u'\text{v}'} = &-\sigma_\text{v}^s\frac{\partial \overline{u}}{\partial y} + f\left(\overline{\text{v}'_A\text{v}'} - \overline{u'_A u'}\right) \\ &- 2\frac{K_s}{h^2}\overline{u'\text{v}'} - \frac{\partial}{\partial y}\left(\overline{u'\text{v}'\text{v}'}\right) \ .\end{aligned} \tag{11.41}$$

Three of the terms in (11.40) can be estimated on basis of the data in Table 1. They are:

$$-\sigma_\text{v}^2\frac{\partial \overline{u}}{\partial y} \simeq 4 \cdot 10^{-4} \text{ m}^2/\text{sec}^3 \ , \tag{11.42}$$

$$\frac{g^*}{T}\overline{\theta' u'} \simeq 3 \cdot 10^{-4} \text{ m}^2/\text{sec}^3 \ , \tag{11.43}$$

$$\frac{K_s}{h^2}\overline{u'\text{v}'} \simeq 10^{-5} \text{ m}^2/\text{sec}^3 \ . \tag{11.44}$$

As in the heat-flux equation, the surface friction term of (11.40) is insignificant.

Since the flow field is in approximate geostrophic balance, the Coriolis terms in (11.40) approximately balance the pressure-gradient terms. If the net difference were small enough, and if the flux-divergence term could be ignored, it would be possible to use the following (very crude!) approximation to (11.40):

$$\sigma_\text{v}^2\frac{\partial \overline{u}}{\partial y} \simeq \frac{g^*}{T}\overline{\theta' u'} \ . \tag{11.45}$$

This is an appealing simplification. We have seen above that $\overline{\theta' u'}$ is likely to be positive south of the jet axis ($\partial\bar{u}/\partial y > 0$) and that it must be negative north of it ($\partial\bar{u}/\partial y < 0$). The sign reversal of $\overline{\theta' u'}$ thus should coincide with that of $\partial\bar{u}/\partial y$, which is exactly what (11.45) indicates. Moreover, this picture agrees with the slopes of the trough and ridge axes on either side of the jet (Fig. 11.6).

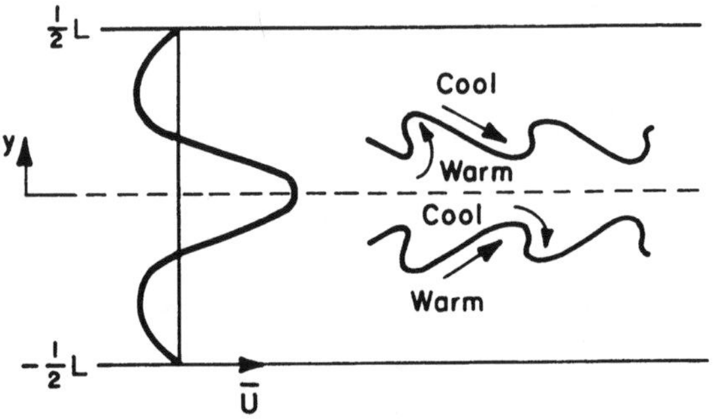

Fig. 11.6: Troughs and ridges on both sides of the jet stream. The slopes of the trough and ridge axes are such that $\overline{u' v'} > 0$ where $\partial\bar{u}/\partial y > 0$, and vice versa. South of the jet, warm air moves eastward (relative to the mean flow), while cold air moves westward. North of the jet axis, warm air slows down, while cooler air accelerates toward the east. The sign of $\overline{\theta' u'}$ thus is linked to those of $\overline{u' v'}$ and $\partial\bar{u}/\partial y$.

In terms of unbalanced wind components, the maintenance of $\overline{u' v'}$ appears to be fairly complicated. The principal source term is the difference between the two Coriolis terms in (11.41); this must make up for the losses caused by the mixing term $\sigma_v^2 \partial\bar{u}/\partial y$ (note that the latter tends to destroy the countergradient momentum flux). Since it is hard to imagine how the delicate lack of balance between $\overline{v'_A v'}$ and $\overline{u'_A u'}$ depends upon the other properties of the flow field, it will be hard to make further progress before the results of general circulation experiments with this model are available. The ultimate goal, of course, is to find a parameterization of (11.40) and (11.41) which does justice to the energetics of the general circulation.

In the budget for $\overline{u' v'}$, the major source term contains the zonal eddy heat flux. The maintenance equation for $\overline{\theta' u'}$ reads

$$
\begin{aligned}
\frac{\partial}{\partial t}\overline{\theta' u'} = 0 = &-\overline{u'\mathrm{v}'}\frac{\partial\overline{\theta}}{\partial y} - \overline{\theta'\mathrm{v}'}\frac{\partial\overline{u}}{\partial y} - \frac{K_s}{h^2}\overline{\theta' u'} \\
&- \frac{\partial}{\partial y}\left(\overline{\theta' u'\mathrm{v}'}\right) - \overline{\frac{\theta'}{\rho}\frac{\partial p'}{\partial x}} + f\overline{\theta'\mathrm{v}'} \ .
\end{aligned}
\tag{11.46}
$$

Introducing the unbalanced wind fluctuations, we can write (11.46) in the form

$$
\begin{aligned}
\frac{\partial}{\partial t}\overline{\theta' u'} = 0 = &-\overline{u'\mathrm{v}'}\frac{\partial\overline{\theta}}{\partial y} - \overline{\theta'\mathrm{v}'}\frac{\partial\overline{u}}{\partial y} - \frac{K_s}{h^2}\overline{\theta' u'} \\
&- \frac{\partial}{\partial y}\overline{\theta' u'\mathrm{v}'} + f\overline{\theta'\mathrm{v}'_A} \ .
\end{aligned}
\tag{11.47}
$$

The orders of magnitude of four of the terms are

$$
-\overline{u'\mathrm{v}'}\frac{\partial\overline{\theta}}{\partial y} \simeq 3\cdot 10^{-5}\ \mathrm{m\ K/sec^2}\ , \tag{11.48}
$$

$$
-\overline{\theta'\mathrm{v}'}\frac{\partial\overline{u}}{\partial y} \simeq 10^{-4}\ \mathrm{m\ K/sec^2}\ , \tag{11.49}
$$

$$
\frac{K_s}{h^2}\overline{\theta' u'} \simeq 10^{-5}\ \mathrm{m\ K/sec^2}\ , \tag{11.50}
$$

$$
f\overline{\theta'\mathrm{v}'_A} \simeq \frac{1}{10}f\overline{\theta'\mathrm{v}'} \simeq 10^{-4}\ \mathrm{m\ K/sec^2}\ . \tag{11.51}
$$

The last of these is probably an overestimate; our analysis has not suggested anywhere that θ' should be correlated with v'_A. Also, the difference between (11.48) and (11.49) is probably not significant, because order-of-magnitude estimates of fluxes and gradients are compounded here. Surface friction effects are fairly small compared to the leading terms; they will be neglected.

South of the jet-stream axis, $\overline{u'\mathrm{v}'}$ and $\partial\overline{u}/\partial y$ are positive (Fig. 11.4); they reverse sign on the other side of the jet. The poleward heat flux $\overline{\theta'\mathrm{v}'}$ is positive everywhere, and the mean temperature gradient $\partial\overline{\theta}/\partial y$ is negative. The term $-\overline{u'\mathrm{v}'}\ \partial\overline{\theta}/\partial y$ thus is positive south of the jet axis, while the term $-\overline{\theta'\mathrm{v}'}\ \partial\overline{u}/\partial y$ is negative there. The roles of the two terms are reversed on the poleward flank of the jet. The distributions of (11.48) and (11.49) are sketched in Fig. 11.7. The two curves are strikingly similar, except for the way in which they approach the boundaries of the zonal channel. It is very tempting to ignore the Coriolis term in (11.47) — after all, there is no evidence that θ' and v'_A

should be correlated — and to postulate that the principal balance mechanism in (11.47) is:

$$\overline{u'v'}\frac{\partial\overline{\theta}}{\partial y} \simeq -\overline{\theta'v'}\frac{\partial\overline{u}}{\partial y} . \tag{11.52}$$

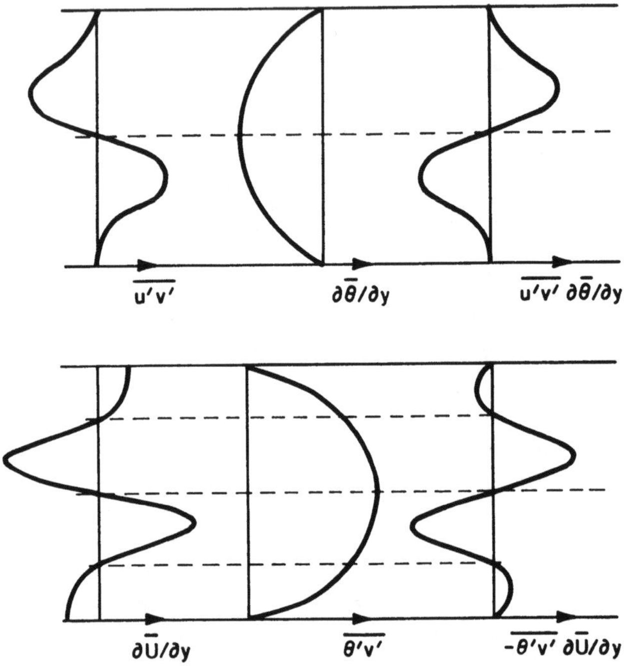

Fig. 11.7: The meridional distributions of the first two terms in the budget equation for the zonal eddy heat flux, $\overline{\theta'u'}$. Apart from the immediate vicinity of the boundaries, the two curves look quite similar; numerical estimates show that the values of $\overline{\theta'v'}$ $\partial\overline{u}/\partial y$ and $\overline{u'v'}$ $\partial\overline{\theta}/\partial y$ are comparable.

This balance may be off by a factor of three, as indicated by (11.48) and (11.49). Nevertheless, (11.52) implies that

$$\frac{\overline{u'v'}}{\partial\overline{u}/\partial y} \simeq -\frac{\overline{\theta'v'}}{\partial\overline{\theta}/\partial y} . \tag{11.53}$$

These ratios define the meridional eddy exchange coefficients for momentum and heat. We find

$$K_M \simeq -K_H . \tag{11.54}$$

Since $\overline{\theta' v'} \simeq 10$ m K/sec and $\partial\bar{\theta}/\partial y \simeq 3 \cdot 10^{-6}$ K/m (Table 1), the eddy heat diffusivity is of order $3 \cdot 10^6$ m^2/sec. Since $\overline{u'v'} \simeq 10$ m^2/sec^2 and $\partial\bar{u}/\partial y \simeq 10^{-5}$ sec^{-1}, the value of K_M must be estimated as -10^6 m^2/sec. This seems to suggest that $-K_M < K_H$, but the accuracy of our estimates is not sufficient to warrant such a conclusion. The value $K_H \simeq 3 \cdot 10^6$ m^2/sec agrees with those used in mean-motion models of the general circulation (Green, 1969).

All of this may contribute to a possible explanation of negative eddy-viscosity phenomena in the general circulation (Starr, 1968), but it does not mean that a negative eddy viscosity will do justice to the complicated energetics of the general circulation. The Coriolis term in (11.47) may not be negligible, and the flux-divergence term must be significant too, at least near the boundaries of the zonal channel. We must refrain from further speculations until the results of numerical experiments become available.

7. Enstrophy Maintenance and Cascade Dynamics

We now turn to a further discussion of the vorticity dynamics of the model. The enstrophy of the eddy motion is governed by the following equation:

$$\begin{aligned} \frac{\partial}{\partial t}\left(\frac{1}{2}\sigma_\zeta^2\right) = 0 = &-\left(\beta + \frac{\partial \xi}{\partial y}\right)\overline{v'\zeta'} + \frac{g^*}{T}\overline{\zeta'\frac{\partial \theta'}{\partial x}} \\ &- \frac{\partial}{\partial y}\left(\frac{1}{2}\overline{\zeta'\zeta' v'}\right) - \frac{K_s}{h^2}\sigma_\zeta^2 - \chi \end{aligned} \tag{11.55}$$

Here, χ represents viscous destruction of enstrophy, based on the presumed existence of an enstrophy cascade. We recall that the fluid is assumed to be almost inviscid and almost nonconducting, so that all viscous terms and heat diffusion terms in the system of equations are negligible, except for those related to the spectral cascades of enstrophy and temperature variance.

The first term in (11.55) depends upon the poleward gradient of the mean absolute vorticity $\xi_a = f + \xi$. The flow is barotropically stable everywhere if the sign of $\partial\xi_a/\partial y = \beta + \partial\xi/\partial y$ is the same all through the channel. If this is the case (there is no reason to expect otherwise), there is a very rapid loss of enstrophy to the mean zonal flow in the center of the channel (Fig. 11.4), because $\overline{v'\zeta'}$ and $\partial\xi/\partial y$ both have maxima there. There are, however, small gains near the two walls. The overall picture, of course, is consistent with a reverse energy cascade; it is worth noting, however, that the geometry of the situation is such that a reverse enstrophy cascade, away from the wavenumbers of greatest baroclinic forcing, is not excluded. This contradicts one of the conclusions of the spectral theory of two-dimensional turbulence (which does not account for the fact that $\partial\xi/\partial y$ may be different from zero).

The second term of (11.55) represents baroclinic forcing; it is the primary source of eddy enstrophy in the system. Since the eddy enstrophy suffers losses at both extremes of the spectrum, it could not survive without the baroclinic source term. Barotropic flow with friction would not last long.

The orders of magnitude of some of the terms in (11.55) are (Table 11.1):

$$\left(\beta - \frac{\partial\xi}{\partial y}\right)\overline{v'\zeta'} \simeq 10^{-16}\text{sec}^{-3} , \tag{11.56}$$

$$\frac{K_s}{h^2}\sigma_\zeta^2 \simeq 10^{-16}\text{sec}^{-3} , \tag{11.57}$$

$$\chi \simeq \frac{1}{3}\sigma_\zeta^2 \simeq 3 \cdot 10^{-16}\text{sec}^{-3} . \tag{11.58}$$

The factor 1/3 in the estimate for χ is based on the assumption that the enstrophy cascade is fairly inefficient. We have to keep in mind, however, that the estimate for χ involves raising σ_ζ to the third power; the number given in (11.58) can easily be off by a factor of ten.

With the aid of the thermal-wind equations, two of the terms in (11.55) can be put in the following alternative form:

$$-\beta\overline{v'\zeta'} + \frac{g^*}{T}\overline{\zeta'\frac{\partial\theta'}{\partial x}} = -\beta\overline{v_A'\zeta'} - f\,\overline{\zeta'(\nabla\cdot v_A')}. \tag{11.59}$$

The first of the two terms on the r.h.s. is of order 10^{-17} sec^{-3} if ζ' is not better correlated with v_A' than with v' (recall that $v' \simeq 10v_A'$); we will neglect it. The second term is estimated by taking the convergence of the thermal wind to be of order 10^{-6} sec^{-1} and assuming that the correlation coefficient between ζ' and $\overline{\nabla\cdot v_A'}$ is about one-third. This yields

$$-f\,\overline{\zeta'(\nabla\cdot v_A')} \simeq 3 \cdot 10^{-16}\ \text{sec}^3 . \tag{11.60}$$

The baroclinic generation of eddy enstrophy thus is able, at least to within an order of magnitude, to balance the losses due to the enstrophy cascade. The flux-divergence term in (11.55) is of order 10^{-16} sec^{-3}; it can handle the redistribution needed to cope with the large loss of eddy enstrophy to the mean

TABLE 11.1

Orders of magnitude of the major parameters and variables in the system of equations.

Parameters	Eddy Fluxes	Velocities	Vorticities	Temperatures
β - 10^{-11} m^{-1} sec^{-1}	$\overline{u'v'}$ - 10 m^2/sec^2	$\overline{u}$ - 10 m/sec	$\overline{\zeta}$ - 10^{-5} sec^{-1}	σ_θ - 3 K
f - 10^{-4} sec^{-1}	$\overline{v'\zeta'}$ - 10^{-5} m/sec^2	σ_u - 10 m/sec	σ_ζ - 10^{-5} sec^{-1}	$\partial\overline{\theta}/\partial y$ - $3 \cdot 10^{-6}$ k/m
g^* - 10^{-2} m/sec^2	$\overline{\theta'v'}$ - 10 mK/sec	σ_v - 6 m/sec	$\partial\overline{\zeta}/\partial y$ - 10^{-11} m^{-1} sec^{-1}	
h - 10^4 m	$\overline{\theta'u'}$ - 10 mK/sec	$\partial\overline{u}/\partial y$ - 10^{-5} sec^{-1}		
K_s - 10^2 m^2/sec				
$\overline{R}$ - 10^{-5} K/sec				
T - 300 K				

zonal flow in the center of the channel. It appears that the entire system of equations is dynamically consistent, at least at the level of accuracy obtainable without support of hard data from numerical experiments.

The source and sink terms in (11.59) do pose other problems, however. An inertial subrange with a conservative enstrophy cascade can exist only if there are no sources or sinks of enstrophy at scales of motion that are small compared to those containing most the kinetic energy, or if at each wavenumber the various sources and sinks balance each other exactly. The first alternative is ruled out here because the surface-friction term $(K_s/h^2)\sigma_\zeta^2$ does not discriminate among scales of motion. If, for example, the kinetic energy spectrum has a k^{-3} range, then surface friction removes enstrophy at the same rate in each octave of that range. In that case a conservative enstrophy cascade can be maintained only if the spectrum of the baroclinic forcing term in (11.59) exactly balances that of the frictional losses and if the enstrophy removal by the mean vorticity gradient becomes negligible at small scales. The first condition implies that the contributions to $\overline{\zeta' \, \partial\theta'/\partial x}$ made by each octave of the spectrum must be independent of wavenumber; the second means that the contributions to $\overline{v'\zeta'}$ associated with small scales of motion must be negligible.

The linear stability analysis in Chapter 3 suggests that the growth rate of small waves my be roughly independent of wavenumber. If that conclusion could be carried over to the nonlinear stability characteristics of the system, the spectrum of $\overline{\zeta' \, \partial\theta'/\partial x}$ would have the shape necessary to balance frictional losses through the inertial subrange. A qualitative balance, however, is not good enough; also, there is no justification for assuming that the behavior of turbulence is similar to that of linearized waves.

The spectrum of $\overline{v'\zeta'}$, on the other hand, probably will behave in the desired manner. In a k^{-3} range, the contribution to $\overline{v'\zeta'}$ made by each octave of the spectrum is proportional to the central wavelength of that octave if the coherence between v' and ζ' is independent of wavenumber. If there is any tendency toward small-scale isotropy, however, the coherence will decrease as the wavelength decreases ($\overline{v'\zeta'} = 0$ in isotropic turbulence). The spectrum of $\overline{v'\zeta'}$ thus may be expected to decrease quite rapidly with increasing wavenumbers, and it is not unlikely that the reverse spectral enstrophy flux associated with $\overline{v'\zeta'}(\beta + \partial\xi\partial y)$ will be quite small at small scales of motion.

The issues raised here clearly deserve further analysis. In two-dimensional turbulence with thermal forcing and surface friction, will there be any small-scale isotropy at all? Will the energy spectrum exhibit a k^{-3} subrange? Will the temperature variance exhibit a spectrum similar to that of enstrophy? Are there any ranges in which the spectral fluxes of enstrophy and/or energy are negligible? Are there ranges in which nonzero spectral fluxes violate conservation principles? At this stage we have no answers to these questions; we hope that they will be studied by people better versed in the spectral theory of two-dimensional turbulence.

We have assumed that enstrophy and temperature variance are cascaded toward the smallest scales of motion, where they are annihilated by molecular friction and heat conduction. In numerical integrations of the primitive equation, the molecular diffusivities can be replaced by the numerical diffusion coefficients which are needed to avoid discontinuities at scales comparable to the mesh size of the integration grid. This should not affect the behavior of the system significantly if the mesh size is sufficiently small, because the cascade rates of enstrophy and temperature variance are determined by the large-scale dynamics of the flow field. The diffusivities involved determine the smallest scale of motion; they do not significantly affect the large-scale dynamics of the system at sufficiently large Reynolds numbers.

The very concept of spectral cascades of enstrophy and temperature variance, however, implies that the largest vorticity and temperature gradients will be associated with scales of motion that are just large enough to avoid rapid attenuation of gradients by diffusion. In other words, the system will generate "fronts" characterized by appreciable vorticity and temperature changes, and most of the dissipation of enstrophy and temperature variance will occur in frontal zones. This line of reasoning adds another dimension to the study of frontogenesis and frontolysis. Will frontogenesis be rapid enough to ensure that enstrophy and temperature variance are destroyed at the required rates? What is the life cycle of a typical frontal zone in our model? Will the midlatitude cyclones in the model develop cold and warm fronts similar to those associated with storms in the earth's atmosphere? Will the presence of frontal zones affect the shape of the various spectra at large wavenumbers? We emphasize that these are questions related to the maintenance of the general circulation; a statistically steady regime cannot be maintained without adequate sinks for enstrophy and temperature variance.

8. Conclusions

We have explored the general circulation of a one-level model of atmospheric motion equipped with a thermodynamic energy cycle which faithfully imitates several features of the thermodynamics of atmospheric motion. The model employs two basic equations, a vorticity equation and a Boussinesq approximation to the first law of thermodynamics. In complexity, our model thus is comparable to a two-level baroclinic system of equations. The model obviously is at a disadvantage compared to multilevel baroclinic models, because it cannot represent the interactions between the vertical and horizontal fields of motion that help to shape the general circulation in the earth's atmosphere. On the other hand, it seems logical to expect that barotropic models might perform much better if they were equipped with a simulated thermodynamic cycle of the kind explored in this chapter. Since the additional computer time involved would not be large, it appears that some experimentation in this direction would be useful.

The equations studied here illustrate the energetics of the general circulation in a particularly simple way. A reverse energy cascade is needed to maintain the mean zonal flow against surface friction, eddy kinetic energy has to be maintained by baroclinic effects, the simulated thermodynamics is necessary to maintain a poleward heat flux, and so on. The model appears to be energetically consistent in all features that have been investigated here, and it shows clearly that the maintenance of the general circulation depends in part upon the spectral cascades of enstrophy and temperature variance. Because the model has fairly realistic energy cycles, it may help to give new momentum to the study of the spectral theory of two-dimensional turbulence. No longer is it necessary to restrict that theory to isotropic fields with artificial forcing at pre-assigned wavenumbers. Our equations also show which covariances are involved in the maintenance of the flow field; this provides a background for attempts to develop mean-motion models of the general circulation.

At this stage, it is not clear whether the problems we have encountered will seriously affect the applicability of the model. Of particular concern is the absence of a high-wavenumber cut-off in the baroclinic forcing term. Without numerical experiments, we have no way of knowing whether the nonlinear characteristics of the model will concentrate most of the baroclinic effects in a narrow range of wavenumbers. This issue is related to the absence or presence of isotrophy at small scales; here too, our insight is inadequate at this time.

Batchelor, G. K., 1969: *Phys. Fluids*, **12**, Supplement II, 233.

Blackmon, M. L., J. M. Wallace, N. C. Lau, and S. L. Mullen, 1977: *J. Atmos. Sci.*, **34**, 1040–1053.

Fjørtoft, R., 1953: *Tellus*, **5**, 225.

Fleagle, R. G., 1955: *Tellus*, **7**, 168.

Gall, R. L., 1976: *J. Atmos. Sci.*, **33**, 349.

Green, J. S. A., 1969: *Quart. J. Roy. Meteor. Soc.*, **96**, 157.

Holton, J. R., 1972: *An Introduction to Dynamic Meteorology*. Academic Press, New York.

Kraichnan, R. H., 1967: *Phys. Fluids*, **10**, 1417.

Kraichnan, R. H., 1976: *J. Atmos. Sci.*, **33**, 1521.

Starr, V. P., 1968: *The Physics of Negative Eddy Viscosity Phenomena*. McGraw-Hill Book Co., New York.

Tennekes, H., and J. L. Lumley, 1972: *A First Course in Turbulence*. MIT Press, Cambridge, Mass.

MAGNETOHYDRODYNAMIC TURBULENCE

Preface

Magnetohydrodynamic (MHD) turbulence is more difficult to present in lectures than fluid turbulence is, for several reasons: (1) less is known about it, theoretically and experimentally; (2) there is not yet a consensus among the practitioners on either a mathematical description or a central core of problems worth solving; and (3) one can never assume that any audience is knowledgeable about both fluid mechanics and plasma physics. Any attempt to "teach" the subject is bound to be in large part a rather chaotic effort to respond to unpredictable questions that originate in very different experiences and perspectives, and often the answers are less than satisfying. The NCAR summer school was no exception. What remains on the page is a pale substitute for the animated and jagged trajectories that the classes followed. MHD turbulence is a new subject trying to be born, fun to think about, but not ready to write textbooks about. These lectures may stand, at least, as a kind of material culture exhibit of what seemed to be worth introducing in the summer of 1987.

David Montgomery
Dartmouth College
27 October 1988

Chapter 1

MAGNETOFLUID TURBULENCE

In this series of chapters, we will review some aspects of the theory of turbulence in magnetofluids. To begin, we define a magnetofluid as any fluid which conducts electric current reasonably well. Such fluids can create and maintain internal magnetic fields which interact nonlinearly with the internal fluid velocity field, and possibly with external magnetic fields. The study of such phenomena is termed *magnetohydrodynamics*, or *MHD* for short. Some of the environments where one would find a magnetofluid are: remote astrophysical settings (stars, the interstellar medium, and exotic objects such as pulsars and binaries); nearby astro- and geophysical settings (the solar surface and corona, the solar wind, the earth's magnetotail and even the earth's oceans); laboratory settings (mainly fusion confinement devices and, more recently, in MHD power generation and liquid metallurgy). In these chapters we will not treat electrostatic plasma turbulence, i.e., phenomena with significant ion-electron charge separation and relatively minor magnetic effects (as would be the case in inertial confinement schemes and in the $\boldsymbol{E} \times \boldsymbol{B}$ - drifting ionospheric plasma.)

At this point it would be useful to write down some qualitative features of real magnetofluids. The most striking quality is the enormous range of physical parameters one encounters. Number densities n [particles/cm^3] range from $n \simeq 1$ (in the solar wind and the interstellar medium) to $n \simeq 10^{14} - 10^{18}$ (in laboratory pinches) up to $n \simeq 10^{26}$ (in dense astrophysical objects). Length scales L_0 can be measured in light years (galaxies) down to millimeters (laboratory plasmas). Also, in laboratory settings alone, temperatures can vary between $1/40$ eV ($300K$) and 20 keV ($2 \times 10^8 K$). Although magnetofluid behavior generally seems very removed from our everyday experience (particularly in comparison with Navier-Stokes phenomena), indirect evidence of such behavior is to be found in the terrestrial magnetic field, which is thought to be due to MHD processes within the earth's core.

Our understanding of magnetofluid turbulence through experimental measurement is, unfortunately, fragmentary. Astronomical observations of remote astrophysical plasmas reveal only their large-scale behavior and do not yield any knowledge of small-scale, turbulent behavior. Typical fusion research plasmas are very hot ($T_0 \geq 1keV$), and will burn up any internal probes one might use to measure turbulent magnetic and velocity fields; diagnostics here are limited to external probes measuring, for example, magnetic fields at the edge and internal density and temperature variations (by laser scattering). Although cooler operating regimes could be created and maintained in order to allow for probe insertion and systematic study of magnetofluid turbulence, this opportunity has not been sufficiently exploited. Another possible source of data, *in situ* space-plasma probes (e.g., particle detectors and antennae on spacecraft) often modify local properties of the plasma (e.g., electrons forming sheaths around solar wind probes) and thereby confuse accurate small scale measurements.

In addition to experimental difficulties, there are also theoretical ones. As a result of the enormous variations in the parameters encountered in magnetofluids, assumptions (namely, that the plasma is collision dominated) upon which the resulting mathematics are derived are often poorly met: mean-free-path lengths are too long, collision frequencies too small, and impurity ions too numerous, to name a few. For example, the mean free path in the solar wind is approximately 1 AU (10^{13} cm). Somewhat fortuitously, however, experience has shown that the MHD description of these systems is still relevant, to a degree that no one has satisfactorily explained.

Three basic regimes of magnetofluid turbulence can be identified by consideration of the fluid's turbulent kinetic and magnetic energies. To define these regimes, we will assume that any mean velocity field has been transformed away (i.e., by a Galilean transformation); a mean magnetic field, however, cannot be similarly removed. The *kinetic regime* is one in which the (turbulent) energy in the flow field is much greater than that in the magnetic field. This partitioning of energy is seen at the solar surface and in the earth's small-scale oceanic motions. The magnetic field has vanishing effect on the flow field, and is easily pushed around. The *magnetic regime* is one where a large mean magnetic field is present; the energy in the mean magnetic field is much greater than that in either the turbulent magnetic or velocity fields. Examples of this occur in fusion plasmas and in pulsars; here, the dynamics of the flow field are severely modified by the mean magnetic field. The third regime is the *equipartition regime*, with no dominating mean magnetic field and with the turbulent kinetic and magnetic energies being comparable. The earth's magnetotail and the solar wind in its local rest frame are examples. In practice, these regimes are the operating domains of different research communities, with different approximations being applied to the putative equations and with differing technical terminology.

Features of magnetofluid behavior that differ from those typical of Navier-Stokes turbulence are tendencies toward inhomogeneity (e.g., density and field gradients encountered in magnetic confinement devices), anisotropy (due to mean magnetic fields), and critical interaction with material boundaries (such as plasma-wall interactions in fusion research devices.) The experimental and mathematical study of MHD turbulence is still somewhat immature, perhaps comparable to the state of hydrodynamics a century ago. Foundations have been laid, however, and it is to a description of these that we now address ourselves.

The MHD Equations

The equations for a continuous magnetofluid may be determined by applying Newton's laws and Maxwell's equations to a collection of charged particles (i.e., ions and electrons), utilizing phase space (i.e., ($\boldsymbol{x}$,v) space) distribution functions and then averaging the discrete particle equations with respect to velocity. Newton's second law and Maxwell's equations (presented here in cgs units) are

$$m\frac{d\mathbf{v}}{dt} = \boldsymbol{F} \,, \tag{1.1}$$

$$\nabla\cdot\boldsymbol{B} = 0 \,, \qquad \nabla\times\boldsymbol{B} = \frac{1}{c}\frac{\partial\boldsymbol{E}}{\partial t} + \frac{4\pi}{c}\boldsymbol{j} \,, \tag{1.2}$$

$$\nabla\cdot\boldsymbol{E} = 4\pi\rho_e \,, \qquad \nabla\times\boldsymbol{E} = -\frac{1}{c}\frac{\partial\boldsymbol{B}}{\partial t} \,. \tag{1.3}$$

Here ρ_e is the electric charge density, $\boldsymbol{j}$ is the conduction current, and $1/c\partial\boldsymbol{E}/\partial t$ is the displacement current. $\boldsymbol{F} = e(\boldsymbol{E} + \mathbf{v}\times\boldsymbol{B}/c)$ is the electromagnetic force on a particle of charge-to-mass ratio e/m. The speed of light is c.

The distribution function $f_j(\boldsymbol{x}, \mathrm{v}, t)$ represents the probability of any individual particle of species j being found in the phase space volume $d\boldsymbol{x}d\mathbf{v}$ centered on $(\boldsymbol{x}, \mathrm{v})$ at time t. In terms of f_j, where the charge per particle is e_j and mean number density is n_{0j},

$$\rho_e = \sum_j n_{0j}e_j\int f_j(\boldsymbol{x},\mathbf{v},t)d\mathbf{v} = \sum_j \rho_j \,, \tag{1.4}$$

$$\boldsymbol{j} = \sum_j n_{0j}e_j\int \mathbf{v}f_j(\boldsymbol{x},\mathbf{v},t)d\mathbf{v} \,, \tag{1.5}$$

$$\rho = \sum_j n_{0j}m_j\int f_j(\boldsymbol{x},\mathbf{v},t)d\mathbf{v} \,. \tag{1.6}$$

These equations represent, respectively, the charge, current, and mass densities which appear in the dynamical equations. Note that they are obtained by averaging over velocities; also note that $n_j = n_{0j} \int f_j d\mathbf{v}$ is the *number density* for the jth species.

The evolution of the f_j is determined by kinetic Boltzmann-like equations

$$\frac{\partial f_j}{\partial t} + \frac{\partial}{\partial \boldsymbol{x}} \cdot f_j \mathbf{v} + \frac{e_j}{m_j} \frac{\partial}{\partial \mathbf{v}} \cdot \left[f_j \left(\boldsymbol{E} + \frac{1}{c} \mathbf{v} \times B \right) \right] = \frac{1}{\epsilon} \sum_{\ell} C_{j\ell} \left(f_j, f_\ell \right) = \frac{1}{\epsilon} C_j \left(f_j \right), \quad (1.7)$$

where $\epsilon \simeq$ [mean free path (or time)]/[macroscopic length (or time)] is a formal Chapman-Enskog expansion parameter which is treated as of much less than unity; the expansion for f_j is written as: $f_j = f_j^{(0)} + \epsilon f_j^{(1)} + \cdots$. For a normal fluid, to 0th order, we have the Euler equations, and to first order, the dissipative fluid equations. The collision term $C_{j\ell}(f_j, f_\ell)$ is the Boltzmann collision integral

$$C_{jl}(f_j, f_l) = n_{0l} \int d^3 \mathrm{v}_1 \int b db \int d\phi |\mathbf{v}_1 - \mathbf{v}| \left[f_j(\mathbf{v}') f_l(\mathbf{v}_1') - f_j(\mathbf{v}) f_l(\mathbf{v}_1) \right], \qquad (1.8)$$

where the two unprimed velocities scatter to produce the primed ones; b is the impact parameter; and ϕ is the azimuthal angle. If small-angle collisions are assumed to be dominant, which is reasonable for single-collision events in a plasma, then the resultant velocities $\mathbf{v}_1'$ and $\mathbf{v}'$ may be written as $\mathbf{v}' = \mathbf{v} + \Delta\mathbf{v}$, $\mathbf{v}_1' = \mathbf{v}_1 + \Delta\mathbf{v}_1$, and the distribution function f_j may be expanded to second order in a Taylor series in $\Delta\mathbf{v}$ and $\Delta\mathbf{v}_1$. This results in a Fokker-Planck equation for f_j [Df_j/Dt stands for the left-hand side of (1.7)]:

$$\frac{Df_j}{Dt} = -\frac{\partial}{\partial \mathbf{v}} \cdot \left(\boldsymbol{F} f_j(\mathbf{v}) \right) + \frac{1}{2} \frac{\partial^2}{\partial \mathbf{v} \partial \mathbf{v}} : \left(\underline{\underline{T}}\, f_j(\mathbf{v}) \right). \qquad (1.9)$$

The integral defining the friction term $\boldsymbol{F}$ and the diffusion tensor $\underline{\underline{T}}$ both diverge in the short and long spatial limits due to the nature of the Coulomb interaction. This divergence may be patched up if, for example, the effects of Debye shielding are accounted for; the interested reader is referred to Landau (1936). Later, more elaborate kinetic theories which yield minor corrections to Landau's results were developed. See, e.g., Montgomery and Tidman (1964), Braginskii (1965), Montgomery (1971), or Ferziger and Kaper (1972).

In review, we start with a Boltzmann-type equation, where the volume force is given by the (local mean) Lorentz force, and the collision integral contains mainly the effects of finite-distance, small-angle scattering events. Next, we perform a formal expansion of f in the small parameter ϵ, which then results in the general macroscopic moment equations underlying MHD. In using these general macroscopic moment equations, the following approximations are necessary.

First, assume that electrons can respond to local electric field variations much faster than the field changes itself. Formally, this is equivalent to demanding

$$\left|\frac{1}{c}\frac{\partial \boldsymbol{E}}{\partial t}\right| \ll \left|\frac{4\pi}{c}\boldsymbol{j}\right| , \tag{1.10}$$

the essential "low-frequency" limit of any good conductor. Neglecting the time derivative in Maxwell's equations (giving the "pre-Maxwell" equations) effectively removes radiation processes from MHD and is, in fact, its essential approximation. (Unfortunately, real life contains many situations in which the radiation field has to be put back in.)

Second, assume near charge neutrality; that is, throughout the plasma the net charge density in a small volume is approximately zero. Formally, this "quasi-neutrality" assumption can be stated as $\sum_s \rho_s(\boldsymbol{x},t) \ll |\rho_j|$, for any species j. The quasi-neutral assumption has the effect of eliminating Poisson's equation for the electric potential from dynamic consideration, but it does *not* mean that $\nabla \cdot \boldsymbol{E} = 0$, or that the electrostatic potential vanishes.

A third approximation is that ions, because of their mass, carry most of the momentum, and that electrons, because of their mobility, carry most of the electric current.

$$\text{momentum density}: \rho\mathbf{v} = \sum_j n_{0j} m_j \int \mathbf{v} f_j d\mathbf{v} \simeq n_{0i} m_i \int \mathbf{v} f_i d\mathbf{v} \tag{1.11}$$

$$\text{current density}: \boldsymbol{j} = \sum_j n_{0j} e_j \int \mathbf{v} f_j d\mathbf{v} \simeq n_{0e} e_e \int \mathbf{v} f_e d\mathbf{v}. \tag{1.12}$$

This allows a one-fluid description to be used, in which the independent vector fields $\mathbf{v}$ and $\boldsymbol{j}$ are on an equal footing.

A fourth basic approximation is to suppose the existence of an Ohm's law: $\boldsymbol{j} = \underline{\underline{\sigma}} \cdot \left(\boldsymbol{E} + \frac{1}{c}\mathbf{v} \times \boldsymbol{B}\right)$ where $\underline{\underline{\sigma}}$ is the conductivity tensor. Since the exact tensor structure is difficult to ascertain (and may only possess secondary importance), the further assumption of a scalar isotropic conductivity is also usually invoked. Thus, with this constitutive relation between $\boldsymbol{E}$ and $\boldsymbol{j}$, the electric field may be removed entirely from the dynamic equations, though it should be kept in mind that it does all the electromagnetic work on the charges.

Finally, but not necessarily least significantly, the assumption is made that the Chapman-Enskog expansion parameter ϵ is much less than unity. This implies that the 0th order term $f_j^{(0)}$ in the expansion represents local thermodynamic equilibrium, and that $f_j^{(0)}$ can be written as a local Maxwellian

$$f_j^{(0)} = \frac{n_j(\mathbf{x},t)}{n_{0j}} \left(\frac{m_j}{2\pi kT}\right)^{3/2} exp\left(\frac{-m_j(\tilde{\mathbf{v}})^2}{2kT}\right) , \tag{1.13}$$

where $T = T(\mathbf{x},t)$ and $\tilde{\mathbf{v}}$ = [particle velocity] - [fluid velocity]. For the interested reader, a fuller description of the kinetic theory calculations and the approximations leading to MHD may be found in Braginskii (1965) and Montgomery and Tidman (1964). The book of Cowling (1957) is an excellent introduction to the basic physical approximations of MHD, as is Shercliff (1965).

In regard to the transport tensors, the usual procedure is to assume that the viscosity $\underline{\underline{\mu}}$ and the resistivity $\underline{\underline{\eta}} \sim \sigma^{-1}$ are isotropic (i.e., essentially scalars). Thus, the dissipative terms can be simply approximated as

$$\nabla \cdot \underline{\underline{\mu}} \cdot \nabla \mathbf{v} \simeq \nabla \cdot \rho\nu\nabla\mathbf{v} + \nabla\left(\rho\frac{\nu}{3}\nabla\cdot\mathbf{v}\right) ,$$

$$\nabla \times \left(\underline{\underline{\eta}} \cdot \nabla \times \boldsymbol{B}\right) \simeq \nabla \times (\eta\nabla \times \boldsymbol{B}) .$$

(ν = the "kinematic" viscosity $\sim \lambda_{imfp} \mathrm{v}_{i_{th}}$)

(λ_{imfp} is the ion mean-free path and v_{ith} is the ion thermal speed),

$$\eta = \text{"the resistivity"} \sim \frac{\omega_{pe}^2}{\nu_{ei}} . \tag{1.14}$$

(ω_{pe} is the electron plasma frequency and ν_{ei} is the electron-ion collision frequency). Additionally, the pressure is often assumed to satisfy, $p = p(\rho)$, (i.e., an adiabatic or isothermal equation of state), such that

$$\left(\frac{\partial}{\partial t} + \mathbf{v}\cdot\nabla\right)\left(\frac{p}{\rho^\gamma}\right) = 0 , \ \gamma \geq 1 . \tag{1.15}$$

Here,

$$p\rho^{-\gamma} = \text{const} , \tag{1.16}$$

defines the equation of state, for any fluid element.

These assumptions yield a closed system of MHD equations

$$\frac{\partial\rho}{\partial t} + \nabla\cdot(\rho\mathbf{v}) = 0 , \tag{1.17a}$$

$$\rho\left(\frac{\partial \mathbf{v}}{\partial t} + \mathbf{v}\cdot\nabla\mathbf{v}\right) = -\nabla p + \frac{\boldsymbol{j}}{c}\times\boldsymbol{B} + \nabla\cdot\left[\rho\nu\nabla\mathbf{v} + \underline{1}\frac{\rho\nu}{3}\nabla\cdot\mathbf{v}\right] , \qquad (1.17b)$$

$$\frac{\partial \boldsymbol{B}}{\partial t} = \nabla\times(\mathbf{v}\times\boldsymbol{B}) - \nabla\times\left((c^2\eta/4\pi)\nabla\times\boldsymbol{B}\right) , \qquad (1.17c)$$

$$\nabla\cdot\boldsymbol{B} = 0 \ , \nabla\times\boldsymbol{B} = 4\pi\boldsymbol{j}/c \ , \qquad (1.17d)$$

and $p\rho^{-\gamma}$ = const. It is also usually assumed that ν and η are constants, often taken as the unmagnetized ν and η when the ion/electron gyro-radii are large compared to λ_{mfp}; when local heating is significant, they are not constant, but depend on the local temperature.

Now, in order to arrive at a non-dimensional set of equations, we will measure length in units of L_0, the macroscopic *length* scale, and times in terms of $T_0 = L_0/U_0$, the macroscopic *time* scale. Here, U_0 is a characteristic velocity, which may be a flow speed *or* an Alfvén speed $C_A \equiv B_0(4\pi\rho_0)^{-1/2}$; B_0 is a characteristic magnetic field strength. We measure pressure in units of $p_0 \sim (\rho_0/m_i)kT_0$ or $p_0/\rho_0 \sim \partial p/\partial\rho = C_S^2$ = (sound speed)^2, where ρ_0 and T_0 are characteristic mass density and temperature, respectively.

Using these characteristic values, along with the assumption that the viscosity $\rho\nu$ = const, we arrive at the following dimensionless MHD equations, the continuity equation, and the equation of state

$$\frac{\partial\rho}{\partial t} + \boldsymbol{\nabla}\cdot(\rho\mathbf{v}) = 0 \ , \quad \left(\frac{\partial}{\partial t} + \mathbf{v}\cdot\nabla\right)(p\rho^{-\gamma}) = 0 \ , \qquad (1.18)$$

the momentum equation

$$\left(\frac{\partial}{\partial t} + \mathbf{v}\cdot\nabla\right)\mathbf{v} = -\frac{\beta}{\rho}\nabla p + \frac{C_A^2}{U_0^2}\left(\boldsymbol{B}\cdot\nabla\boldsymbol{B} - \nabla\frac{B^2}{2}\right) + R^{-1}\left[\nabla^2\mathbf{v} + \frac{1}{3}\nabla(\nabla\cdot\mathbf{v})\right] , \qquad (1.19)$$

and the magnetic induction equation

$$\frac{\partial\boldsymbol{B}}{\partial t} = \nabla\times(\mathbf{v}\times\boldsymbol{B}) + S^{-1}\nabla^2\boldsymbol{B}, \qquad (1.20)$$

where

$$\beta = \frac{C_S^2}{U_0^2} \sim (\text{Mach no.})^{-2} \text{ or } \frac{\text{thermal pressure}}{\text{magnetic pressure}} , \qquad (1.21)$$

$$R = \frac{U_0 L_0}{\nu} \sim \frac{\text{inertial terms}}{\text{viscous terms}} , \qquad (1.22)$$

$$S = U_0 \frac{L_0}{\eta} \sim \frac{\text{advective term in } \partial \boldsymbol{B}/\partial t \text{ eq}}{\text{resistive term in } \partial \boldsymbol{B}/\partial t \text{ eq}} . \tag{1.23}$$

In these equations, there are three relevant dimensionless numbers: β, R, and S; the quantity C_A^2/U_0^2 is conventionally set to either 0 or 1, depending upon whether $C_A^2 \ll U_0^2$ or $C_A^2 \sim U_0^2$. The problem's natural "regimes" are:

kinetic regime:	$U_0 \sim$ a flow speed ($\gg C_A$)
magnetic regime:	$U_0 \sim$ an Alfvén speed ($\equiv C_A$) $\gg$ a flow speed
equipartition regime:	$U_0 \sim$ either one (C_A and flow speed comparable).

The dimensionless numbers R and S thus have different possible identities:

R = "R":	Reynolds number, if $U_0 \sim$ a flow speed
R = "M":	viscous Lundquist number, if $U_0 \sim$ an Alfvén speed
S = "R_M":	magnetic Reynolds number, if $U_0 \sim$ a flow speed
S = "S":	Lundquist number, if $U_0 \sim$ an Alfvén speed.

In order to determine the incompressible limit, we write

$$p = p_0 + \beta^{-1} p_1 + \beta^{-2} p_2 + \cdots . \tag{1.24}$$

Placing this into the MHD equations and taking the limit $\beta \to \infty$ yields $p \to p_0 = \text{const}$, and, using $p\rho^{-\gamma} \to const$, $\rho \to$ const. If ρ = const, the continuity equation becomes the incompressibility condition, $\nabla \cdot \mathbf{v} = 0$. The pressure gradient does not, however, disappear from the momentum equation; instead, we have $\nabla \beta p \to \nabla p_1$. If we now take the divergence of the momentum equation, we see that the pressure in the incompressible case is determined by a Poisson's equation

$$\nabla^2 p^* = -\boldsymbol{\nabla} \cdot [\rho(\mathbf{v} \cdot \nabla \mathrm{v}) - B \cdot \nabla B], \tag{1.25}$$

$$p^* = p_1 + \frac{B^2}{2}. \tag{1.26}$$

Henceforth, when discussing the incompressible case, we will drop the subscript from p_1 and call it p. The whole area of compressible MHD turbulence has only just begun to be explored.

Assuming incompressibility ($\rho \equiv 1$), the MHD equations take the following forms in the different regimes:

"magnetic" or "equipartition" regimes

$$\left(\frac{\partial}{\partial t} + \mathbf{v}\cdot\nabla\right)\mathbf{v} = -\nabla\left(p + \frac{B^2}{2}\right) + \boldsymbol{B}\cdot\nabla\boldsymbol{B} + (R \text{ or } M)^{-1}\nabla^2\mathbf{v}\,, \tag{1.27}$$

$$\frac{\partial \boldsymbol{B}}{\partial t} = \nabla\times(\mathbf{v}\times\boldsymbol{B}) + (R_M \text{ or } S)^{-1}\nabla^2\boldsymbol{B}\,. \tag{1.28}$$

"kinetic" regime (***B*** "goes along for the ride," and does not influence **v**)

$$\left(\frac{\partial}{\partial t} + \mathbf{v}\cdot\nabla\right)\mathbf{v} = -\nabla p + R^{-1}\nabla^2\mathbf{v}\,, \tag{1.29}$$

$$\frac{\partial \boldsymbol{B}}{\partial t} = \nabla\times(\mathbf{v}\times\boldsymbol{B}) + R_M^{-1}\nabla^2\boldsymbol{B}\,. \tag{1.30}$$

In both regimes, $\nabla\cdot\boldsymbol{B} = 0$ and $\nabla\cdot\mathbf{v} = 0$. Note that the Navier-Stokes equations are recovered when $\boldsymbol{B} \equiv 0$.

In addition to the equations of motion, the MHD equations, we must also prescribe boundary conditions at all relevant surfaces. If we are studying homogeneous flows, we usually expand in Fourier series and thereby have periodic boundary conditions which are free in the sense that the field variables are not required to have any particular value. On the other hand, we can prescribe more realistic boundary conditions for investigating, say, MHD in actual plasma confinement devices ($\boldsymbol{n}$ is the normal to the boundary); the following, then, are possible conditions that a magnetofluid might satisfy on a bounding surface

$$\mathbf{v}\cdot n = 0 \quad \text{"free–slip"}\,, \tag{1.31}$$

$$\boldsymbol{n}\cdot\nabla(\boldsymbol{n}\times\mathbf{v}) = 0 \quad \text{"stress–free"}\,, \tag{1.32}$$

$$\mathbf{v} = 0 \quad \text{"no–slip"}\,, \tag{1.33}$$

$$\boldsymbol{B}\cdot\boldsymbol{n} = 0 \quad \text{perfect conductor}\,, \tag{1.34}$$

$$\boldsymbol{n}\times(\nabla\times\boldsymbol{B}) = 0 \quad \text{perfect conductor.} \tag{1.35}$$

The material boundary of an actual confinement device tends to be much messier than the above conditions indicate: there are steep temperature and density gradients, a large un-ionized component, chemical interaction with the wall, etc. Furthermore, for any practical fusion machine, slots and slits must be cut in the wall to permit external ***B*** and ***E*** fields to penetrate, creating more complications. It is no exaggeration to say that the most unsatisfactory part of the foundations of MHD is the lack, in many cases, of convincing boundary conditions. It is also true that we have virtually no systematic turbulence theory that applies to the important cases of compressible magnetofluids and variable transport coefficients.

Braginskii, S. I., 1965: "Transport Processes in a Plasma," Vol. I of *Reviews of Plasma Physics*, ed. M. A. Leontovich (New York: Consultant's Bureau).

Cowling, T. G., 1957: *Magnetohydrodynamics* (New York: Interscience).

Ferziger, J. M., and M. G. Kaper, 1972: *Mathematical Theory of Transport Processes in Gases* (Amsterdam: North-Holland).

Landau, L. D., 1936: *JETP*, **7**, 203; 1937: *Phys. Soviet Un.*, **10**, 154.

Montgomery, D., and D. A. Tidman, 1964: *Plasma Kinetic Theory* (New York: McGraw-Hill).

Montgomery, D., 1971 *Theory of the Unmagnetized Plasma* (New York: Gordon and Breach).

Shercliff, J. A., 1965: *A Textbook of Magnetohydrodynamics* (Oxford, Pergamon Press).

Chapter 2

THE ORIGIN OF TURBULENCE
PART I: SHEAR FLOW AND CURRENT GRADIENTS

Where does MHD turbulence come from? Like Navier-Stokes turbulence, it comes from gradients, e.g., gradients in pressure, velocity (shear flows), vorticity, temperature, magnetic field, current density, etc. Nature abhors a gradient, and will only support mild ones in a laminar state. Above thresholds, violent motions arise to smooth gradients out. When boundary conditions obstruct such smoothing, steady-state turbulence results; all steady-state turbulence is driven through boundary conditions.

A Prototype Example: Plane Poiseuille Flow

This flow is confined by rigid, smooth horizontal boundaries situated at $\pm a$, and is driven by a constant horizontal pressure gradient $-\nabla p_0$. The Navier-Stokes equations admit a steady-state horizontal velocity field $\mathbf{v}^{(0)} = U_0(a^2-y^2)\mathbf{e}_x$, where U_0 is proportional to the imposed pressure gradient. The stability of this solution is investigated by introducing small perturbations (as depicted in Fig. 2.1) to the velocity and pressure fields:

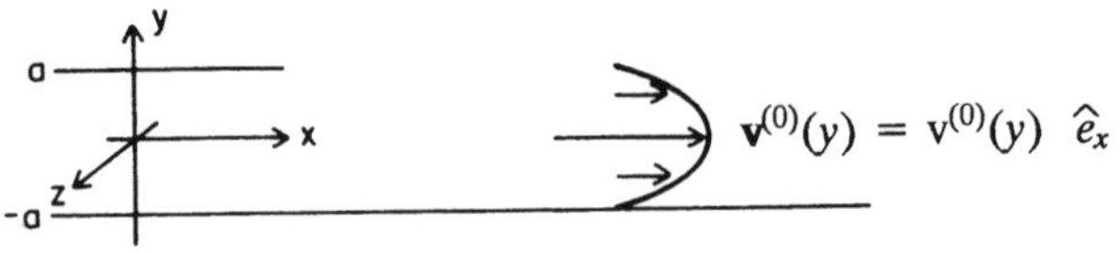

Figure 2.1: Flow direction and geometry for plane Poiseuille flow.

$$\mathbf{v} = \mathbf{v}^{(0)} + \mathbf{v}^{(1)} \ , \quad p = p^{(0)} + p^{(1)}. \tag{2.1}$$

With these expressions substituted for $\mathbf{v}$ and p, and with products of the perturbation fields neglected, the incompressible Navier-Stokes equations read

$$\frac{\partial \mathbf{v}^{(1)}}{\partial t} + \mathbf{v}^{(1)} \cdot \nabla \mathbf{v}^{(0)} + \mathbf{v}^{(0)} \cdot \nabla \mathbf{v}^{(1)} = -\nabla p^{(1)} + \frac{1}{R}\nabla^2 \mathbf{v}^{(1)} \ , \quad \nabla \cdot \mathbf{v}^{(1)} = 0. \tag{2.2}$$

Units have been chosen so that a and the fluid density are both unity, and the resulting Reynolds number is proportional to the imposed pressure gradient.

The perturbation fields may be assumed to have the normal mode form

$$\mathbf{v}^{(1)} = \mathbf{v}(y) \times \exp(i\alpha x + i\beta z - i\omega t) \ , \quad p^{(1)} = p(y) \times \exp(i\alpha x + i\beta z - i\omega t) \ , \quad (2.3)$$

where ω may be complex.

Straightforward manipulations now lead to the Orr-Sommerfeld equation

$$\left(\frac{d^2}{dy^2} - \alpha^2 - \beta^2\right)^2 \mathrm{v} = iR\left[(\alpha U_0(y) - \omega)\left(\frac{d^2}{dy^2} - \alpha^2 - \beta^2\right)\mathrm{v} - \alpha \mathrm{v}\left(\frac{d^2 U_0}{dy^2}\right)\right] , \quad (2.4)$$

where v is the y-component of $\mathbf{v}$, $\mathbf{v}^{(0)} = U_0(1 - y^2)\ \hat{e}_x$ and $\mathrm{v}(\pm 1) = 0$, $d\mathrm{v}(\pm 1)/dy = 0$.

Solutions $\mathrm{v}(y)$ satisfying the boundary conditions exist only for certain sets of values of α, β, ω, and U_0. Squire's theorem reveals that as R increases from zero, the first unstable mode (i.e., when $Im\ \omega > 0$) to appear will have $\beta = 0$, so that for the purpose of determining stability thresholds, only the case $\beta = 0$ need be considered.

The problem has been solved numerically (Orszag, 1971) by means of a polynomial expansion

$$\mathrm{v}(y) = \sum_{n=0}^{\infty} a_n T_n(y) \ , \quad (2.5)$$

where the $T_n(\cos\theta) \equiv \cos(n\theta)$ are the Chebyshev polynomials. When this expansion is placed into the Orr-Sommerfeld equation, we can use the orthogonality properties of the Chebyshev polynomials to uncover an infinite set of coupled, linear equations in the coefficients a_n; we can reduce these to a finite set of $N - 4$ equations in the coefficients $a_0, a_1, \cdots a_{N-1}$; at this point, we use the boundary conditions to produce four more equations. The result is a set of N linear algebraic equations for the N unknowns $a_0, \cdots a_{N-1}$. (This is the "tau" method of incorporating boundary conditions, and is due to C. Lanczos, 1956; e.g., Gottlieb and Orszag, 1971.) The equations may be written in the matrix form

$$M\chi = \omega\chi \ , \quad (2.6)$$

where M is an $N \times N$, non-Hermitian matrix and the vector χ contains the coefficients a_n. Solutions for the eigenvalues ω are obtained numerically, and N is increased until convergence is achieved (generally in the range $20 \leq N \leq 50$).

It is found that the first unstable mode has $\alpha_c = 1.02056$, $\beta = 0$, and occurs at the critical Reynolds number $R_c = 5772.22$. Carefully constructed laboratory experiments (see M. Nishioka et al., 1975) have verified this result, but early experiments consistently gave values of R_c in the range 1000 to 1200. Unless great pains are taken, these lower values on the laboratory thresholds will result.

The discrepancy has been explained as follows. There exists a finite-amplitude, two-dimensional, x-periodic perturbation to the original flow which is an equilibrium state or a quasi-equilibrium (decaying very slowly) for $R \gtrsim 1000$. For $R \gtrsim 1000$, this two-dimensional state is itself linearly unstable to three-dimensional perturbations. Early lab experiments actually tested the stability, not of the plane Poiseuille flow, but of something resembling the two-dimensional quasi-equilibrium state, which was superimposed on the original flow by uncontrolled fluctuations in the entry region of the channel. Nishioka et. al. (1975) were able to suppress these fluctuations and thus approximate closely the "ideal" plane Poiseuille flow. But under anything like normal engineering conditions, the actual transition in shear-flow turbulence can be expected to occur not far above $R = 1000$ to 1200. (See Orszag and Kells, 1980, and Orszag and Patera, 1983.) Over 50 years were required to sort out these subtleties.

Chapter 2 (continued)

THE ORIGIN OF TURBULENCE PART II: MHD ANALOGUE

The MHD "analogue" of the Poiseuille flow problem is considered now. Consider two rigid, perfectly conducting boundaries at $y = +1, -1$ confining a plasma, and suppose the current is in the z direction. producing a magnetic field in x direction, i.e.,

$$\boldsymbol{j}^{(0)} = j_0(y)\,\hat{e}_z\,, \qquad \boldsymbol{B}^{(0)} = B_0(y)\,\hat{e}_x\,,\ j_0(y) = -B_0'(y). \tag{2.7}$$

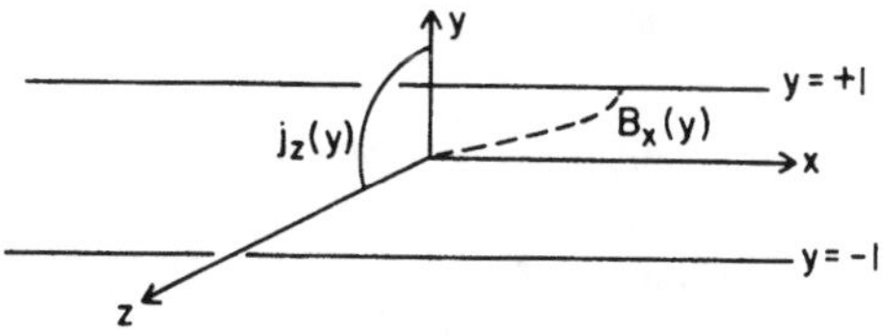

Figure 2.2: Geometry of the magnetic field and electric current density for the "sheet pinch."

The governing equations are

$$\frac{\partial \mathbf{v}}{\partial t} + \mathbf{v}\cdot\nabla\mathbf{v} = -\nabla p^* + \boldsymbol{B}\cdot\nabla\boldsymbol{B} + \frac{1}{M}\nabla^2\mathbf{v}\,, \tag{2.8a}$$

$$\frac{\partial \boldsymbol{B}}{\partial t} + \mathbf{v}\cdot\nabla\boldsymbol{B} = \boldsymbol{B}\cdot\nabla\mathbf{v} + \frac{1}{S}\nabla^2\boldsymbol{B}. \tag{2.8b}$$

For the purely magnetic problem without a steady flow

$$\mathbf{v} = 0, \qquad \nabla p^* = 0, \tag{2.9}$$

so that

$$\frac{\partial \boldsymbol{B}^{(0)}}{\partial t} = \frac{1}{S}\nabla^2\boldsymbol{B}^{(0)}\,, \tag{2.10}$$

implying

$$\frac{\partial \boldsymbol{j}^{(0)}}{\partial t} = \frac{1}{S}\nabla^2\boldsymbol{j}^{(0)}. \tag{2.11}$$

It follows from this equation that the current resistivity decays, and we cannot have a true steady state as in the hydrodynamic Poiseuille problem.

Since the magnetic field decays according to the equation

$$\boldsymbol{B}^{(0)}(t) = \exp\left(\frac{t\nabla^2}{S}\right)\boldsymbol{B}^{(0)}(y,t=0)\ , \tag{2.12}$$

it is not possible to study the stability in a way that is completely analogous to that used for the Poiseuille problem. However, one can follow one of the following procedures to make the problem closely parallel:

(1) "Freeze" $\boldsymbol{B}^{(0)}$ and calculate stability anyway.

(2) Cut slits and apply external electric field, thus losing the one-dimensionality.

(3) Imagine a dielectric-coated boundary which permits $\boldsymbol{E}_t \neq 0$, and consider a spatially dependent resistivity

$$\boldsymbol{E} = E_0\,\hat{e}_z = \text{const}\ , \tag{2.13}$$

$$\eta(y)j_z^{(0)}(y) = \text{const}\ . \tag{2.14}$$

The approach (1) is followed in the paper by R. B. Dahlburg et al. (1983). It is expected to give reliable results if the relevant time scales in the stability problem are small compared to the decay time. So, unlike the Poiseuille problem, where the steady state to be perturbed is unique, one can consider perturbing any given field $\boldsymbol{B}^{(0)}$. For perturbations of the form

$$\sim e^{i\alpha x - i\omega t} f(y)\ , \tag{2.15}$$

the MHD analogue of the Orr-Sommerfeld equation is

$$\begin{aligned}(D^2-\alpha^2)^2\mathrm{v} &= -\,i\omega M(D^2-\alpha^2)\mathrm{v} \\ &\quad -\,i\alpha MB_0(D^2-\alpha^2)b + i\alpha M(D^2B_0)b\ ,\end{aligned} \tag{2.16}$$

$$(D^2-\alpha^2+i\omega S)b = -i\alpha SB_0\mathrm{v}, \tag{2.17}$$

where

$$D \equiv \frac{d}{dy}. \tag{2.18}$$

In the same notation, the Orr-Sommerfeld equation would have been

$$(D^2-\alpha^2)^2\mathrm{v} = i\alpha R\left[\left(U_0-\frac{\omega}{\alpha}\right)\left(D^2-\alpha^2\right)\mathrm{v}-\left(D^2U_0\right)\mathrm{v}\right]. \tag{2.19}$$

The MHD boundary conditions are

$$\left.\begin{array}{ll} \mathrm{v} & = 0 \\ D\mathrm{v} & = 0 \\ b & = 0 \end{array}\right\} \text{ at } y = \pm 1. \tag{2.20}$$

The problem is solved by expanding in Chebyshev series, e.g.,

$$B_0(y) = \sum_{n=0}^{N} B_n T_n(y) \, , \tag{2.21}$$

$$\mathrm{v}(y) = \sum_{n=0}^{\infty} \mathrm{v}_n T_n(y). \tag{2.22}$$

The stability analysis parallels Orszag's, with a few minor variations. The $B_0(y)$ profiles studied by Dahlburg et al. (1983) are:

$$B_0^I(y) = y - y^3/3 \, ,$$

$$B_0^{II}(y) = \tan^{-1}\gamma y - \gamma y(\gamma^2 + 1)^{-1} \, ,$$

$$B_0^{III}(y) = y - y^{21}/21 \, ,$$

$$B_0^{IV}(y) = \sinh^{-1}\gamma y - \gamma y(\gamma^2 + 1)^{-1/2} \, . \tag{2.23}$$

These fields all correspond to "sheet pinch" geometries, as can be seen from field and current distribution for $B_0^{II}(y)$ (Fig. 2.3).

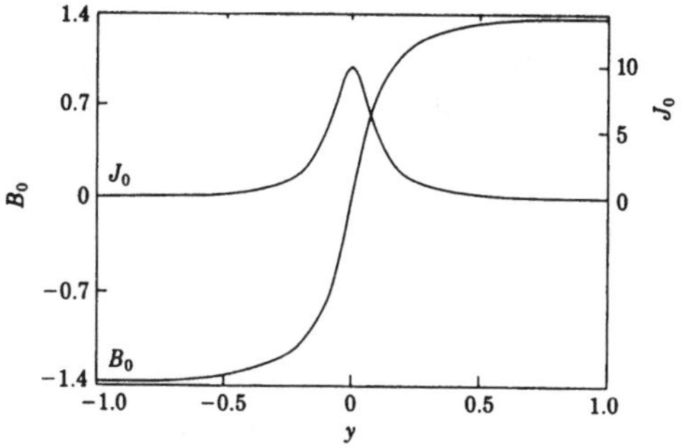

Figure 2.3: Current density and magnetic field profiles as functions of y for $B_0^{II}(y)$.

It appears from the numerical study that the system is stable unless j_0 has an inflection point (see Figs 2.4a and 2.4b). It has not yet been possible to prove this result analytically.

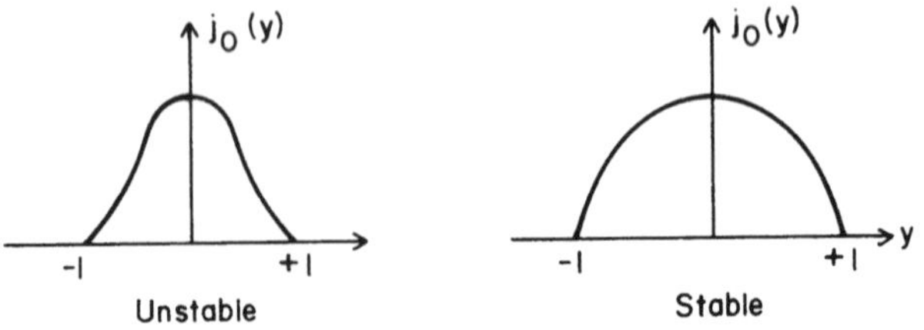

Figure 2.4: Conjectured unstable (a) and stable (b) current profiles (based only upon a limited numerical exploration).

The neutral stability curve $\omega = 0$ in the α, S plane for different values of M for B_0^{II} with $\gamma = 10$ is shown below in Fig. 2.5.

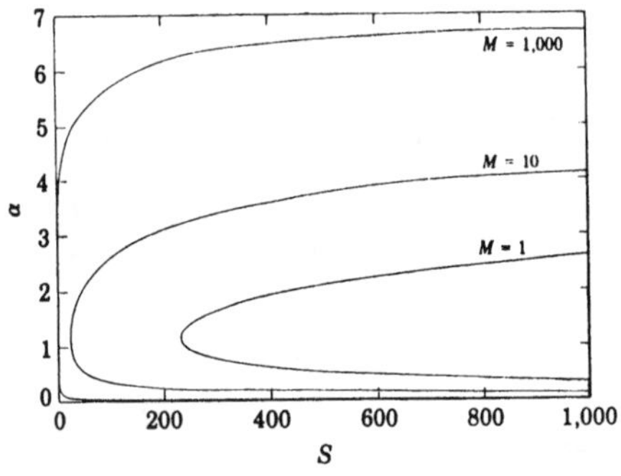

Figure 2.5: Neutral stability curve ($\omega = 0$) in α, S plane for different M values for $B_0^{II}(y)$ with $\gamma = 10$.

The critical values of the numbers S, M separating regions of stability from regions of instability are shown in the plot of Fig. 2.6. The eigenmodes are of such a character that the field lines in the xy plane look like those in Fig. 2.7. It is clear that the eigenmodes have "magnetic islands," which are a common feature of the "tearing mode" type of instabilities, and this general geometrical structure persists in the general nonlinear problem.

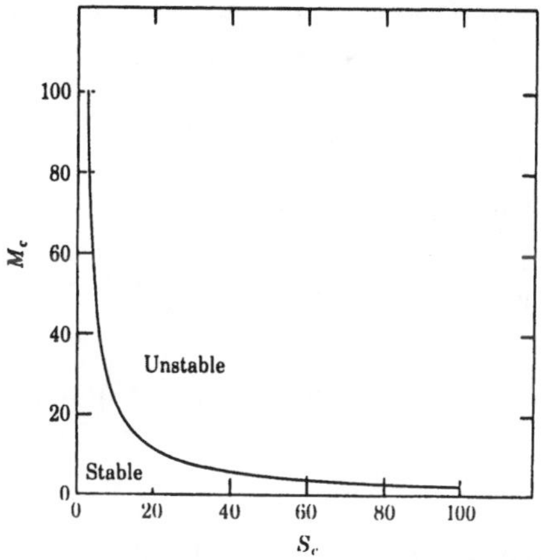

Figure 2.6: Critical curve in S-M plane separating the stable and unstable regions for $B_0^{II}(y)$ with $\gamma = 10$.

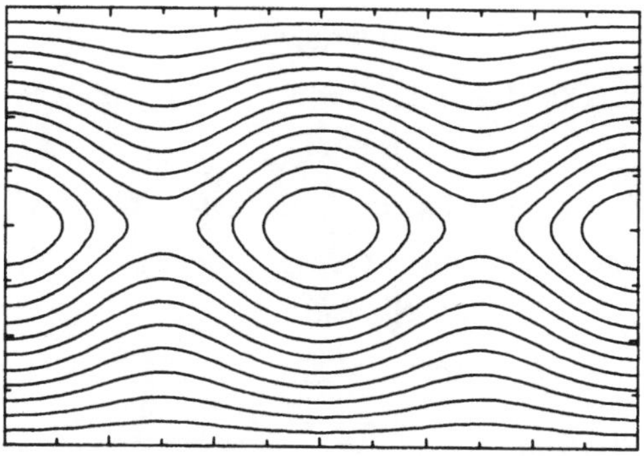

Figure 2.7: Magnetic field lines for linearly unstable configuration (quasi-equilibrium plus 5% eigenmode).

The reader is reminded that only two-dimensional perturbations have been considered here. It was seen in the Orr-Sommerfeld problem that the introduction of 3D instabilities can make the system unstable for Reynolds numbers substantially lower than the critical Reynolds number obtained from a purely 2D analysis. It is conceivable that 3D perturbations would have a similar effect upon the analogous MHD problem and would diminish the region of stability in the parameter space (see Montgomery, et al., 1987).

In a recent 3D numerical solution of the incompressible dissipative MHD equations, initial conditions were chosen so that, except for a slight amount of 3D noise, they were identical to the periodic conditions for the sheet pinch that has been extensively studied in 2D. Sharp departures from the 2D results were observed, centering around a much faster (Alfvénic) break-up of the ordered pattern, with a much faster deterioration of the characteristic x-point configuration. The results suggest that 3D MHD fluids, like 3D Navier-Stokes fluids, are much more prone to the onset of "subcritical" turbulence than 2D ones, in the absence of any stabilizing dc magnetic field (see Montgomery et al., 1987).

A Note on References

The study of the stability problem within the ideal MHD framework ($\nu = 0, \eta = 0$) began within a few years of the development of MHD. Some of the useful references on MHD stability are:

T. G. Cowling, *Magnetohydrodynamics* (Interscience, 1957).

G. Bateman, *MHD Instabilities* (Cambridge: MIT Press, 1978).

I. Bernstein, E. Frieman, M. Kruskal, and R. Kulsrud, *Proc. Roy. Soc.* (London), **A244**, 17 (1958).

M. Kruskal and M. Schwarzschild, *Proc. Roy. Soc.*, **A223**, 348 (1954).

W. Mannheimer and C. Lashmore-Davies, *MHD Instabilities in Simple Plasma Configurations* (Washington: US Naval Research Laboratory, 1984).

V. D. Shafranov, *Sov. Phys.: Technical Phys.*, **15**, 175 (1970).

Cowling's small volume is still a good introduction to MHD for people without any previous background. The Kruskal and Schwarzschild (1954) paper is one of the first papers to study the stability of a "pinch" configuration. Bernstein et al. (1958) systematically developed an energy principle for *ideal* MHD stability problems.

The distinctive features of resistive instabilities were not realized in the early years of plasma physics. Only when people started building plasma devices was it concluded that many configurations which would have been stable according to the ideal MHD analysis were actually not stable. One of the first studies of the resistive instabilities is the rather difficult paper (which has in one sense given rise to the whole subject): Furth, H. P., J. Killeen, and M. N. Rosenbluth, 1963: *Phys. Fluids*, **6**, 459.

Dahlburg, R. B., T. A. Zang, D. Montgomery, and M. Y. Hussaini, 1983: Proc. Nat'l. Acad. Sci. (USA), **80**, 5798.

Gottlieb, D., and S. A. Orszag, 1971: *Numerical Analysis of Spectral Methods: Theory and Applications*, NSF-CBMS Monograph no. 26, *Soc. Ind. App. Math.*, Philadelphia, Penn.

Lanczos, C. 1956: "Tau-method," in *Applied Analysis*, (Englewood Cliffs, N. J.: Prentice-Hall).

Montgomery, D., J. P. Dahlburg, and R. B. Dahlburg, 1987: *Bull. Am. Phys. Soc.*, Ser. II, **32** (No. 9), 1819.

Nishioka, M., S. Iida, and Y. Ichikawa, 1975: *J. Fluid Mech.*, **72**, 731.

Nishioka, M., S. Iida, and S. Kanbayashi, 1978: in *Proc. 10th Turbulence Symposium of the Institute of Aeronautical and Space Sciences*, Tokyo Univ., p. 55 ff.

Orszag, S. A., 1971: *J. Fluid Mech.*, **50**, 689.

Orszag, S. A., and L. C. Kells, 1980: *J. Fluid Mech.*, **96**, 159.

Orszag, S. A., and A. T. Patera, 1983: *J. Fluid Mech.*, **128**, 347.

Representative *non-ideal* references:

Dahlburg, R. B., T. A. Zang, and D. Montgomery, 1986: *J. Fluid Mech.*, **169**, 71.

Matthaeus, W. H., and S. L. Lamkin, 1986: *Phys. Fluids*, **29**, 2513.

Chapter 3

IDEAL MHD STABILITY, INCOMPRESSIBILITY CONDITIONS, AND IDEAL INVARIANTS

Much of the early MHD literature was concerned with ideal equilibria and their stability properties. An ideal MHD plasma is in equilibrium if the pressure and the $\boldsymbol{j} \times B$ force in the momentum equation balance each other, so that the plasma is not accelerated in any direction. This happens when

$$\nabla p = \boldsymbol{j} \times B \,, \tag{3.1}$$

a condition which can, of course, only be realized if $\nabla \times (\boldsymbol{j} \times \boldsymbol{B}) = 0$. By taking the scalar product of both sides of (3.1) with $\boldsymbol{B}$ and $\boldsymbol{j}$, we see, respectively, that

$$\boldsymbol{B} \cdot \nabla p = 0 \,, \tag{3.2a}$$

$$\boldsymbol{j} \cdot \nabla p = 0 \,, \tag{3.2b}$$

indicating that both the magnetic field lines and the current density vector lie on nested surfaces of constant pressure. This highly idealized construction has exerted extraordinary influence on thinking about magnetofluids.

In cylindrical geometry, with $p = p(r)$, we have

$$\boldsymbol{j} = (0, j_\theta(r),\ j_z(r)) = \nabla \times \boldsymbol{B} \,, \tag{3.3a}$$

$$\boldsymbol{B} = (0, B_\theta(r),\ B_z(r)) \,, \tag{3.3b}$$

which, when inserted into (3.1), gives

$$\frac{dp(r)}{dr} = j_\theta B_z - j_z B_\theta \,. \tag{3.4}$$

In the MHD literature many different kinds of equilibria satisfying (3.4) have been studied. It is not difficult at all to find ideal MHD equilibria which obey such relations, but the presence of even small amounts of dissipation make the problem much more difficult. An important historical example is the Bennett pinch (Bennett, 1934) in which $\boldsymbol{B} = B_\theta(r)\, \boldsymbol{e}_\theta$ only. Inserting this expression for $\boldsymbol{B}$ in (3.4) and making use of the relation between $\boldsymbol{j}$ and $\boldsymbol{B}$ in (3.3), we get the equilibrium equation

$$p(r) + \frac{B_\theta^2}{2} + \int_0^r \frac{B_\theta^2}{r'}dr' = \text{const} . \tag{3.5}$$

After having determined an equilibrium state of the plasma, the next question has been whether or not this equilibrium is stable. The hope in magnetic confinement physics has been to find a *stable equilibrium* and then set it up experimentally. The relevance of this paradigm has seldom been questioned. A successful method for studying the ideal stability properties of a plasma equilibrium is based on an *energy principle*. The total energy, E, of an ideal MHD plasma, obeying a polytrope law for the pressure, integrated over the whole plasma volume, is a sum of the kinetic, the thermal, and the magnetic energy, as expressed by

$$E = \int (\frac{1}{2}\rho v^2 + \frac{p}{\gamma - 1} + \frac{B^2}{2})d^3x = \text{const.} \tag{3.6}$$

The plasma is now perturbed slightly about an equilibrium with no flow, so that a small fluid element originally at the equilibrium position $\boldsymbol{r}_0$ is displaced by the Lagrangian displacement vector $\boldsymbol{\xi}(\boldsymbol{r}_0, t)$. This displacement cannot be completely arbitrary, since it must satisfy (3.6), which states that E is constant under the plasma motion. If E at equilibrium is a minimum with respect to displacements $\boldsymbol{\xi}$, there are no possible energy-conserving stable positions, and the plasma is locked into a stable $\boldsymbol{\xi} = 0$ position.

Expanding all the field quantities $\boldsymbol{B}$, j, ρ and p to first order in terms of $\boldsymbol{\xi}$ and its derivatives leads to

$$\rho(\boldsymbol{r}_0)\frac{d^2\boldsymbol{\xi}(\boldsymbol{r}_0, t)}{dt^2} = \boldsymbol{F}(\xi), \tag{3.7}$$

where

$$\begin{aligned} \boldsymbol{F}(\xi) = \nabla(\rho_0 \cdot \nabla p_0 + \gamma p_0 \nabla \cdot \xi) + (\nabla \times B_0) \times (\nabla \times (\xi \times B_0)) \\ + (\nabla \times [\nabla \times (\xi \times B_0)]) \times B_0 . \end{aligned} \tag{3.8}$$

(The zero subscripts indicate equilibrium quantities.) The function $\boldsymbol{F}(\xi)$ is self-adjoint since for reasonable boundary conditions

$$\int \boldsymbol{\xi} \cdot \boldsymbol{F}(\eta) d^3x = \int \boldsymbol{\eta} \cdot \boldsymbol{F}(\xi) d^3x. \tag{3.9}$$

The question is now whether the perturbations ξ will grow or decay with time. Instead of solving the differential equation (3.7), it suffices to look for solutions $\xi(r_0, t) = \xi(r_0)\exp(-i\omega t)$ which, when inserted into (3.7), give the time-independent eigenvalue equation

$$-\omega^2 \xi \rho_0 = \boldsymbol{F}(\xi), \tag{3.10}$$

and to ask if the eigenvalue ω^2 can be greater than 0 (stable perturbation) or less than 0 (unstable perturbation).

It can be shown that the sign of ω^2 is determined by the sign of the functional

$$\delta W = -\frac{1}{2}\int \xi \cdot \boldsymbol{F}(\xi) d^3x, \tag{3.11}$$

which represents the change in the potential energy introduced by the perturbation. If $\delta W > 0$ for all ξ, the plasma is stable. However, if $\delta W < 0$ for any physically allowable ξ, the equilibrium is unstable.

It turns out that the simple, cylindrical Bennett pinch described above is unstable to various kinds of perturbations. However, there do exist some "stabilized" pinches, which in their toroidal versions can be classified as "tokamaks," characterized by $B_z \gg B_\theta$ and $\beta \equiv p/(B^2/8\pi) \ll 1$, or "screw pinch" where $B_z \sim B_\theta$ and $\beta \sim 1$; or "theta pinch," where $B_\theta \gg Bz$ and $\beta \sim 1$. We note that additional toroidal effects must be included in the equilibrium and stability analyses of these devices, and there is no limit to the complexity of possible stability analyses for various boundary conditions and equilibrium profilers.

Both the tokamak and the screw pinch will confine a plasma. However, both have been seen to have a certain level of MHD turbulence, and both have ill-understood pathologies that prevent their use as a fusion reactor at present. We shall not even scratch the surface of the many hundreds of stability analyses of their putative equilibria that have been carried out.

Let us now turn to the question of the conditions under which the MHD equations can be simplified by the assumption of incompressibility. The basic set of MHD equations is as follows:

$$\begin{aligned} \frac{\partial \boldsymbol{B}}{\partial t} &= \nabla(\mathbf{v} \times \boldsymbol{B}) + \eta \nabla^2 B \\ &= \nabla \times [\mathbf{v} \times \boldsymbol{B} - \eta \nabla \times B], \end{aligned} \tag{3.12}$$

$$\frac{\partial \mathbf{v}}{\partial t} + \mathbf{v}\cdot\nabla\mathbf{v} = -\frac{1}{\rho}\nabla p + \frac{1}{\rho}\boldsymbol{j}\times\boldsymbol{B} + \nu\left[\nabla^2\mathbf{v} + \frac{1}{3}\nabla(\nabla\cdot\mathbf{v})\right], \tag{3.13}$$

$$\nabla\cdot\boldsymbol{B} = 0\,, \tag{3.14}$$

$$\frac{\partial \rho}{\partial t} + \nabla\cdot(\rho\mathbf{v}) = 0, \tag{3.15}$$

and

$$\boldsymbol{j} = \nabla\times\boldsymbol{B}. \tag{3.16}$$

To see when the incompressibility assumption is valid, we follow the variation in density of a fluid element as it moves around:

$$\frac{\delta\rho}{\rho} \sim \frac{\delta p}{\rho\frac{dp}{d\rho}} = \frac{\delta p}{\rho c_s^2}\,, \tag{3.17}$$

where, in the first step, we have assumed that pressure and density are approximately related to each other via some equation of state $p = p(\rho)$, and used the definition of sound speed, i.e., $c_s^2 = dp/d\rho$. The pressure variation seen by a fluid element can be obtained rather simply by balancing terms in the equation of motion, i.e.,

$$\frac{\delta p}{\rho} \sim \delta(\mathrm{v}^2) \text{ or } \frac{\delta(B^2)}{\rho}\,, \tag{3.18}$$

whichever is greater.

By $\delta(\ldots)$, we mean the maximum variation seen by a fluid element which moves with velocity $\mathbf{v}$, *not* the maximum variation from place to place in the fluid. (If the turbulence is fully developed, there may be no distinction; but in MHD, and in particular for confined magnetofluids, the turbulence may be quite local, and fluid elements may take a long time to sample regions with very different B.)

Substituting (3.18) into (3.17), we have

$$\frac{\delta\rho}{\rho} \sim \frac{\delta(\mathrm{v}^2) \text{ or } \delta(C_A^2)}{C_s^2}. \tag{3.19}$$

The incompressibility assumption is justified only if $\delta\rho/\rho \ll 1$. C_A^2 is the square of the Alfvén speed.

In fact, for many magnetofluids other than liquid metals, the assumption of incompressibility is hard to justify. However, in the presence of a strong dc magnetic field, fluid elements may be slow to move away from their initial locations even in the presence of a significant level of fluctuations.

The accuracy of the incompressibility assumption depends strongly on the specific properties of the turbulence under consideration. From a "turbulence theory" point of view, incompressibility is desirable if one wants to take advantage of the knowledge of Navier-Stokes turbulence that exists. However, it is fair to point out that compressible turbulence theory is undeveloped, even for neutral fluids.

Having stated the conditions needed for incompressibility to be satisfied, we shall henceforth assume $\nabla \cdot \mathbf{v} = 0$. The pressure can then be determined from

$$\nabla \cdot \left(\frac{1}{\rho}\nabla p\right) = \nabla \cdot \left[\frac{\boldsymbol{j} \times \boldsymbol{B}}{\rho} - \mathbf{v} \cdot \nabla \mathbf{v}\right] + \text{boundary conditions} . \tag{3.20}$$

This expression simplifies if we assume constant density (= 1, say, in appropriate dimensionless units). In that case, we may write

$$\nabla^2 p^* = \nabla \cdot [\boldsymbol{B} \cdot \nabla \boldsymbol{B} - \mathbf{v} \cdot \nabla \mathbf{v}] , \tag{3.21}$$

where

$$p^* \equiv p + \frac{B^2}{2}. \tag{3.22}$$

In Fourier space, the relation (3.21) can be written in the presence of periodic boundary conditions as

$$-k^2 p^*(\boldsymbol{k}) = -\boldsymbol{k}\,k : \left[(\boldsymbol{B}\,\boldsymbol{B})_{\boldsymbol{k}} - (\mathbf{v}\,\mathbf{v})_{\boldsymbol{k}}\right] \tag{3.23}$$

or

$$p^*(\boldsymbol{k}) = \frac{\boldsymbol{k}\boldsymbol{k}}{k^2} : \sum_{\boldsymbol{l}} [\boldsymbol{B}(\boldsymbol{k}-\boldsymbol{l})\boldsymbol{B}(\boldsymbol{l}) - \mathbf{v}(\boldsymbol{k}-\boldsymbol{l})\mathbf{v}(\boldsymbol{l})] . \tag{3.24}$$

We see that pressure disappears from the equations altogether when the expression (3.24) is inserted back into (3.13).

Using the vector potential $\boldsymbol{A}$, we may pull off the curl in (3.12), so

$$\frac{\partial \boldsymbol{A}}{\partial t} = \mathbf{v} \times \boldsymbol{B} - \eta \boldsymbol{j} - \nabla \varphi , \tag{3.25}$$

where we have noted that the emerging equation is undetermined up to the gradient of a scalar. If we were to invoke the Coulomb gauge, for example, φ could be determined from

$$\nabla^2\varphi = -\nabla \cdot (\mathbf{v} \times \boldsymbol{B}). \tag{3.26}$$

$\boldsymbol{A}$ is related to the magnetic flux through Stokes's theorem:

$$\int_{Surface} \boldsymbol{B} \cdot d\boldsymbol{s} = \oint_{Perimeter} \boldsymbol{A} \cdot d\boldsymbol{l}\,. \tag{3.27}$$

If $\eta = 0$, this flux is a conserved quantity

$$\frac{D}{Dt}\int_{Surface} \boldsymbol{B} \cdot d\boldsymbol{s} = 0 = \frac{D}{Dt}\oint_{Perimeter} \boldsymbol{A} \cdot d\boldsymbol{l}\,. \tag{3.28}$$

Turbulence is regulated, to an as yet unknown extent, by pointwise ideal invariants (i.e., invariants which follow from the assumption $\nu = 0 = \eta$). It is useful to draw an analogy between hydrodynamics and MHD in both 2D and 3D:

Navier-Stokes	*MHD*
3D	3D
$\frac{D}{Dt}\oint \mathbf{v} \cdot d\boldsymbol{l} = 0$	$\frac{D}{Dt}\oint \boldsymbol{A} \cdot d\boldsymbol{l} = 0$
(Kelvin's theorem)	(Alfvén's theorem)
2D	2D
$\mathbf{v} = (v_x,\ v_y,\ 0)$	$\boldsymbol{B} = \left(B_x,\ B_y, 0\right) = \nabla \times \boldsymbol{A}$
$\frac{\partial}{\partial z} = 0$	$\frac{\partial}{\partial z} = 0$
$\frac{D}{Dt}\oint \mathbf{v} \cdot d\boldsymbol{l} = 0$	$\boldsymbol{A} = (0,\ 0,\ a)$
$\Downarrow$	$\frac{D}{Dt}\oint \boldsymbol{A} \cdot d\boldsymbol{l} = 0$
$\left(\frac{\partial}{\partial t} + \mathbf{v} \cdot \nabla\right)\omega = 0$	$\Downarrow$
$\nabla \times \mathbf{v} = (0,\ 0,\ \omega)$	$\left(\frac{\partial}{\partial t} + \mathbf{v} \cdot \nabla\right)a = 0$
	$\boldsymbol{j} = (0,\ 0,\ j) = -\nabla^2\boldsymbol{A}.$

We note that the line contour involved in the 2D version of Kelvin's theorem lies in the (x,y) plane, while the line contour involved in the 2D version of Alfvén's theorem must have some part out of the (x,y) plane in order to give a non-trivial result. A is, of course, unvarying along the part of the contour $\perp$ the x,y plane.

Much has been made of "frozen-in" magnetic flux and moving MHD fluids. The flux is frozen-in until $\eta\nabla^2$ becomes large; then "reconnection" occurs, and a plasma can readily move across a magnetic field. The extent to which small scales develop is crucial to the question of how small-scale development occurs. Small scales develop very readily in Navier-Stokes fluids, and until now, the analogous subject of "vortex reconnection" has played a negligible role in elucidating their dynamics.

We have a pretty good picture of how the small-scale features develop in 2D, and the extent to which they modify the macroscopic features in the case of random, periodic initial conditions. Again, the analogy between hydrodynamics and MHD is useful:

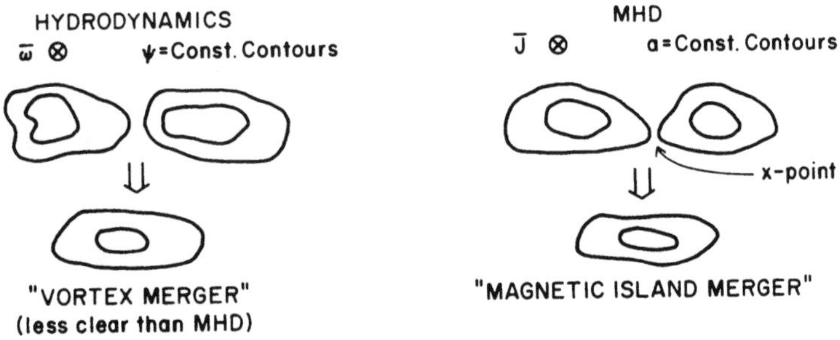

Figure 3.1: The "slightly non-ideal" event for 2D Navier-Stokes fluids is the merger of contours of constant $\omega = -\nabla^2\psi$, but the analogous MHD event is the merger of contours of constant a, for which $\nabla^2 a = -j$.

What we do not know are :

1) How important "topological invariant violations" are in real life (solar wind penetration into magnetosphere; tokamak disruptions due to "magnetic island" merger), or indeed whether topological invariants are invariant enough to be worth considering.

2) The significant differences between 2D and 3D. There may be 3D analogies of 2D coherent structures that we have not conceived of yet. This is one of the principal motivations for doing 3D simulations.

There are also "global" invariants in 2D and 3D for both the Navier-Stokes equations and MHD. These are given by:

Energy $= \int_{volume} \left(\frac{1}{2}\mathrm{v}^2 + \frac{1}{2}B^2\right) dV$, (2D, 3D)

Helicity $= \begin{cases} \frac{1}{2}\int \mathbf{v}\cdot\boldsymbol{\omega} dV & \text{(3D NS)} \\ \frac{1}{2}\int \boldsymbol{A}\cdot\boldsymbol{B} dV & \text{(3D MHD)} \end{cases}$

"Cross helicity" (2D, 3D)

or "Cross energy $= \frac{1}{2}\int \mathbf{v}\cdot\boldsymbol{B} dV$, (MHD)

Enstrophy $= \int \omega^2 dV$, (2D NS)

"Mean square vector potential" $= \int a^2 dV$. (2D MHD)

Other global invariants, but less useful or well-understood, also exist:

$$\int \omega^n dV \,, \qquad (2D\ NS)$$

and

$$\int a^n dV \,. \qquad (2D\ MHD)$$

The role of various ideal invariants in constraining turbulence is one of the least well understood and most controversial in the whole subject. References on MHD equilibria and their stability are very numerous. Some useful publications are:

Bateman, G., 1978: *MHD Instabilities* (Cambridge: MIT Press).

Bennett, W., 1934: *Phys. Rev.*, **45**, 890.

Bernstein, I., E. Frieman, M. Kruskal, and R. Kulsrud, 1958: *Proc. Roy. Soc.* **A244**, 17.

Cowling, T. G., 1957: *Magnetohydrodynamics* (New York: Interscience).

Freidberg, J. P., 1982: *Rev. Mod. Phys.*, **54**, 801.

Kruskal M., and M. Schwarzschild, 1954: *Proc. Roy. Soc.*, **A223**, 348.

Mannheimer, W., and C. Lashmore-Davies, 1984: *MHD Instabilities in Simple Plasma Configurations* (Washington: U.S. Naval Research Laboratory). *non-ideal* Shafranov, V. D., 1970: *Sov. Phys.: Technical Phys.*, **15**, 175.

Chapter 4

SOME CONNECTIONS WITH STATISTICAL MECHANICS

The incompressible MHD equations, in familiar dimensionless units, are

$$\left(\frac{\partial}{\partial t} + \mathbf{v}\cdot\nabla\right)\mathbf{v} = -\nabla\left(p + \frac{B^2}{2}\right) + \boldsymbol{B}\cdot\nabla\boldsymbol{B} + \nu\nabla^2\mathbf{v}\ , \qquad (4.1)$$

$$\frac{\partial \boldsymbol{B}}{\partial t} = \nabla\times(\mathbf{v}\times\boldsymbol{B}) + \eta\nabla^2\boldsymbol{B}\ , \qquad (4.2)$$

$$(\rho \equiv 1) \qquad \nabla\cdot\mathbf{v} = 0\ . \qquad (4.3)$$

Here, ν and η are dimensionless and represent inverse Reynolds numbers characteristic of the particular flow of interest. The Navier-Stokes equations may be recovered by setting $\boldsymbol{B} \equiv 0$. For the case of turbulence, the fields $\mathbf{v}$ and $\boldsymbol{B}$ are highly disordered.

Now, we wish to attempt to connect MHD turbulence with mainstream statistical mechanics. To do this, let us set $\boldsymbol{B} \equiv 0$ and consider the Navier-Stokes equations; these provide a simpler set of equations than the full MHD set and can serve as a general example of both. To begin, we expand the velocity $\mathbf{v}(x,t)$ in orthogonal functions, for example, in a Fourier series,

$$\mathbf{v}(\boldsymbol{x},t) = \sum_{\boldsymbol{k}} \mathbf{v}(\boldsymbol{k},t)e^{i\boldsymbol{k}\cdot\boldsymbol{x}}. \qquad (4.4)$$

Henceforth, the argument of $\mathbf{v}$ will indicate whether we mean the physical ($\boldsymbol{x}$-space) values $\mathbf{v}(x)$ or the Fourier ($\boldsymbol{k}$-space) values $\mathbf{v}(k)$; also, the time will be omitted from the argument, when this causes no difficulty. Putting this expansion for $\mathbf{v}(x)$ into the incompressibility condition $\nabla\cdot\mathbf{v}(x) = 0$ yields the algebraic relationship

$$\boldsymbol{k}\cdot\mathrm{v}(k) = 0\ , \qquad (4.5)$$

while putting it into the Navier-Stokes equation yields a set of ordinary differential equations

$$\frac{d\mathbf{v}(k)}{dt} + i\boldsymbol{k}\cdot\sum_{\boldsymbol{q}}\mathbf{v}(k-q)v(q) = -i\boldsymbol{k}p(\boldsymbol{k}) - \nu k^2\mathbf{v}(k). \qquad (4.6)$$

Now, if we take a dot product of $i\boldsymbol{k}$ with these equations and use $\boldsymbol{k} \cdot \mathrm{v}(k) = 0$, we get

$$k^2 p(\boldsymbol{k}) = -\boldsymbol{kk} : \sum_{\boldsymbol{q}} \mathrm{v}(k-q)\mathrm{v}(q). \tag{4.7}$$

Next, put this expression for $\mathrm{p}(\boldsymbol{k})$ back into the $\boldsymbol{k}$-space evolution equations for $\mathrm{v}(\boldsymbol{k})$; after some algebra, we have,

$$d\frac{\mathrm{v}_\alpha(\boldsymbol{k})}{dt} = -\nu k^2 \mathrm{v}_\alpha(\boldsymbol{k}) + M_{\alpha\beta\gamma}(\boldsymbol{k}) \sum_{\boldsymbol{q}} \mathrm{v}_\beta(\boldsymbol{k}-q)\mathrm{v}_\gamma(\boldsymbol{q}), \tag{4.8}$$

$$M_{\alpha\beta\gamma}(\boldsymbol{k}) = -\frac{i}{2}\left[k_\gamma P_{\alpha\beta}(\boldsymbol{k}) + k_\beta P_{\alpha\gamma}(\boldsymbol{k})\right], \tag{4.9}$$

$$P_{\alpha\beta}(\boldsymbol{k}) = \delta_{\alpha\beta} - \frac{k_\alpha k_\beta}{k^2}. \tag{4.10}$$

We can deal with the statistics of both the MHD and Navier-Stokes cases by considering a generic form which represents (4.8). Order the real and imaginary parts of the Fourier coefficients into a long column vector whose i^{th} component is X_i. Then

$$\frac{d}{dt}X_i = -\nu_i X_i + \sum_{j,k} C_{ijk} X_j X_k \qquad (X_i \equiv real,\ i,j,k \equiv \text{integers}) \tag{4.11}$$

can stand for (4.8), with coupling coefficients C_{ijk}, which are related to the $M_{\alpha\beta\gamma}$ $(\boldsymbol{k})$. Here $\nu k^2 \rightarrow \nu_i$. Now, we set $\nu_i \equiv 0$ and restrict the possible indices of X_i to some finite but large set; thus, we have a truncated, conservative model:

$$X_i \equiv \frac{d}{dt}X_i = \sum_{j,k} C_{ijk} X_j X_k. \tag{4.12}$$

Equation (4.12) can be shown to conserve *finitely truncated* representations of ideal invariants which are quadratic in the expansion coefficients, like the energy.

A *phase space* whose coordinates are the X_i can be introduced and used to develop an equilibrium statistical mechanics, since (4.12) describes a dynamics that obeys the Liouville equation. In particular, the conventional Gibbs-Boltzmann arguments can be used to construct canonical ensembles from the energy.

There is a second quadratic invariant that may be formed for 3D Navier-Stokes flows, namely the helicity H:

$$H = \int \mathbf{v} \cdot \nabla \times \mathbf{v} d\boldsymbol{x} = \int \mathbf{v} \cdot \boldsymbol{\omega} d\boldsymbol{x} = \tag{4.13}$$

$$\sum_{\boldsymbol{k}} \mathbf{v}^*(\boldsymbol{k}) \cdot [i\boldsymbol{k} \times \mathbf{v}(\boldsymbol{k})] = i \sum_{\boldsymbol{k}} \boldsymbol{k} \cdot [\mathbf{v}(\boldsymbol{k}) \times \mathbf{v}^*(\boldsymbol{k})] . \tag{4.14}$$

Using this invariant in the derivation of the canonical ensemble demands the introduction of another Lagrange multiplier, or equivalently, another temperature. The equilibrium probability distribution D_{eq} then becomes

$$D_{eq} \propto exp(-\alpha E - \beta H) . \tag{4.15}$$

As would be expected, the case with no mean helicity ($\langle H \rangle = 0$, or equivalently $\beta = 0$) is a statement of the previous result. The significance of nonzero mean helicity in the behavior of the representative system is unclear. A more detailed discussion may be found in the references.

In the two-dimensional ideal Navier-Stokes formulation one can recover two quadratic invariants: energy E and enstrophy Ω (helicity H is trivially zero in 2D):

$$E = \frac{1}{2} \int \mathbf{v}^2 d^2 x, \tag{4.16}$$

$$\Omega = \frac{1}{2} \int \omega^2 d^2 x. \tag{4.17}$$

Doing the statistical mechanics with these invariants gives another two-temperature canonical distribution:

$$D_{eq} \propto \exp(-\alpha E - \beta \Omega) . \tag{4.18}$$

Furthermore, one can write down the probability distribution for modal energy

$$f\left(\frac{1}{2}|\mathbf{v}(\boldsymbol{k})|^2\right) \propto \exp\left(-(\alpha + \beta k^2)\frac{|\mathbf{v}(\boldsymbol{k})|^2}{2}\right), \tag{4.19}$$

since the D_{eq} factors into "one-body" Maxwellians, with the result that the expectation for modal energy at wavenumber $\boldsymbol{k}$ is

$$\left\langle |\mathbf{v}(\boldsymbol{k})|^2 \right\rangle = \frac{1}{\alpha + \beta k^2} . \tag{4.20}$$

The two inverse temperatures may be calculated, for a given $\langle E \rangle$ and $\langle \Omega \rangle$, from the defining equations (sum over all polarizations):

$$\sum_{\boldsymbol{k}} \frac{1}{\alpha + \beta k^2} = 2\langle E \rangle, \tag{4.21}$$

$$\sum_{\boldsymbol{k}} \frac{k^2}{\alpha + \beta k^2} = 2\langle \Omega \rangle. \tag{4.22}$$

For a finite-resolution representation of the field, it is easily deduced that $\alpha = \alpha(\langle E \rangle, \langle \Omega \rangle, k_{max})$ and $\beta = \beta(\langle E \rangle, \langle \Omega \rangle, k_{max})$. The temperatures α^{-1} and β^{-1} may be negative, though not both simultaneously; all that is required is that $\alpha + \beta k^2$ be > 0, for all k.

We now ask the question of what would happen to $\langle |\mathbf{v}(\boldsymbol{k})|^2 \rangle = 1/\alpha + \beta k^2$ if we fixed $\langle E \rangle$ and $\langle \Omega \rangle$, but let k_{max} go to infinity. It can be shown (Kraichnan, 1967) that in this case $\beta \to \infty$, $\alpha \to -\infty$, and $|\alpha/\beta| \to k_{min}^2$; that is, all the energy condenses into the fundamental wavelength, and the remaining enstrophy is distributed in infinitesimal increments out to $k_{max} = \infty$. This leads one to suggest that there is a fundamental tendency in two-dimensional ideal flows for the system to be dominated by some box-filling structure.

This idea is also clear in the "selective decay" process in viscous flow. It can be shown, using a variational procedure, that in viscous flow the decay of enstrophy relative to energy is negative; that is, enstrophy decays more rapidly than energy. In terms of the Fourier components of the velocity field,

$$\frac{d}{dt}\frac{\Omega}{E} = -\frac{\nu}{2E^2}\sum_{\boldsymbol{k},\boldsymbol{q}} \left(k^2 - q^2\right)^2 |\mathbf{v}(\boldsymbol{k})|^2 |\mathbf{v}(\boldsymbol{q})|^2 \le 0, \nu \neq 0, \tag{4.23}$$

because all of the terms on the right-hand side of this expression are positive. It should be noted that this expression is a prediction of the time-asymptotic behavior (namely, $\Omega/E \to k_{min}^2$ as $t \to \infty$), but does not give much information about the rate of approach to that limiting state. A lot of interesting behavior can occur on the way, but experience has shown that one eventually approaches the final state (two box-filling, counter-rotating vortices). If, however, energy is distributed between two wavenumbers $\boldsymbol{k}_1$ and $\boldsymbol{k}_2$, then

$$\frac{d}{dt}\frac{\Omega}{E} \simeq -\frac{\nu}{2E^2}\left(k_1^2 - k_2^2\right)^2 E^2 \simeq -\nu\left(\Delta\left(k^2\right)\right)^2, \tag{4.24}$$

and the approach may be slow if $k_1 \sim k_2$.

One can thus see two consequences of this tendency toward long-wavelength spectra in dissipative 2D Navier-Stokes flows: "selective decay," in which there is the asymptotic approach to the extremal ratios of ideal (inviscid) global invariants, and the "inverse cascade," in which band-limited excitations in energy are fed to the spectrum and cascaded to low-wavenumber k, while excitations in enstrophy are cascaded to high-wavenumber k. Inverse cascades will be discussed in greater detail in later chapters.

Kraichnan, R. H., 1967: *Phys. Fluids*, **10**, 1417.

Kraichnan, R. H. and D. Montgomery, 1980: Repts. *Progress in Physics*, **43**, 547.

Seyler, C. E., Y. Salu, D. Montgomery, and G. Knorr, 1975: *Phys. Fluids*, **18**, 803.

Ting, A. C., W. H. Matthaeus, and D. Montgomery, 1986: *Phys. Fluids*, **29**, 3261.

Chapter 5

MORE STATISTICAL MECHANICS

Let us consider once again the fourier transformed Navier-Stokes equations.

$$\frac{\partial v_\alpha(\boldsymbol{k},t)}{\partial t} = M_{\alpha\beta\gamma}\sum_q v_\beta(\boldsymbol{k}-q)v_\gamma(\boldsymbol{q}) - \nu k^2 v_\alpha(\boldsymbol{k},t), \tag{5.1}$$

where $M_{\alpha\beta\gamma}$ is as defined in the previous chapter. At the moment we are interested in the general behavior of this type of equation, so let us consider the model equation (all the X_i are real)

$$\frac{\partial X_i}{\partial t} = \sum_{jk} C_{ijk}X_jX_k - \nu_i X_i. \tag{5.2}$$

We will leave the coupling coefficients C_{ijk} fairly arbitrary for now. Note that for an appropriate choice of the modes, a one-to-one correspondence between this equation and the Fourier-transformed Navier-Stokes or MHD equations can be made.

Let's consider the probability distribution function in the phase space of expansion coefficients $D = D(X_1, X_2, \cdots, t)$. In the absence of friction we can write a Liouville equation $dD/dt = 0$. If we include friction and expand the total time derivative we obtain

$$\frac{\partial D}{\partial t} + \sum_{ijk} C_{ijk}X_jX_k\frac{\partial D}{\partial X_i} = \sum_i \nu_i \frac{\partial}{\partial X_i}(X_iD). \tag{5.3}$$

Though there may be other "meta-stable" attractors, clearly there is an attractor for all of the $X_i = 0$. To avoid having the problem simply decay away, we must introduce some kind of forcing. Care must be taken to avoid having the forcing introduce statistics which are not really intrinsic to equation (5.2).

If we have D we can calculate ensemble averages for various observables:

$$\langle f\rangle \equiv \int Df(X)\,dX, \tag{5.4}$$

where $dX = dX_1dX_2\cdots$. For example, we can form velocity moments such as

$$\begin{gathered}\langle \mathbf{v}(\boldsymbol{x})\rangle \\ \langle \mathbf{v}(\boldsymbol{x})\cdot\mathbf{v}(\boldsymbol{x}+\boldsymbol{r})\rangle \\ \langle \mathbf{v}(\boldsymbol{x})\cdot\nabla\times\mathbf{v}(\boldsymbol{x}+\boldsymbol{r})\rangle\ .\end{gathered} \tag{5.5}$$

Unfortunately, we usually cannot hope to calculate D and its moments directly. What we try to do is to assume that D exists and has moments, and then try to find relations purely among the moments. Customarily some general assumptions, of varying accuracy, are made concerning D to make the moments satisfy simplifying relations.

If we take the ensemble average of the equations of motion, we get an equation for $\langle X_i \rangle$,

$$\frac{\partial \langle X_i \rangle}{\partial t} = \sum_{jk} C_{ijk} \langle X_j X_k \rangle - \nu_i \langle X_i \rangle, \tag{5.6}$$

which involves $\langle X_j X_k \rangle$. So we multiply the equation of motion by X_j, ensemble average, and perform the algebra to get an equation for $\langle X_j X_k \rangle$,

$$\frac{d}{dt} \langle X_j X_k \rangle = \sum_{lm} (C_{jlm} \langle X_k X_l X_m \rangle + C_{klm} \langle X_j X_l X_m \rangle) - (\nu_i + \nu_j) \langle X_i X_j \rangle, \tag{5.7}$$

which involves $\langle X_k X_l X_m \rangle$. Simply going to higher moments does not produce a closed set of equations, and we say that we have a "closure problem."

An analogous series of equations for the reduced probability distributions arises in kinetic theory; they are called the BBGKY hierarchy, and can be solved with perturbation theory if a small parameter can be identified (Brittin et al., 1967). For example, kinetic equations can be derived in the following cases:

i) low density → Boltzmann equation + (Choh-Uhlenbeck, for next order);

ii) weak interaction → Fokker-Planck equation;

iii) many particles per Debye sphere → Balescu-Lenard equation (Coulomb case).

However, in turbulence theory, it is more difficult to identify a small parameter with which to perform perturbation expansions.

Symmetries play important roles in turbulence theory because they can reduce the number of independent variables (Robertson, 1941). This is vital in a problem with as many degrees of freedom as fully developed turbulence. Let us move to the real velocity variables for now, so that we may identify some physical symmetries. First, move to the rest frame of the fluid to eliminate overall translation. In terms of moments this means

$$\mathbf{v}(\boldsymbol{k} = 0) = 0. \tag{5.8}$$

We can also consider an isotropic ensemble of fluids so that

$$\langle \mathbf{v} \rangle = 0. \tag{5.9}$$

Note that each member of the ensemble may be anisotropic. The first interesting moment is

$$\langle \mathrm{v}_i(\boldsymbol{x},t)\mathrm{v}_j(\boldsymbol{x}+\boldsymbol{r},t)\rangle \equiv R_{ij}(\boldsymbol{x},r,t). \tag{5.10}$$

This expression could also have a time separation between the velocities, but we will not consider that here.

We will now make various simplifying assumptions (Batchelor, 1953; Matthaeus and Smith, 1981).

I. *Homogeneity*. We may assume that the statistics are translation-invariant. Then R_{ij} is independent of $\boldsymbol{x}$ and we can take a Fourier transform of the displacement,

$$R_{ij}(\boldsymbol{r}) = \int d\boldsymbol{k} S_{ij}(\boldsymbol{k}) e^{i\boldsymbol{k}\cdot\boldsymbol{r}}. \tag{5.11}$$

Then incompressibility ($\nabla \cdot \mathbf{v} = 0$) gives

$$k_i S_{ij} = k_j S_{ij} = 0. \tag{5.12}$$

II. *Isotropy*. Then $a_i b_j S_{ij}$ is invariant under rotation for any vectors $\boldsymbol{a}$ and $\boldsymbol{b}$.

III. *Stationarity*. Then R_{ij} is independent of t.

It should be mentioned that these conditions are dubious at best in most real fluids. However, if we cannot solve this simple system first, then we may have little hope of solving more complicated real cases.

The most general homogeneous, isotropic second moment of the velocity for the Navier-Stokes equation has the form

$$\langle \mathrm{v}_\alpha(\boldsymbol{k},t)\mathrm{v}_\beta(\boldsymbol{k}',t)\rangle = \delta(\boldsymbol{k}+\boldsymbol{k}')\left(\frac{P_{\alpha\beta}(\boldsymbol{k})E(k)}{4\pi k^2} + i\epsilon_{\alpha\beta l}\frac{k_l F(k)}{8\pi k^4}\right), \tag{5.13}$$

where the energy spectrum $E(k)$ and the helicity spectrum $F(k)$ are functions of the magnitude of $\boldsymbol{k}$ and satisfy

$$\frac{1}{2}\langle \mathrm{v}^2 \rangle \equiv \int_0^\infty E(k)\,dk \tag{5.14}$$

$$\langle \mathbf{v} \cdot \boldsymbol{w} \rangle = \int_0^{\infty} F(k)dk. \tag{5.15}$$

Here $E(k) \geq 0$ and it may be shown that $|F(k)| \leq kE(k)$, but while $E(k)$ is necessarily positive, $F(k)$ can be negative.

Helicity H ($\equiv \int \mathbf{v} \cdot \boldsymbol{\omega} d^3x$) is an ideal invariant, meaning that it is conserved under a rather wide variety of boundary conditions if the viscosity is identically zero. The prototypical flow with helicity (called helical flow) is *linked vortex tubes*.

$$H \equiv \int \mathbf{v} \cdot \boldsymbol{\omega} d^3x = 2\Phi_1\Phi_2 \,, \tag{5.16}$$

$$\Phi_1 = \int \boldsymbol{\omega} \cdot d\boldsymbol{A}_1 \text{ and } \Phi_2 = \int \boldsymbol{\omega} \cdot d\boldsymbol{A}_2 \,, \tag{5.17}$$

where the integrations are made over the cross section of each tube. Of course, as the flow evolves in time, the geometry may become much more complex.

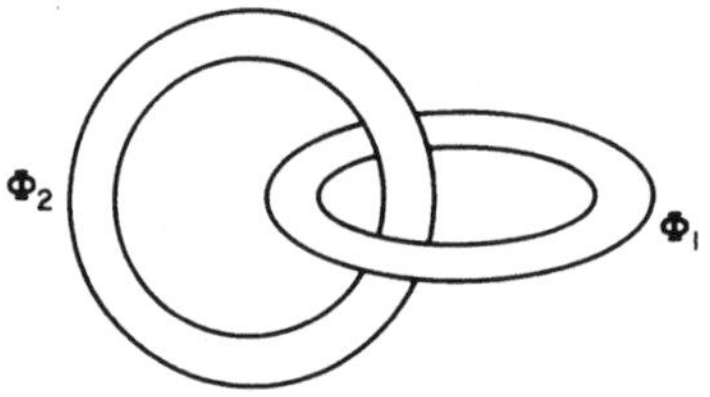

Fig. 5.1: A pair of linked vortex tubes.

One may also define magnetic helicity (Moffatt, 1969; 1978)

$$H_M (\text{or}K) = \int \boldsymbol{A} \cdot \boldsymbol{B} d^3x, \tag{5.18}$$

where $\boldsymbol{A}$ is a vector potential of the magnetic field $\boldsymbol{B} = \nabla \times \boldsymbol{A}$. A common example of a magnetic field with helicity is a field in a toroidal magnetic confinement device with both a toroidal and poloidal magnetic field. H_M measures the amount of linked magnetic flux inside the torus.

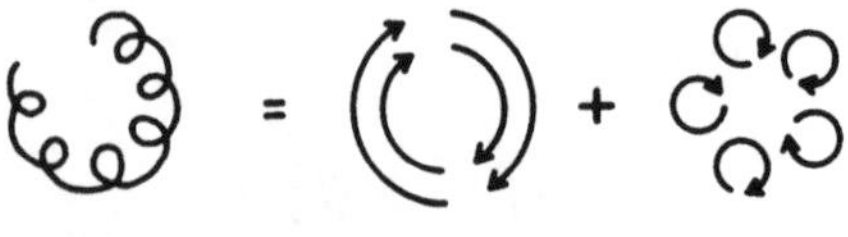

Figure 5.2: An impressionistic decomposition of a helical magnetic field into its "toroidal" and "poloidal" components, which are linked.

Helicity tends to make a flow quiescent. When $\boldsymbol{A}$ and $\boldsymbol{B}$ are primarily lined up, so are $\boldsymbol{B}$ and $\boldsymbol{J}$. When $\boldsymbol{B}$ and $\boldsymbol{J}$ are lined up, the small scales are "force free." This is analogous to the Navier-Stokes case in which the vortex stretching (and thereby the small scales) are suppressed when $\boldsymbol{\omega}$ and $\mathbf{v}$ are lined up in a "Beltrami flow."

Consider the reflection of the coordinates $\boldsymbol{r} \rightarrow -\boldsymbol{r}$. Then $\mathbf{v}\cdot\mathbf{v} \rightarrow \mathbf{v}\cdot\mathbf{v}$, but $\mathbf{v}\cdot\boldsymbol{\omega} \rightarrow -\mathbf{v}\cdot\boldsymbol{\omega}$. Helicity may be thought of as measuring a departure from reflection invariance.

Assume now the flow has no helicity. For a reflection-invariant, isotropic, homogeneous ensemble, $E(k)$ is the only arbitrary function left in the general form of the second moment:

$$\langle \mathrm{v}_\alpha(\boldsymbol{k},t)\mathrm{v}_\beta(\boldsymbol{k}',t)\rangle = \delta(\boldsymbol{k}+\boldsymbol{k}')\frac{P_{\alpha\beta}(\boldsymbol{k})E(k)}{4\pi k^2}. \tag{5.19}$$

We can write this in a form that shows its physics better. A general, second-rank isotropic tensor can be written in the form (Batchelor, 1953, 1970)

$$R_{ij}(\boldsymbol{r}) = F(r)r_ir_j + G(r)\delta_{ij}. \tag{5.20}$$

The incompressibility condition, applied to R_{ij}, gives

$$4F + r\frac{\partial F}{\partial r} + \frac{1}{r}\frac{\partial G}{\partial r} = 0. \tag{5.21}$$

Rewriting to remove scale dependence from the arbitrary functions gives

$$R_{ij}(\boldsymbol{r}) = \mathrm{v}^2(\frac{f-g}{r^2}r_ir_j + g\delta_{ij}), \tag{5.22}$$

when f and g satisfy the condition

$$g = f + \frac{1}{2}\,r\,\frac{df}{dr}, \tag{5.23}$$

and $\mathrm{v}^2 = \frac{1}{3}\langle\mathbf{v}\cdot\mathbf{v}\rangle$ is a mean square velocity component. Here f and g represent the longitudinal and the transverse correlation, respectively. While we cannot calculate f and g analytically, experiments indicate the behavior in Fig 5.3. The *integral scale* $L_p \equiv \int_0^\infty f(r,t)\,dr$ estimates the length scale of the most energetic motions.

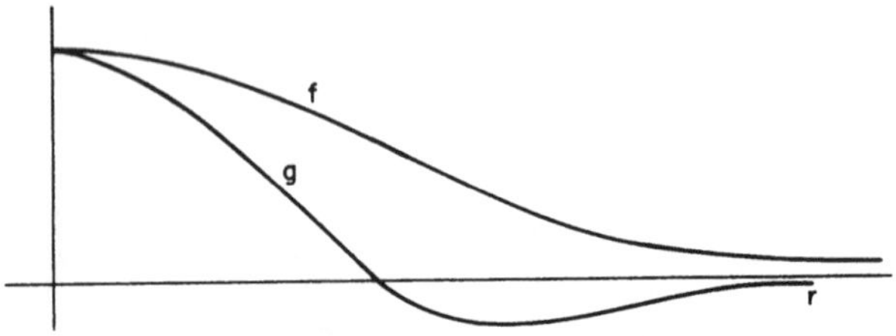

Figure 5.3: Qualitative form of the scalar functions $f(r)$, $g(r)$ in the velocity correlation tensor.

Near $r = 0$, $f(r) \simeq 1 + \frac{r^2}{2}f''(0) + \cdots$. We can define

$$\lambda \equiv -f''(0). \tag{5.24}$$

λ is a length and is called the Taylor microscale, and may also obtained as

$$\frac{1}{\lambda^2} = \frac{1}{5}\frac{\int_0^\infty k^2 E(k)dk}{\int_0^\infty E(k)dk} = \frac{1}{5}\frac{\Omega}{E} = \frac{1}{5}\frac{\text{"enstrophy"}}{\text{"energy"}}, \tag{5.25}$$

which indicates that it is a kind of expectation value for a length scale of the flow. However λ has no sharp location in the energy spectrum, i.e., it represents a length simply somewhere between integral length scale and the Kolmogorov cutoff (Fig. 5.4).

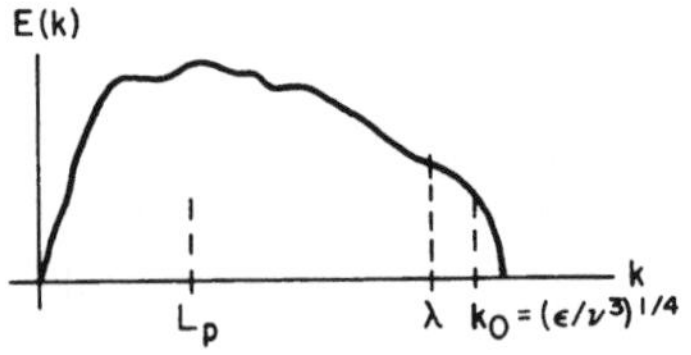

Figure 5.4: Qualitative form of the energy spectrum for a high-Reynolds-number velocity field.

Let us now return to the X_i notation. Suppose the X_i's in Eq. (5.2) were to have a multivariate Gaussian distribution

$$P(X_1, X_2, \cdots) \propto \exp\left[-\sum_{ij} a_{ij}(X_i - \langle X_i\rangle)(X_j - \langle X_j\rangle)\right] \tag{5.26}$$

for some set of a_{ij}. If we offset the definitions of the variables so that $\langle X_i \rangle = 0$, we can integrate over all but one to find the distribution of any single one of the X_i's. Then

$$f_1(X_i) \propto \exp(-a_i X_i^2) \ , \tag{5.27}$$

with a_i a constant. There is good reason to believe that many features of homogeneous turbulence are roughly "Gaussian."

The multivariate Gaussian distribution has the property that its fourth moment may be expressed exactly in terms of second moments:

$$\langle X_i X_j X_k X_l \rangle = \langle X_i X_j \rangle \langle X_k X_l \rangle + \langle X_i X_l \rangle \langle X_j X_k \rangle + \langle X_i X_k \rangle \langle X_j X_l \rangle . \tag{5.28}$$

The assumption that the distribution satisfies this relation at least approximately is central to closure methods. If we were able to expand the third order moments $\langle X_i X_j X_k \rangle$ in terms of the fourth order moments $\langle X_i X_j X_k X_l \rangle$, and the fourth order moments in terms of the second order, we would have closure. This was the hope of the 1950s, and it was thought that one could follow the decay of turbulence in this direction.

However, this simplest closure scheme does not work; when an initially realizable state is evolved in time it produces an energy spectrum that is negative at high wavenumbers (Fig. 5.5). An important logical flaw in the derivation is that if the distribution is Gaussian, then $\langle X_i X_j X_k \rangle \equiv 0$, whereas we have an equation for its evolution that requires that it be nonzero. The deviations from Gaussian statistics are the key to the coupling between the higher-order moments, and cannot be neglected entirely.

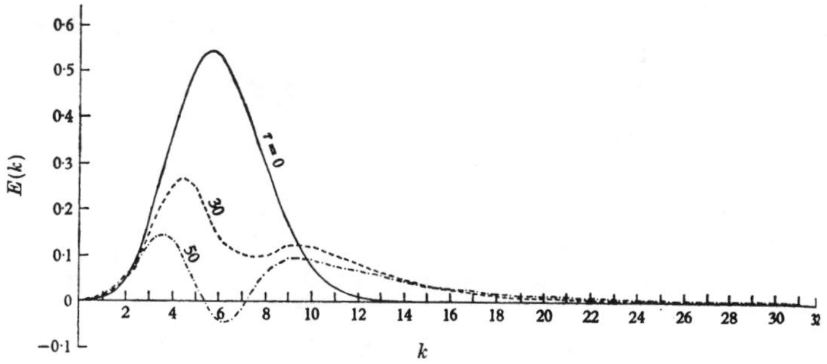

Figure 5.5: A typical negatively developing energy spectrum that results from the straightforward Gaussian closure (see H. Ogura, 1963).

One possible modification is to introduce a non-Gaussian component to the distribution function as a "small parameter" and, by analogy with BBGKY, hope to solve for the perturbations,

$$D = D_G + D_{NG},$$

where D_G is the Gaussian distribution and D_{NG} is the non-Gaussian component. With $D_G \gg D_{NG}$ we can hope that we can solve for D_{NG}. This has not yet been entirely successfully done, but something like it is involved in many of the recently developed closure schemes.

Even if that were capable of implementation, the question remains of how much can be represented in a purely statistical theory. Coherent structures and intermittency are not modeled by any current statistical theory, and it is not clear how to include them at all in the strongly interacting regime of fully developed turbulence.

During the last decades several sophisticated closure schemes have been developed (for example, DIA, LHDIA, TFM, EDQNM, etc). However the situation is still confused, and none of the closure schemes can really give more solid results than are obtained from dimensional analysis. At best they reproduce the dimensional analysis results. We need something better than what has been proposed so far.

A good introduction to closure schemes, which we shall not inquire into much here is by S. A. Orszag (1973).

Batchelor, G. K., 1953: *Theory of Homogeneous Turbulence* (Cambridge Univ. Press).

Batchelor, G. K., 1970: *An Introduction to Fluid Dynamics*, Cambridge Univ. Press, 615 pp.

Brittin, W., A. O. Barut, and M. Guenin, 1967: Kinetic theory, Lectures in *Theor. Physics*, Vol. 9c, ed., W. Brittin, Gordon and Breach, Publishers.

Matthaeus, W. H., and C. Smith, 1981: *Phys. Rev.*, **A24**, 2135.

Moffatt, H. K., 1969: *J. Fluid Mech.*, **35**, 117.

Moffatt, H. K., 1978: *Magnetic Field Generation in Electrically Conducting Fluids* (Cambridge Univ. Press).

Ogura, H., 1962: *Phys. Fluids*, **5**, 395.

Orszag, S. A., 1973: *Proc. Les Houches Summer School*, 1973 ed. by R. Balian and J. L. Peube (New York: Gordon and Breach).

Robertson, H. P., 1941: *Proc. Camb. Phil. Soc.*, **36**, 209.

Chapter 6

SPECTRAL CASCADES AND TURBULENT RELAXATION PROCESSES

We shall assume some familiarity with the idea of direct turbulent cascade. In this chapter, the focus will be mainly on *inverse* cascades.

The historical background for the idea of an inverse cascade (i.e., a spectral transport of some physical quantity from large k-values toward small k-values) can be found in the works of Onsager (1949) and Fjørtoft (1953). Since then, the idea has been elaborated upon considerably.

Following Onsager's ideas, let us consider a two-dimensional box of area A lying in the (x,y) plane containing N discrete, infinitally long parallel line vortices oriented along the z-axis (see Fig. 6.1).

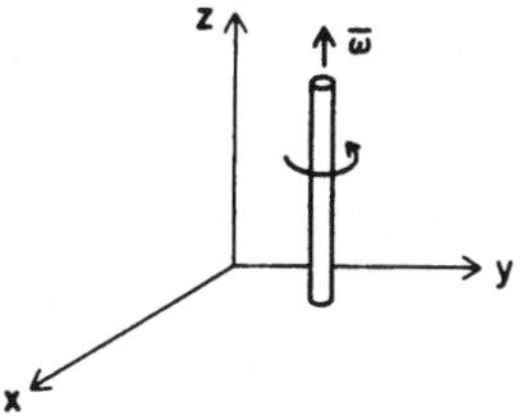

Figure 6.1: Geometry of the two-dimensional line vortex model.

The vorticity field $\boldsymbol{\omega}$(x,y) can then be written as

$$\boldsymbol{\omega} = \sum_{i}^{N} \omega_i \delta^{(2)}(\boldsymbol{x} - \boldsymbol{x}_i)\, \hat{\boldsymbol{e}}_z\,, \tag{6.1}$$

where $\boldsymbol{x}_i$ is the two-dimensional (2D) location of the ith vortex.

The vortices move with the flow field that the others generate (the viscosity is assumed negligible). We note that this 2D hydrodynamic system of discrete line vortices has an electrostatic plasma analogue in a 2D system of line charges in an external B-field, $\boldsymbol{B} = B_0\ \hat{\boldsymbol{e}}_z$. These line charges will create a flow field which, in the limit of vanishing radius, is $\mathbf{v} = c\boldsymbol{E}\times \boldsymbol{B}_0/B_0^2$, where $\boldsymbol{E}$ is determined self-consistently from Poisson's equation. The equations of motion of the line charges are the same as the equations of motion for the line vortices in the hydrodynamic case (Taylor and McNamara, 1971).

Introducing the coordinates

$$(q_i,\ p_i) \equiv (x_i,\ y_i \mathrm{sgn}\omega_i), \tag{6.2}$$

we find that q_i and p_i (which are Hamiltonian variables) satisfy the Hamilton equations

$$q_i = \frac{\partial H}{\partial p_i}\ , \tag{6.3a}$$

$$p_i = -\frac{\partial H}{\partial q_i}\ , \tag{6.3b}$$

$$\frac{\partial H}{\partial t} = 0 = \frac{dH}{dt}\ , \tag{6.3c}$$

where

$$H = K\sum_{i<j}^{N} \omega_i\omega_j \ln(|\boldsymbol{x}_i - \boldsymbol{x}_j|) = K_0 \sum_{i<j}^{N} (\pm \ln r_{ij})\ . \tag{6.4}$$

In the presence of a container with rigid walls, the Coulomb potentials get replaced by Ewald potentials (see Seyler, 1976), but the Hamiltonian formalism survives.

In (6.4), K and K_0 are constants and

$$r_{ij} = \sqrt{(x_i - x_j)^2 + (y_i - y_j)^2}. \tag{6.5}$$

Note that all the line vortices have been assumed to have the same magnitude $|\omega_i|$ in the last sum in (6.4). Furthermore, note that the Hamiltonian H only depends upon the positions of the line vortices, not on their velocities. Assuming equilibrium statistical mechanics in its most familiar form, we can write the probability density D_{eq} (q,p) in the form of a Gibbs distribution (canonical distribution),

$$D_{eq}(q,p) = \exp(-E/kT), \tag{6.6}$$

where the total energy E is equal to the Hamiltonian H. The corresponding partition function Z can be written as:

$$\begin{aligned} Z = \int e^{-E(q,p)/kT} dq dp &= \int \Omega(E) e^{-E/kT} dE \\ &= \int e^{(\ln[\Omega(E)] - E)/kT} dE\ , \end{aligned} \tag{6.7}$$

where we have introduced the phase space volume per unit energy, or the structure function $\Omega(E)$, commonly called the "density of states," which measures the number of states in the phase space having the same energy E. Although the Gibbs factor $\exp(-E/kT)$ in (6.7) is a decreasing function of E, the structure function $\Omega(E)$ usually increases dramatically with E (see Fig. 6.2), so that the integrand in (6.7) is fairly sharply localized around the expectation value of the energy, E_0. We can therefore conventionally make the following Taylor expansion around E_0:

$$Z \simeq e^{\ln\Omega(E_0)-E_0/kT} \int \exp\left[-C(E-E_0)^2/2\right] d\left[E-E_0\right], \tag{6.8}$$

where

$$C \equiv \left|\frac{d^2}{dE_0^2}\left[\ln\Omega(E_0)\right]\right| .$$

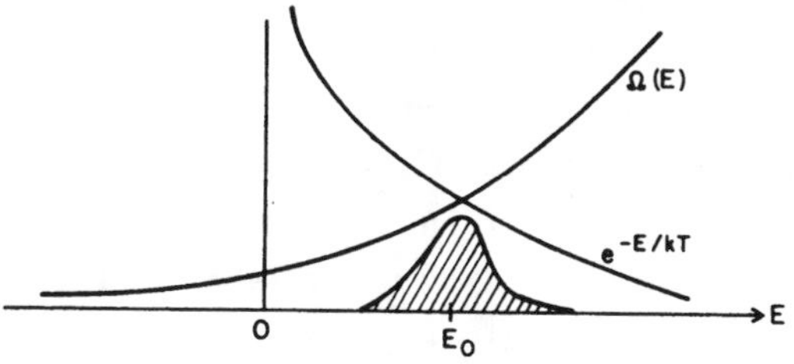

Figure 6.2: The probability distribution in E is effectively confined to the shaded area near E_0, where the maximum occurs.

Introducing the Helmholtz free energy F defined by

$$F = E_0 - TS , \tag{6.9}$$

where S is the entropy, and making use of the thermodynamic relation between Z and F,

$$Z = e^{-F/kT} = e^{S/k-E_0/kT} , \tag{6.10}$$

we can determine the relation between $\Omega(E_0)$ and S by comparing (6.8) and (6.10). We see that

$$S = k\ln[\Omega(E_0)] , \tag{6.11}$$

incidentally coinciding with Boltzmann's celebrated epitaph.

Since we are considering the equilibrium statistical properties of a system within a fixed volume, we can use the thermodynamic relation

$$\frac{1}{T} = \left(\frac{\partial S}{\partial E}\right)_V \tag{6.12}$$

to write

$$\frac{1}{kT} = \frac{\Omega'(E_0)}{\Omega(E_0)} \,. \tag{6.13}$$

Let us now consider the total volume of phase space $\Phi(E)$ up to the energy E defined by

$$\Phi(E) = \int_{-\infty}^{E} \Omega(E')dE'. \tag{6.14}$$

As seen from (6.4), $E \to -\infty$ as all the vortices move close together in oppositely signed pairs, while large positive E corresponds to the vortices of different signs being far apart. However, since the box we are considering has a finite area, $\Phi(E)$ has to level off as shown in Fig. 6.3a. Now, since $\Omega(E) = d\Phi/dE$, we see that $\Omega(E)$ cannot continue to increase monotonically as for most physical systems, but must have a negative slope for sufficiently high E, as shown in Fig. 6.3b. Due to the relation in (6.13), the region of negative $d\Omega/dE$ is called the "negative temperature regime" and as seen from (6.12), the entropy in this regime must decrease with increasing E! A higher level of organization must necessarily characterize the higher energy states.

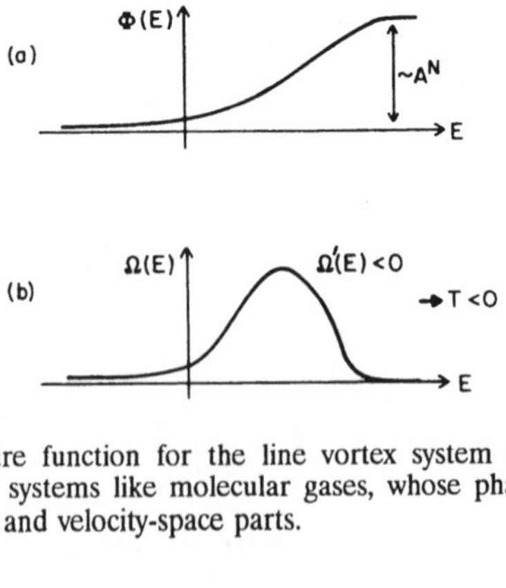

Figure 6.3: The structure function for the line vortex system differs qualitatively from those which characterize systems like molecular gases, whose phase spaces have independent configuration-space and velocity-space parts.

This result has been supported by numerical simulations of the dynamics of a discrete vortex system (Joyce and Montgomery, 1973; Montgomery and Joyce, 1974). In these simulations the eventual configuration of a vortex system which initially was away from equilibrium with an energy E corresponding to the negative temperature regime, was found to consist of two large vortices of different signs displaced as far from each other as was permitted by the dimensions of the box. Similar results were confirmed by Seyler's Monte Carlo evaluations of $\Omega(E)$ (Seyler, 1976). See also the work of Lundgren and Pointin (1977).

As mentioned in the beginning of this chapter, one other piece of historical background for the inverse cascade is due to Fjørtoft. His arguments, proceeding from the continuum approximation, were based on the existence of two rugged invariants in the 2D inviscid Navier-Stokes equations, namely, the energy,

$$E = \sum_k E(\boldsymbol{k}) = \sum_k |\mathbf{v}(\boldsymbol{k})|^2 = \text{constant} , \tag{6.15a}$$

and the "enstrophy,"

$$\Omega = \sum_k \Omega(\boldsymbol{k}) = \sum_k k^2 |\mathbf{v}(\boldsymbol{k})|^2 = \text{constant} . \tag{6.15b}$$

If energy E and enstrophy Ω are to be transferred among the various modes under these two constraints, at least three modes must be active in any transfer. Furthermore, if the modes are ordered $k_1 < k_2 < k_3$, energy transfer out of mode k_2 must go to *both* k_1 and k_3. Although the transfer up the spectrum must be accompanied by transfer down, the relative strengths of the up and down transfers can be widely different. Note that these arguments rely on the existence of two invariants of the system, as is the case for two-dimensional Navier-Stokes (NS) and MHD turbulence. In three dimensions, however, when E is the only invariant, spectral transfer of energy can be unidirectional.

By considering 2D NS turbulence with dissipation included and the system being forced by injection of energy and enstrophy within a narrow band in k-space, Kraichnan (1967) conjectured that enstrophy should be cascaded toward higher k and dissipated there, whereas the energy should be inversely cascaded toward low k as sketched in Fig. 6.4. See also Batchelor (1969), Leith (1968), and Lilly (1969). Of course (6.15) shows that for any finite k, one quantity implies the finiteness of the other, so the existence of a "dual cascade" in its pure form (unidirectional transfer for both E and Ω) must involve some very subtle limiting processes.

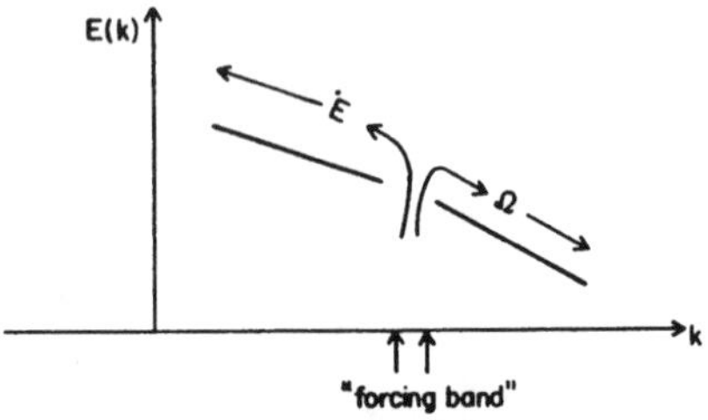

Figure 6.4: Schematic of a conjectured "dual cascade" spectrum for energy and enstrophy, with an "injection band" and two inertial subranges.

The slopes of the energy spectrum in the two regions below and above the forcing band can be determined by dimensional arguments. For the lower part of the spectrum where energy is being inversely cascaded, we note that $E(k)$ has the dimensionality (always referred to unit mass)

$$E(k) \sim L^3/T^2, \tag{6.16}$$

where L denotes a dimension corresponding to a length while T denotes a time dimension. Similarly, the rate of supply of energy ϵ in the forcing band has the dimension

$$\epsilon \sim L^2/T^3. \tag{6.17}$$

Assuming that

$$E(k) = C\epsilon^\alpha k^\beta , \tag{6.18}$$

where C is a dimensionless constant, we find, by matching exponents of L and T, that

$$\alpha = 2/3 \quad \text{and} \quad \beta = -5/3, \tag{6.19}$$

which is the usual Kolmogorov result. (The experimental value for the constant C seems to be something less than 1.5 for the direct cascade in three dimensions.)

Considering the upper part of the spectrum where enstrophy is being cascaded, we have

$$\Omega(k) \sim L/T^2, \tag{6.20}$$

while the supply rate of enstrophy has the dimensionality

$$\eta \equiv \langle \frac{d\omega^2}{dt} \rangle \sim T^{-3}. \tag{6.21}$$

Assuming $\Omega(k) = C'\eta^{\alpha}k^{\beta}$ with C' a dimensionless constant, we find $\alpha = 2/3$ and $\beta = -1$, or in terms of an energy spectrum,

$$E(k) = \frac{\Omega(k)}{k^2} = C'\eta^{2/3}k^{-3} . \tag{6.22}$$

Although the inverse energy cascade in 2D NS turbulence has not been rigorously proved theoretically or error bars attached to the relevant exponent, its existence has been demonstrated in numerical simulations by several authors, for example by Fyfe et al. (1977). As shown in Fig. 6.5a, their simulations of forced 2D NS turbulence with periodic boundary conditions showed that $\Omega(t)$ levels off at a nearly constant value where the injection of enstrophy is balanced by dissipation at high k, whereas the kinetic energy in the system continues to increase during the simulation indicating the absence of an effective energy dissipation mechanism. The developed spectrum for the simulation is shown in Fig. 6.5b, and we see that the slopes are consistent with the dimensional analysis presented above. This is an early low-resolution calculation which has since been improved upon (Frisch and Sulem, 1984; Herring and McWilliams, 1985).

In the case of 2D dissipative MHD turbulence with narrow band forcing in k-space, Fyfe and Montgomery (1976) argued that energy should cascade to higher wavenumbers, while the vector potential should be inversely cascaded toward lower wavenumbers. Following the same kind of dimensional arguments as above, we can estimate the slope of the k-spectrum in the low k-range, where there is a transfer of vector potential. We have, in Alfvénic units,

$$A^2 \sim B^2L^2 \sim L^4/T^2 , \tag{6.23}$$

$$E_A(k) \sim A^2 \cdot L \sim L^5/T^2, \tag{6.24}$$

and the supply rate of vector potential ϵ_A,

$$\epsilon_A \sim A^2/T \sim L^4/T^3 . \tag{6.25}$$

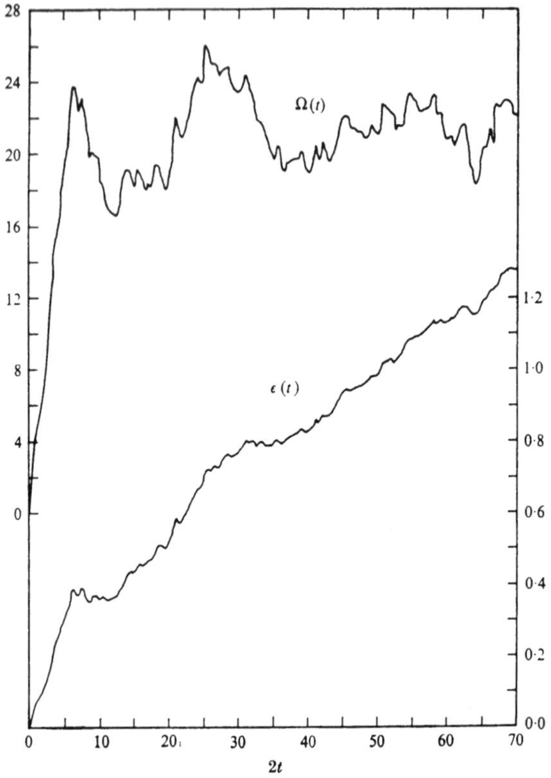

Figure 6.5a: Global ideal invariants for forced dissipative turbulence for two dimensions and periodic boundary conditions as functions of time (from Fyfe et al., 1977).

Assuming

$$E_A(k) \sim C'' \epsilon A^{\alpha} k^{\beta}, \tag{6.26}$$

we find

$$\alpha = 2/3 \qquad \text{and} \qquad \beta = -7/3, \tag{6.27}$$

or in terms of a magnetic spectrum,

$$E_B(k) \sim k^2 E_A(k) = C'' \epsilon_A^{2/3} k^{-1/3}. \tag{6.28}$$

The direct cascade prediction for the high k-numbers requires assuming something about **v** and ***B*** at small scales. The "Alfvén effect," described by Kraichnan (1965), in which higher-k **v**(k) and ***B***(k) are treated as small-amplitude

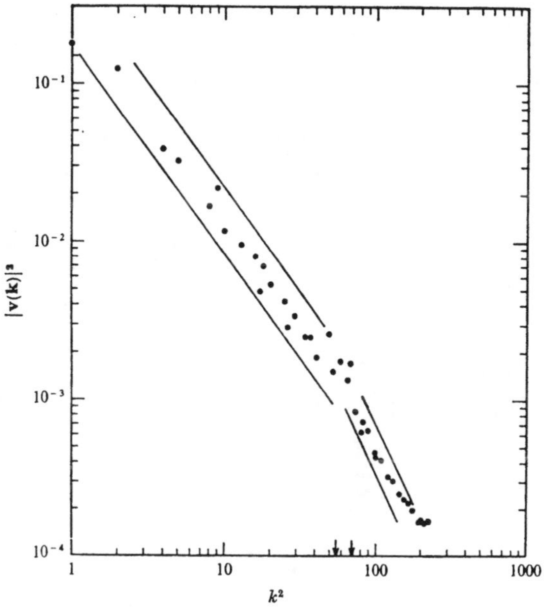

Figure 6.5b: Time-averaged modal energy spectra corresponding to the situation in Fig. 6.5a.

perturbations on a spatially uniform state of lower-k components, tends to create equipartition of magnetic and kinetic energy:

$$|\mathbf{v}(\boldsymbol{k})|^2 \simeq |\boldsymbol{B}(\boldsymbol{k})|^2 . \tag{6.29}$$

To a rough approximation, the relation (6.29) usually seems to have been fulfilled in numerical computations for viscosities and magnetic diffusivity not too far apart.

Assuming this equipartition, Kolmogorov's arguments can be repeated for the high-k end of the spectrum to give

$$E_B(k) \sim E_{\mathrm{v}}(k) \sim k^{-5/3}. \tag{6.30}$$

Kraichnan (1965) has argued instead for a $k^{-3/2}$ dependence in place of of Eq. (6.30).

Some results of a numerical simulation of the 2D dissipative MHD turbulence with narrow k-band forcing by Fyfe et al. (1977) are shown in Fig. 6.6. In

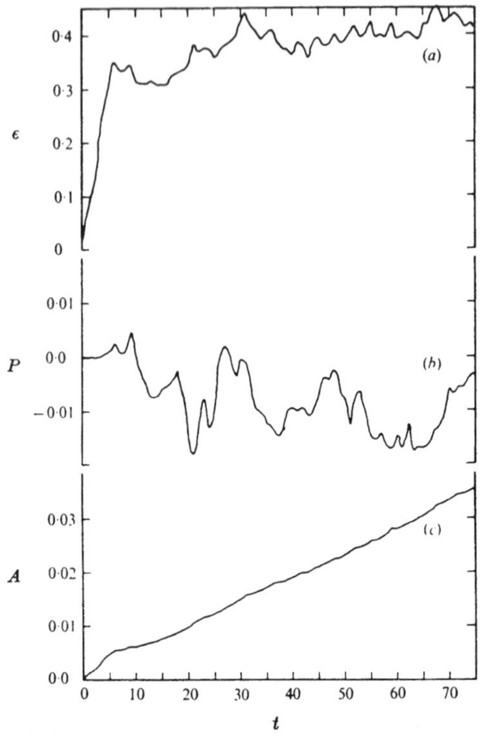

Figure 6.6a: Global ideal invariants for forced 2D MHD turbulence for periodic boundary conditions (from Fyfe et al., 1977). Saturation of the total energy (ϵ) and cross helicity (P) simultaneously with the growth of mean-square-vector potential (A) is consistent with an inverse cascade of A.

Fig. 6.6a, the temporal evolution of energy, cross helicity ($\int \mathbf{v}\cdot \boldsymbol{B} d^2x$, an MHD ideal invariant) and vector potential squared are shown. We see that these quantities behave approximately as expected from the description above. Fig. 6.6b shows the time-averaged spectrum $E_B(k)$, and a line showing the $-2/3$ power law for the lower part of the spectrum is also drawn. In general, we note that the MHD simulations are much more noisy than the corresponding NS simulations. Although equipartition between magnetic and kinetic energy was assumed for higher k, the simulations indicate that $E_B(k)$ is slightly larger than $E_v(k)$ for values of the magnetic Prandtl number close to one. The magnetic Prandtl number (ratio of viscosity to magnetic diffusivity) can control some interesting effects, especially when $\nu/\eta \gg 1$. An analogous program for three dimensions

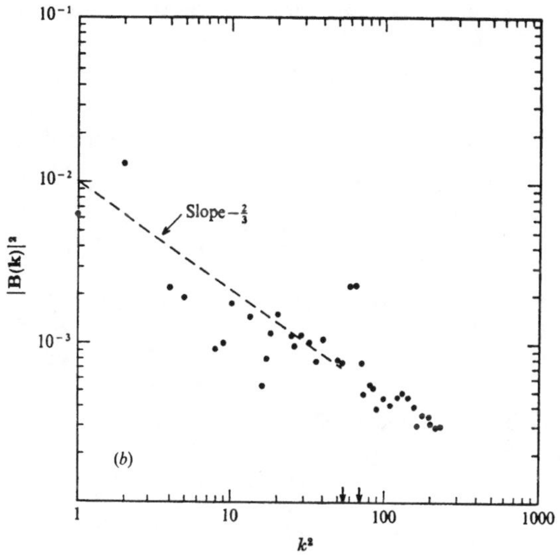

Figure 6.6b: Modal magnetic energy spectrum (late time average) for the situation in Fig. 6.6a. The "forcing band" is between the two arrows. The dotted line has a slope appropriate to that given by the prediction of an inverse cascade of A.

was carried out (Meneguzzi et al., 1981) where the inversely cascading quantity was magnetic helicity.

Summarizing, the possible direct and inverse cascades are shown in Table 6.1.

Table 6.1

	2D	3D
NAVIER-STOKES	Ω up E down	E up, if H = mechanical helicity = 0. If $H \neq 0$, still controversial
INCOMPRESSIBLE MHD	E up A down H_c: it depends	E up H_m down H_c: it depends

The absolute equilibrium ensemble for these cases can be written as:

Table 6.2

$$D_{eq} \propto \begin{pmatrix} \exp(-\alpha E - \beta \Omega) & \text{2D, NS} \\ \exp(-\alpha E - \beta H) & \text{3D, NS} \\ \exp(-\alpha E - \beta A - \gamma H_C) & \text{2D, MHD} \\ \exp(-\alpha E - \beta H_m - \gamma H_C) & \text{3D, MHD} \end{pmatrix}$$

In the two tables above, the following variables are used:

$$\text{Total energy}: \quad E = \int \left(\mathrm{v}^2 + B^2\right) \qquad (6.31a)$$

$$\text{Mean mechanical helicity}: \quad H = \int \boldsymbol{\omega} \cdot \mathbf{v} \qquad (6.31b)$$

$$\text{Mean magnetic helicity}: \quad H_m = \int \boldsymbol{A} \cdot B \qquad (6.31c)$$

$$\text{Mean cross helicity}: \quad H_C = \int \mathbf{v} \cdot \boldsymbol{B} \qquad (6.31d)$$

$$\text{Mean enstrophy}: \quad \Omega = \int \omega^2 \qquad (6.31e)$$

$$\text{Mean square vector potential}: \quad A = \int \boldsymbol{A}^2 \qquad (6.31f)$$

$$\text{Inverse "temperatures"}: \quad \alpha,\ \beta,\ \gamma \qquad (6.31g)$$

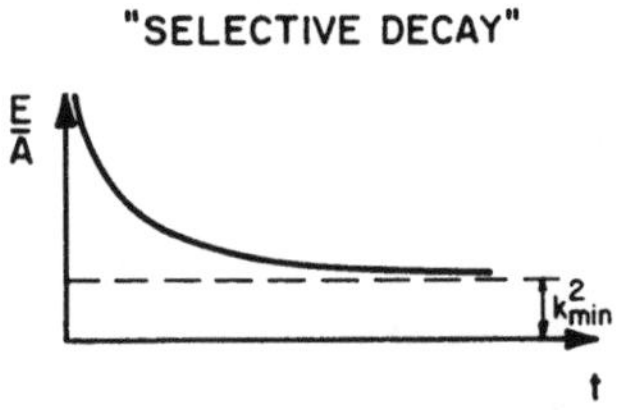

Figure 6.7: Time decay of E/A (schematic), a "selective decay" which is characteristic of 2D MHD.

We note that Ω and A only play a role in 2D equilibria. Let us now consider the turbulent relaxation of a 2D MHD system without any forcing. In this case, the energy decays faster than the mean square vorticity, a phenomenon known as "selective decay" (see Fig. 6.7). The final 2D minimum energy state is all magnetic; everything gets crowded into the longest wavelength, and more interesting geometrical patterns can result from less simple-minded boundary conditions.

At this point we would like to mention a qualitatively different behavior in 2D MHD and 2D NS relaxation. As sketched in Fig. 6.8a, the mean square vorticity density is a monotonically decreasing function of time in the 2D NS case, while Fig. 6.8b indicates that the mean square current density can experience a local maximum for some $t > 0$. This maximum becomes higher for increasing Reynolds number and is coincident with the generation of intense current filaments connected with "magnetic island" merging (see, e.g., Matthaeus and Montgomery, 1980).

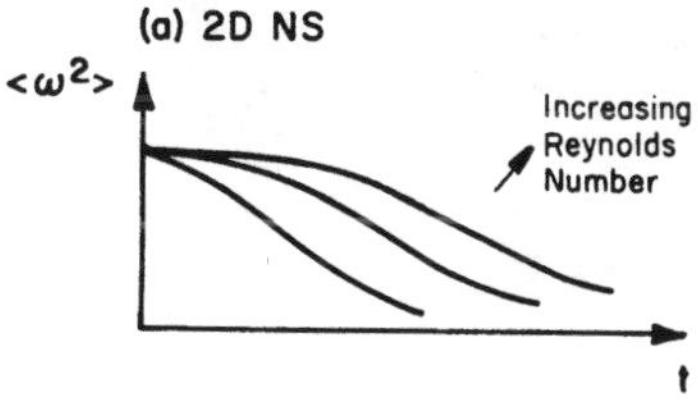

Figure 6.8a: Typical mean enstrophy decay, characteristic of the 2D NS case.

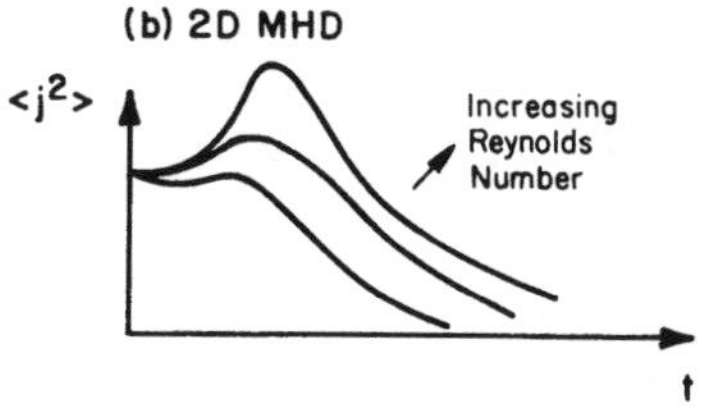

Figure 6.8b: $\langle j^2 \rangle$, which plays a role in 2D MHD that is analogous to $\langle \omega^2 \rangle$ for the 2D NS case, eventually decays, but not monotonically.

The turbulent relaxation process, in 2D MHD, is complicated when there is a non-vanishing mean cross helicity H_C, as defined by (6.31d). A feeling of how these complications arise can be obtained by considering the non-dissipative MHD equations:

$$\frac{\partial \mathbf{v}}{\partial t} + (\mathbf{v} \cdot \nabla)\mathbf{v} = -\nabla p^* + (\boldsymbol{B} \cdot \nabla)\boldsymbol{B} \,, \tag{6.32a}$$

$$\frac{\partial \boldsymbol{B}}{\partial t} + (\mathbf{v} \cdot \nabla)\boldsymbol{B} = (\boldsymbol{B} \cdot \nabla)\mathbf{v} \,, \tag{6.32b}$$

$$p^* \equiv p + \frac{1}{2}B^2 \,. \tag{6.32c}$$

Introducing the Elsässer variables

$$Z^{\pm} \equiv \mathbf{v} \pm \boldsymbol{B} \; (\nabla \cdot Z^{\pm} = 0) \tag{6.33}$$

in (6.33), we obtain

$$\frac{\partial Z^{\pm}}{\partial t} + (Z^{\mp} \cdot \nabla) Z^{\pm} = -\nabla p^{*}(Z^{+}, Z^{-}). \tag{6.34}$$

We see that all modal transfer stops in (6.34) if either $\mathbf{Z}^{+}$ or $\mathbf{Z}^{-}$ becomes zero. These situations, which are called "Alfvénic states" and are characterized by $\mathbf{v} \simeq \pm \boldsymbol{B}$, can often be observed in the solar wind. In 2D MHD computations, it is observed that the relative amount of energy $\zeta^{\pm}$ in either $\mathbf{Z}^{+}$ or $\mathbf{Z}^{-}$, defined by

$$\zeta^{\pm} \equiv \frac{\int |\mathbf{Z}^{\pm}|^2}{\int \left(|\mathbf{Z}^{+}|^2 + |\mathbf{Z}^{-}|^2 \right)}, \tag{6.35}$$

evolves as indicated in Fig. 6.9, e.g., the majority species wins (see Ting et al., 1986, and references therein).

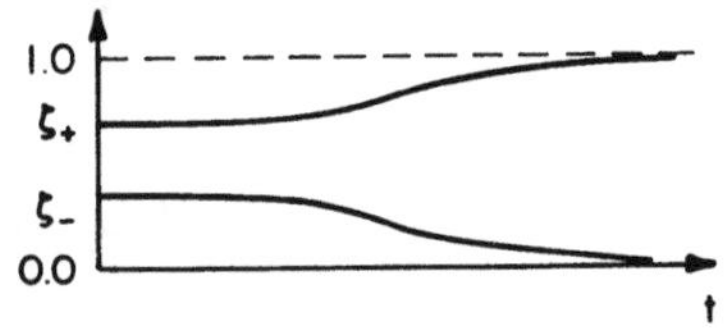

Figure 6.9: Eventual dominance by either the Z^{+} or Z^{-} Elsässer field.

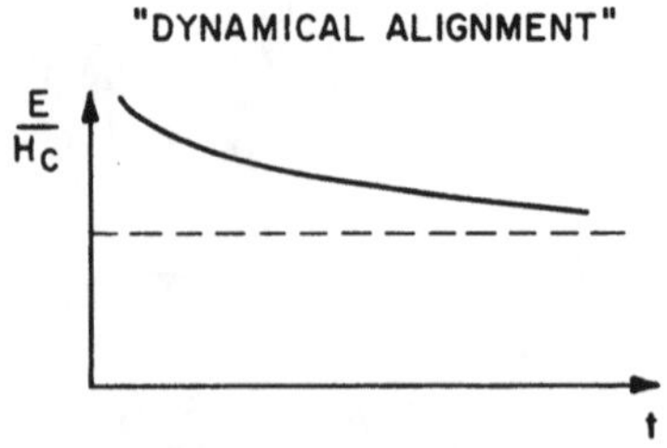

Figure 6.10: Typical numerically calculated evolution of E/H_C for the "dynamic alignment" process.

Similarly to the case of selective decay, it can be shown that in circumstances in which there is a lot of initial cross helicity, E decays faster than H_C as shown in Fig. 6.10.

The process shown in Fig. 6.10 is known as "dynamic alignment" and is very different from the selective decay process. While the selective decay process tends to decay $|\boldsymbol{V}|^2$ relative to $|\boldsymbol{B}|^2$, the dynamic alignment process tends to make $|\boldsymbol{V}| \simeq |\boldsymbol{B}|$ as in the Alfvénic state. The complicated behavior of turbulent relaxation under the simultaneous action of these two competing processes has been studied by Ting, Matthaeus, and Montgomery (1986). This work identified at least four regions of parameter space of initial conditions where different dynamical behavior occurs.

Batchelor, G. K., 1969: *Phys. Fluids Suppl.*, **12**, II–233.

Fjørtoft, R., 1953: *Tellus*, **5**, 225.

Frisch, U., and P. L. Sulem, 1984:, *Phys. Fluids*, **27**, 1921.

Fyfe, D. and D. Montgomery, 1976: *J. Plasma Phys.*, **16**, 181.

Fyfe, D., D. Montgomery, and G. Joyce, 1977: *J. Plasma Phys.*, **17**, 369.

Herring, J., and J. McWilliams, 1985: *J. Fluid Mech.*, **153**, 229.

Joyce, G., and D. Montgomery, 1973: *J. Plasma Phys.*, **10**, 107.

Kraichnan, R., 1965: *Phys. Fluids*, **8**, 1385.

Kraichnan, R., 1967: *Phys. Fluids*, **10**, 1417.

Leith, C., 1968: *Phys. Fluids*, **11**, 671.

Lilly, D. K., 1969: *Phys. Fluids Suppl.*, II, **12**, 240.

Lundgren, T., and Y. Pointin, 1977: *J. Stat. Phys.*, **17**, 323.

Matthaeus, W. H., and D. Montgomery, 1980: *Ann. N. Y. Acad. Sci.*, **357**, 203.

Meneguzzi, M., U. Frisch, and A. Pouquet, 1981: *Phys. Rev. Let.*, **47**, 1061.

Montgomery, D., and G. Joyce, 1974: *Phys. Fluids*, **17**, 1139.

Onsager, L., 1949: *Nuovo Cimento Suppl.*, **6**, 279.

Seyler, C. E., Jr., 1976: *Phys. Fluids*, **19**, 1336.

Taylor, J. B., and B. McNamara, 1971: *Phys. Fluids*, **14**, 1492.

Ting, A. C., W. H. Matthaeus, and D. Montgomery, 1986: *Phys. Fluids*, **29**, 3261.

Chapter 7

ANISOTROPIC EFFECTS IN MHD

A. The Strauss Equations

Most real turbulence is neither isotropic nor homogeneous: in MHD plasma a current profile $\boldsymbol{j}(r)$ breaks homogeneity, while a strong dc magnetic field renders turbulence anisotropic at all scales. Most real MHD turbulence thus is far from satisfying the theoretically popular idealizations and symmetries.

Let us first consider the effect of a strong externally imposed magnetic field, $\boldsymbol{B}_0$, by decomposing the total field as follows:

$$\boldsymbol{B} \sim \frac{1}{\epsilon}\boldsymbol{B}_0 + \boldsymbol{B}^{(0)} + \epsilon \boldsymbol{B}^{(1)} + 0(\epsilon^2) \ ,$$

with

$$\boldsymbol{B}_0 = B_0 \hat{e}_z \ , \ \epsilon \ll 1 \ . \tag{7.1}$$

We are interested in exploring the consequences of an externally imposed dc magnetic field that is large compared to the internally generated field.

By assuming there is negligible Alfvén-wave activity, the MHD equations are written in dimensionless form in terms of variables t, x, y, and $\zeta = \epsilon z$:

$$\frac{\partial \boldsymbol{B}}{\partial t} = \nabla \times (\boldsymbol{V} \times \boldsymbol{B}) + \eta \nabla^2 B \ , \tag{7.2a}$$

$$\rho\left(\frac{\partial \boldsymbol{V}}{\partial t} + \boldsymbol{V}\cdot\nabla\boldsymbol{V}\right) = -\nabla p + \boldsymbol{j}\times\boldsymbol{B} + \nu\nabla^2\boldsymbol{V} \ , \tag{7.2b}$$

$$\nabla\cdot\boldsymbol{V} = 0 \ , \tag{7.2c}$$

$$\nabla\cdot\boldsymbol{B} = 0 \ , \tag{7.2d}$$

$$\boldsymbol{j} = \nabla\times\boldsymbol{B}(\Rightarrow \nabla\cdot\boldsymbol{j} = 0) \ . \tag{7.2e}$$

We introduce expansions for $\mathbf{v}$, p, and $\boldsymbol{j}$ as

$$\left.\begin{array}{lll} \boldsymbol{V} & \sim & \mathbf{v}^{(0)} + \epsilon\mathbf{v}^{(1)} \\ p & \sim & p^{(0)} + \epsilon p^{(1)} \\ \boldsymbol{j} & \sim & \boldsymbol{j}^{(0)} + \epsilon\boldsymbol{j}^{(1)} \end{array}\right\} + 0(\epsilon^2), \tag{7.3}$$

and write

$$\nabla = \hat{e}_x\partial_x + \hat{e}_y\partial_y + \epsilon\hat{e}_z\partial_\zeta$$
$$\equiv \nabla_\perp + \epsilon\,(\hat{e}_z\partial_\zeta). \tag{7.4}$$

The magnetic field B is expressed in terms of a vector potential $\boldsymbol{A}$:

$$\boldsymbol{B} = \frac{1}{\epsilon}B_0\hat{e}_z + \nabla\times\boldsymbol{A} \tag{7.5a}$$

with

$$\boldsymbol{A} \sim \boldsymbol{A}^{(0)} + \epsilon\boldsymbol{A}^{(1)} + 0(\epsilon^2)\,, \tag{7.5b}$$

where the Coulomb gauge is imposed by demanding $\nabla\cdot\boldsymbol{A} = 0$.

Utilizing (7.5a), we can write the magnetic field equation (7.2a) in terms of $\boldsymbol{A}$ as

$$\frac{\partial\boldsymbol{A}}{\partial t} = \boldsymbol{V}\times\left(\frac{1}{\epsilon}B_0\,\hat{e}_z + \nabla\times\boldsymbol{A}\right) - \eta\boldsymbol{j} - \nabla\phi, \tag{7.6}$$

where the gradient of the scalar potential ϕ results from pulling off a curl operator from (7.2a). It is assumed that ϕ has an expansion

$$\phi \sim \frac{1}{\epsilon}\phi_0 + \phi^{(0)} + \epsilon\phi^{(1)} + 0(\epsilon^2). \tag{7.7}$$

The momentum equation (7.2b) becomes

$$\rho\left(\frac{\partial\boldsymbol{V}}{\partial t} + \boldsymbol{V}\cdot\nabla\boldsymbol{V}\right) = -\nabla p + \frac{1}{\epsilon}B_0\boldsymbol{j}\times\,\hat{e}_z + \boldsymbol{j}\times(\nabla\times\boldsymbol{A}) + \nu\nabla^2\boldsymbol{V}\,. \tag{7.8}$$

Considering the leading order contributions from (7.7), (7.8), ($O(1/\epsilon)$) and (7.2c,e) ($O(1)$) we find

$$\boldsymbol{V}^{(0)}\times\hat{e}_z B_0 = \nabla_\perp\phi_0 = 0\,, \tag{7.9a}$$

$$\boldsymbol{j}^{(0)}\times\,\hat{e}_z B_0 = 0\,, \tag{7.9b}$$

$$\nabla_\perp\cdot\boldsymbol{V}^{(0)} = 0\,, \tag{7.9c}$$

$$\boldsymbol{j}^{(0)} = \nabla_\perp\times\boldsymbol{B}^{(0)} = -\nabla_\perp^2\boldsymbol{A}^{(0)}. \tag{7.9d}$$

Setting $\phi_0 \equiv -B_0\psi$, (7.9a) gives

$$\boldsymbol{V}_\perp^{(0)} = \nabla_\perp\psi\times\hat{e}_z, \tag{7.10}$$

while $v_{\parallel}^{(0)}$, the component of $\mathbf{v}^{(0)}$ in the direction of $\boldsymbol{B}_0$ (i.e., along the z-axis) is determined from boundary conditions, i.e., in the absence of mean flow, $v_{\parallel}^{(0)} = 0$.

Now, from (7.9b)

$$\boldsymbol{j}^{(0)} = .\hat{e}_z j_z(x,y,\zeta,t), \tag{7.11}$$

so that (7.9d) implies

$$\boldsymbol{A}^{(0)} = \hat{e}_z A_z(x,y,\zeta,t) \tag{7.12a}$$

and

$$\boldsymbol{B}^{(0)} = (\nabla_\perp A_z) \times \hat{e}_z \ . \tag{7.12b}$$

Considering the O(1) terms in (7.6) we find

$$\begin{aligned}\frac{\partial A_z}{\partial t}\hat{e}_z &= \boldsymbol{V}_\perp^{(1)} \times \hat{e}_z B_0 + \boldsymbol{V}_\perp^{(0)} \times [(\nabla_\perp A_z) \times \hat{e}_z] \\ &\quad - \eta j_z \hat{e}_z - \frac{\partial}{\partial \zeta}\phi_0 \hat{e}_z - \nabla_\perp \phi^{(0)} \ . \end{aligned} \tag{7.13}$$

Note that the effect of a constant external electric field E in the z-direction can be included in (7.13) by allowing $\phi^{(0)}$ to have the form

$$\phi^{(0)} \to \phi^{(0)}(x,y,\rho,t) - \zeta E_{ext} \ , \tag{7.14}$$

then $E = E_{ext}$ would be added to the right-hand side of (7.13).

Also, introducing the vorticity $\omega = \nabla \times \mathbf{v}$, so that (following the above expansion conventions)

$$\omega^{(0)} = \nabla_\perp \times V_\perp^{(0)} = -e_z \nabla_\perp^2 \psi \equiv \omega\, \hat{e}_z \, , \tag{7.15}$$

we can write the O(1) terms of (7.8), after some algebra, as

$$\begin{aligned}&\rho\left(\frac{\partial \omega}{\partial t} + \boldsymbol{V}^{(0)} \cdot \nabla_\perp \omega\right)\hat{e}_z = \\ &\nabla_\perp \times \left(j_z \hat{e}_z \times \boldsymbol{B}^{(0)}\right) + \nabla_\perp \times \left(\boldsymbol{j}^{(1)} \times \hat{e}_z\right) B_0 \\ &+ \nu \nabla_\perp^2 \omega\, \hat{e}_z \ . \end{aligned} \tag{7.16}$$

Finally, considering the z-components of (7.13) and (7.16), we arrive at the pair of equations first derived in a very different way by Strauss (1976) (see also Montgomery, 1982):

$$\frac{\partial A_z}{\partial t} + \boldsymbol{V}_\perp^{(0)} \cdot \nabla_\perp A_z = \eta \nabla_\perp^2 A_z + B_0 \frac{\partial \psi}{\partial \zeta} + E_{ext} \,, \tag{7.17a}$$

$$\rho \left(\frac{\partial \omega}{\partial t} + \boldsymbol{V}_\perp^{(0)} \cdot \nabla_\perp \omega \right) = \boldsymbol{B}^{(0)} \cdot \nabla_\perp j_z + B_0 \frac{\partial j_z}{\partial \zeta} + \nu \nabla^2 \omega . \tag{7.17b}$$

Note that the $O(\epsilon)$ terms of equation (7.2e) allowed us to write

$$\nabla_\perp \cdot \boldsymbol{j}^{(1)} = -\frac{\partial}{\partial \zeta} j_z$$

in (7.17b).

The above system (7.17) is commonly referred to as the Strauss equations of "reduced" MHD. Their turbulence properties (as well as those of various modifications, derived later) have not been entirely investigated, yet. They describe a highly anistropic limit that has no analogy without the strong mean magnetic field $\boldsymbol{B}_0$.

B. Anisotropic $\boldsymbol{k}$-Spectra by Alfvén Wave Effects

Throughout the above discussion the presence of Alfvén waves has been assumed negligible. Their presence would introduce a fast $(t/\epsilon, z)$ time and z dependence in the problem. An initial $O(1)$ population of Alfvén waves would render the above analysis inadequate.

A question that could be posed is: What would be the evolution of an initially isotropic k-spectrum in the presence of a strong $\boldsymbol{B}_0 = B_0 \, \hat{e}_z$? This problem was addressed in a numerical simulation of the 2D incompressible MHD equations in a periodic geometry with $\nu \neq 0$ and $\eta \neq 0$ (Shebalin et al., 1983). It was found (see Fig. 7.1) that the spectrum evolved anisotropically, by transferring energy to modes perpendicular to $\boldsymbol{B}$ far more rapidly than it was transferred to modes with $\boldsymbol{k}$ parallel to $\boldsymbol{B}$. The evolution proceeded in a direction such as to render the "Strauss approximations" valid after awhile, even if they were not so initially.

As a first step toward an elementary understanding of why perpendicular transfer is easier, we first show the existence of Alfvén modes through linearizing the MHD equations about a quiescent state with an externally imposed magnetic field:

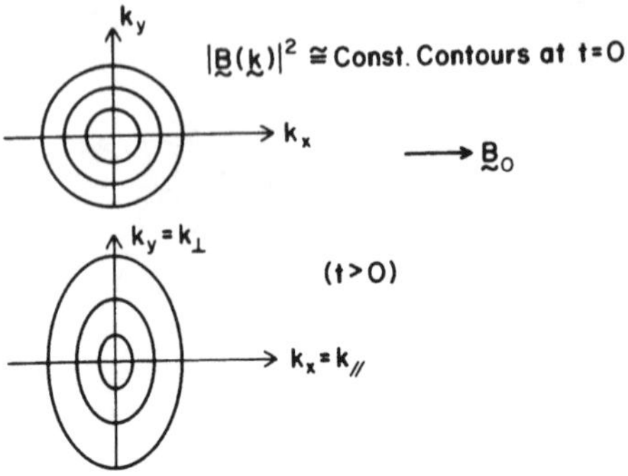

Figure 7.1: Spectral density contours elongate in the perpendicular direction as a consequence of much greater transfer rates in this direction (schematic).

$$B = B_0 + \epsilon B_1 e^{ik\cdot x} , \tag{7.18a}$$

$$\mathbf{v} = \epsilon \mathbf{v}_1 e^{ik\cdot x} , \tag{7.18b}$$

$$p = \epsilon p_1 e^{ik\cdot x} , \tag{7.18c}$$

$$j = \epsilon j_1 = \epsilon i k \times B_1 e^{ik\cdot x} , \tag{7.18d}$$

plus, of course, complex conjugate terms.

Substituting in (7.2) and keeping only terms of $O(\epsilon)$, we get

$$\frac{\partial B_1}{\partial t} = ik \times (\mathbf{v}_1 \times B_0) = i\mathbf{v}_1 (k \cdot B_0) \tag{7.19}$$

and

$$\rho \frac{\partial \mathbf{v}_1}{\partial t} = -ikp_1 - ik(B_1 \cdot B_0) + iB_1 (k \cdot B_0) . \tag{7.20}$$

Taking a dot product of (7.20) with ik we find (since $ik\cdot \mathbf{v}_1 - ik\cdot B_1 = 0$, due to the solenoidal character of **v**, B),

$$p + B_0 \cdot B_1 = 0, \tag{7.21}$$

so that

$$\rho \frac{\partial \mathbf{v}_1}{\partial t} = ik \cdot B_0 B_1 . \tag{7.22}$$

Combining (7.19) and (7.22) we find that $\mathbf{v}_1$, B_1 satisfy similar simple-harmonic-oscillator equations

$$\frac{\partial^2}{\partial t^2}\begin{pmatrix}\mathbf{v}_1\\ \boldsymbol{B}_1\end{pmatrix} + \frac{(\boldsymbol{k}\cdot\boldsymbol{B}_0)^2}{\rho}\begin{pmatrix}\mathbf{v}_1\\ \boldsymbol{B}_1\end{pmatrix} = 0, \tag{7.23}$$

with solutions in the form

$$\begin{pmatrix}\mathrm{v}_1\\ \boldsymbol{B}_1\end{pmatrix} \sim e^{i\boldsymbol{k}\cdot(\boldsymbol{x}\pm B_0 t/\sqrt{\rho})}. \tag{7.24}$$

We notice that perturbations are wavelike, traveling at the Alfvén speed

$$C_a \equiv \frac{B_0}{\sqrt{\rho}} \tag{7.25}$$

in the direction either parallel or antiparallel to $\boldsymbol{B}$. In order to study nonlinear interactions between these Alfvén modes, we introduce the Elsässer variables (setting, for simplicity, $\rho = 1$)

$$z^{\pm} = \mathbf{v} \pm \boldsymbol{B}. \tag{7.26}$$

Substituting

$$\mathbf{v} = \frac{1}{2}(z^+ + z^-)$$

and

$$\boldsymbol{B} = \frac{1}{2}(z^+ - z^-)$$

in the ideal ($\nu = 0,\ \eta = 0$) MHD equations

$$\frac{\partial \boldsymbol{B}}{\partial t} = \nabla\times(\mathbf{v}\times\boldsymbol{B}_0)\ , \tag{7.27a}$$

$$\begin{aligned}\frac{\partial \mathbf{v}}{\partial t} + \mathbf{v}\cdot\nabla\mathbf{v} &= \boldsymbol{j}\times\boldsymbol{B} - \nabla p + \boldsymbol{j}\times B_0\\ &= (\boldsymbol{B}+\boldsymbol{B}_0)\cdot\nabla(\boldsymbol{B}+\boldsymbol{B}_0) - \nabla p^*,\end{aligned} \tag{7.27b}$$

with $p^* = p + (1/2)(B + B_0)^2$, we find, after algebra,

$$\frac{\partial z^{\pm}}{\partial t} + z^{\mp}\cdot\nabla z^{\pm} \pm \boldsymbol{B}_0\cdot\nabla z^{\pm} = -\nabla p^*. \tag{7.28}$$

We notice that if either one of $\boldsymbol{z}^+$ or $\boldsymbol{z}^-$ is set equal to zero, then the other, as given by (7.24), provides an *exact* solution of an Alfvén wave propagating without distortion at speed C_A in the direction of $\boldsymbol{B}_0$, either parallel ($\boldsymbol{z}^+$) or antiparallel ($\boldsymbol{z}^-$). For one mode, $\mathbf{v} = +\boldsymbol{B}$, for the other, $\mathbf{v} = -\boldsymbol{B}$, in the dimensionless units.

We present now a simplified perturbative argument in order to estimate possible energy transfer between modes. We introduce

$$z^{\pm} \sim \epsilon z_1^{\pm} + \epsilon^2 z_2^{\pm} + 0(\epsilon^3) \tag{7.29}$$

and we assume

$$\nabla p^* \sim \epsilon^2 \nabla p_2^* + 0(\epsilon^3). \tag{7.30}$$

Then, $z_1^{\pm}$ are (uncoupled) Alfvén waves of the form (7.24)

$$z_1^{\pm} = \sum_k A_k^{\pm} \exp\left(i\boldsymbol{k} \cdot (\boldsymbol{x} \pm \boldsymbol{B}_0 t)\right) \ . \tag{7.31}$$

Considering the next order, we have

$$\frac{\partial z_2^{\pm}}{\partial t} \pm \boldsymbol{B}_0 \cdot \nabla z_2^{\pm} + \nabla p_2^* = -z_1^{\mp} \cdot \nabla z_1^{\pm} \ . \tag{7.32}$$

Equation (7.32) is of the same structure as the equations for $z_1^{\pm}$, but has an effective driving term on the right, driven by the linear Alfvén waves.

Clearly, resonant interactions, which are the most efficient mechanism for fast energy transfer between modes, will occur among triads of modes with wavenumbers $\boldsymbol{k}_1$, $\boldsymbol{k}_2$, and $\boldsymbol{k}_3$ related by the wavenumber and frequency matching conditions

$$\boldsymbol{k}_1 = \boldsymbol{k}_2 + \boldsymbol{k}_3 \ , \tag{7.33a}$$

$$\boldsymbol{k}_1 \cdot \boldsymbol{B}_0 = \pm \boldsymbol{k}_2 \cdot \boldsymbol{B}_0 \pm \boldsymbol{k}_3 \cdot \boldsymbol{B}_0 \ , \tag{7.33b}$$

with $\pm$ signs uncorrelated. However, (7.32) shows that there is no interaction between right-traveling pairs and left-traveling ones.

Reflection on the content of Eqs. (7.33) quickly reveals that to satisfy them simultaneously requires that at least one member of the triad ($\boldsymbol{k}_2$, say) must have zero component for the wavenumber, if opposite-traveling pairs only are allowed to interact. Thus, in a triad interaction, no $|k_z|_{max}$ can ever increase, but $|\boldsymbol{k}_\perp|_{\max}$ can, giving us an indication of the origin of the anisotropic evolution of the spectrum.

Higher-order processes will eventually transfer energy in the k_z-direction as well, unless they are interrupted by dissipation. Therefore it is fair to say that, although the perpendicular transfer process is an ideal one, dissipation is crucial to its maintenance and the smaller ν and η are the higher the anisotropy of the long-term state (Shebalin et al., 1983).

For given ν and η the anisotropy is not increased by increasing $\boldsymbol{B}_0$, since that increases the frequency of the waves but not the transfer rate between modes (to lowest order). The $\boldsymbol{B}_0$ field is simply responsible for creating the Alfvén modes, while their nonlinear interactions, in combination with dissipation, are responsible for the creation of the anisotropy.

We conjecture that in a driven, direct cascade, the spectrum should be of Kolmogorov type in $k_\perp$ (with a $k_\perp^{-5/3}$ dependence) while showing exponential fall-off in $k_\parallel$. This idea is based upon the guess that a quasi-2D direct cascade can be set up for each k_z band on a time scale short relative to the characteristic parallel transfer time scale.

C. MHD Turbulence in Tokamaks

The tokamak is the most-studied laboratory MHD plasma. Its usual mode of operation is partially quiescent, and the turbulence observed in it is not fully developed. The MHD turbulence that has been observed in tokamaks can roughly be divided into three kinds.

1. Broad-band, low-level noise

This kind of turbulence, or noise, exists in the low-temperature regime with rather low levels of magnetic field fluctuations ($\delta B/B < 10^{-3}$). This noise spans 2 to 3 orders of magnitude in frequency, and is the only kind of MHD turbulence that has been measured locally inside a tokamak. In the higher-temperature turbulence regimes, local probe measurements cannot be performed without destroying the probe and/or the plasma.

The probe measurements present some difficulties when one tries to determine the $\boldsymbol{k}$-spectrum of the noise, since the probe measures the temporal magnetic fluctuations at a fixed location. However, these measurements clearly indicate that the $\boldsymbol{k}$-spectrum is anisotropic with $k_\perp \gg k_\parallel$, so that the correlation distance perpendicular to $\boldsymbol{B}$ is short as compared to the correlation distance parallel to $\boldsymbol{B}$.

Internal probe measurements of low-level MHD turbulence in tokamaks have more or less stopped in the present experimental fusion program, but it would be desirable to have more systematic data on this kind of turbulence in order to obtain a better insight into the physical mechanisms responsible for its behavior.

2. Internal Disruptions, "Sawteeth"

The signature of this kind of turbulence as detected by external X-ray diagnostics is shown in Fig. 7.2.

Figure 7.2: Characteristic shape of a "sawtooth" X-ray signal from a tokamak, thought to be primarily a local temperature diagnostic.

Since the X-ray diagnostics measure density fluctuations and not magnetic fluctuations, no information is provided on the magnitude of $\delta B/B$.

The internal disruptions seem to involve relative few toroidal Fourier modes (m,n) (where m refers to the poloidal wavenumber, and n to the toroidal). The slow increase of the sawtooth signal seems to be on a "resistive" time scale and is thought to be due to heating of the plasma center, which increases the local electrical conductivity and thereby the current density. When the current localized in the central part of the plasma exceeds a certain limit, the magnetic field lines are disrupted on a fast Alfvén time scale, causing heat flux to be transferred to the walls, thereby lowering the central temperature. The whole process is then repeated again. This quasi-periodic, nonlinear process is not a fully developed turbulent phenomenon, but is reminiscent of "transition" behavior in such situations as Rayleigh-Benard convection.

It should be noted that the rising phase has some high-frequency internal structure indicating some turbulent activity. Furthermore, it should be pointed out that a detailed understanding of the mechanisms involved in the internal disruptions has not yet been obtained.

3. Major Disruptions

These disruptions are far more serious than the internal disruptions. A major disruption is characterized by a very rapid loss of confinement, so that the plasma is suddenly transferred to the walls. Since these disruptions most often occur at high temperatures ($\geq 100\ eV$, corresponding to > $10^7\ K$), they can produce severe damage to the walls. A major disruption is an unthinkable event in the imaginable operation of a reactor producing commercial fusion power.

It seems to be possible to more or less avoid the major disruptions by running the tokamak in a "safe" parameter regime. However, this operating regime is rather restrictive and a better understanding of the mechanisms leading to a major disruption is one of the most important issues of the magnetic fusion program. One way of obtaining further insight into this problem seems to be to study major disruptions experimentally at lower temperatures, where the disruption are still be achievable, but with less severe consequences.

It should be mentioned that MHD turbulence is not the only kind of turbulence found in tokamaks. An important example is the so-called "drift-wave" turbulence, which is thought to be of primarily electrostatic origin. Although the electrostatic types of turbulence do not disrupt magnetic field lines, their presence can have a strong impact on various transport properties. While clearly important, electrostatic turbulence lies outside the scope of these chapters.

As mentioned under the description of internal disruptions, the value of the current in the central part of the plasma can determine the stability properties of the plasma equilibrium. In order to discuss this behavior in more precise terms, we introduce the so-called "safety factor" q, which measures how many times a helical magnetic field line has to go around the tokamak in the toroidal direction ("the long way around") before completing one revolution in the poloidal direction ("the short way around"). The safety factor is thus given by

$$q(r) = \frac{r}{R} \frac{B_\phi(r)}{B_\theta(r)} , \tag{7.34}$$

where R is a major radius, r is a minor (interior) radius, and θ and ϕ are the poloidal and toroidal angles, respectively.

Considering an "unwrapping" of the torus into a cylinder, so that the toroidal axis is mapped onto the z-axis with a periodicity length L, we can introduce a "cylindrical safety factor"

$$q_{cyl}(r) = \frac{2\pi r}{L} \frac{B_z(r)}{B_\theta(r)} . \tag{7.35}$$

The $q(r)$ profile depends on the current profile $j_\phi(r)$ (or $j_z(r)$ in the cylindrical case), and for typical tokamak equilibria the $q(r)$ profile has the shape shown in Fig. 7.3.

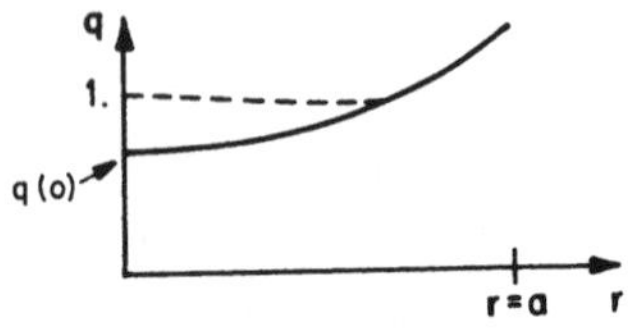

Figure 7.3: Conjectured schematic shape of the $q(r)$ profile of an operating tokamak, with an assumed axisymmetric current and magnetic field distribution (schematic).

The $q(r)$ profile typically has at least a local minimum at $r = 0$ where $q(0) \propto B_z(0)/j_z(0)$.

The so-called "tearing-mode" cylindrical instabilities at high S (low η, ν; sometimes $\nu = 0$), which are characterized by "magnetic island" formation, are associated with "mode rational surfaces," i.e., surfaces where $q(r)$ equals a ratio of integers. This is an involved subject with an enormous literature which cannot be done justice to here.

The "sawtooth" behavior described above may be due to localized nonlinearities (not fully developed turbulence) associated with $q = 1, 2, 3 \ldots$ surfaces having penetrated inside the plasma.

In general, the situation still is somewhat confused. Much of the existing information is numerical, but numerics always fall short of experimental complexities. At this moment, we have far more conjectures about the internal behavior of MHD fields in tokamaks than we do measurements of them.

D. Numerical Investigations of the Strauss Equations

Numerical studies of some turbulent properties of the Strauss equations (17 a,b) were performed by Dahlburg et al. (1985; 1986). In these papers the Strauss equations were solved by a pseudo-spectral method in three dimensions with periodic boundary conditions. A strong external magnetic field $\boldsymbol{B}_0$ was imposed in the z-direction and perfectly conducting, free-slip boundaries were assumed on the sides of a square cross section in the x-y plane:

$$j \times \hat{n} = 0 \, ,$$

$$\boldsymbol{B} \cdot \hat{n} = 0 \, ,$$

$$\mathbf{v} \cdot \hat{n} = 0 \, .$$

In all the computations the electrical resistivity η was finite, whereas the kinematic viscosity ν was either zero or finite. Some of the runs modelled initial-value decays with η = const., while others modelled electric-field-supported situations with $\eta(\boldsymbol{x}) \cdot j_z(\boldsymbol{x})$ = const.

Some results of the decay simulations are shown in Figs. 7.4 to 7.7.

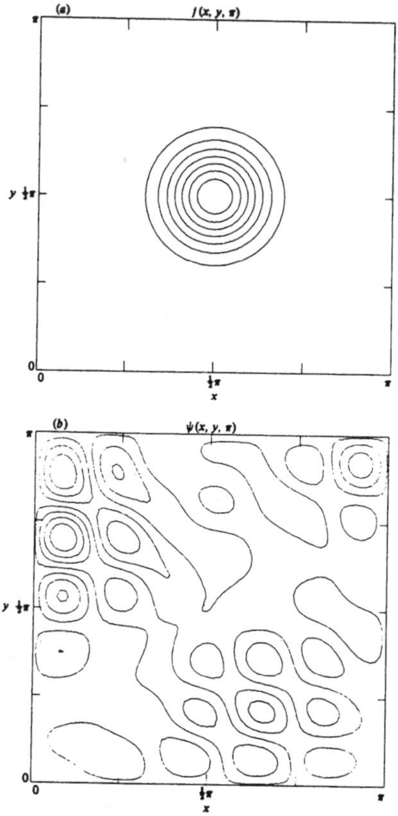

Figure 7.4: Initial current distribution j and stream-function distribution ψ (very small kinetic energy) for the simulations of Dahlburg et al. (1986). Contours are drawn in a poloidal plane, z = const.

Figure 7.4 shows the initial distribution at the midplane of the cylinder of: a) the current density, and b) the stream-function. A small amount of random noise was added to the system in order to trigger the turbulence reasonably fast.

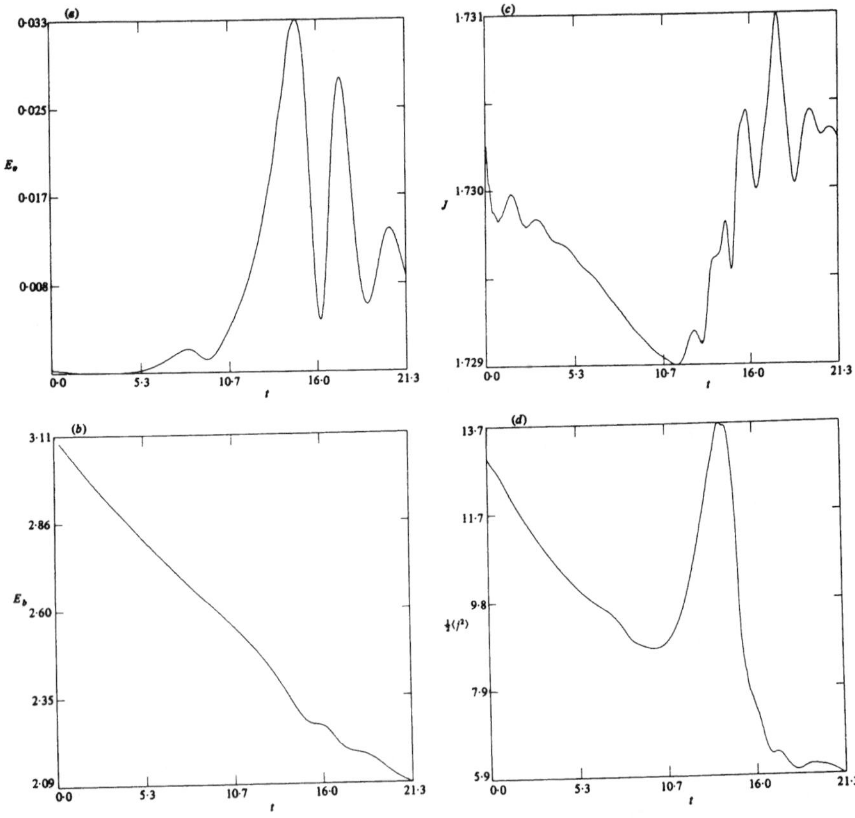

Figure 7.5: Temporal evolution of total kinetic energy (a), poloidal magnetic energy (b), total toroidal current (c), and ohmic dissipation (d) for the simulations of Dahlburg et al. (1986).

Although the magnetic energy decreases more or less monotonically throughout the run, the kinetic energy displays several "bursts," the first and largest of which is accompanied by a strong burst of mean-square current density. This

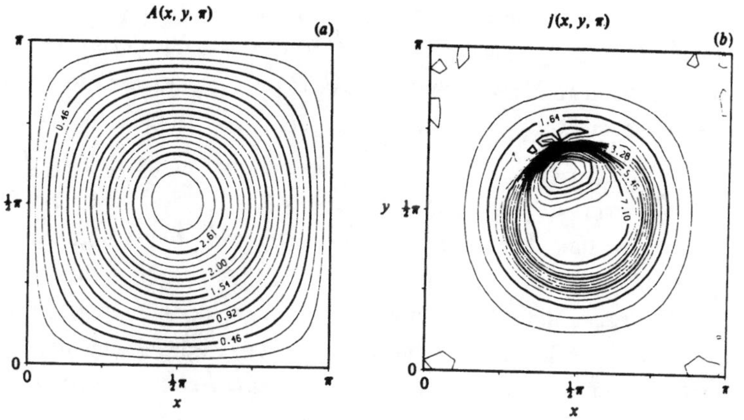

Figure 7.6: Poloidal contours of A (magnetic field lines) and current j for time near maximum disruption (from Dahlburg et al., 1985).

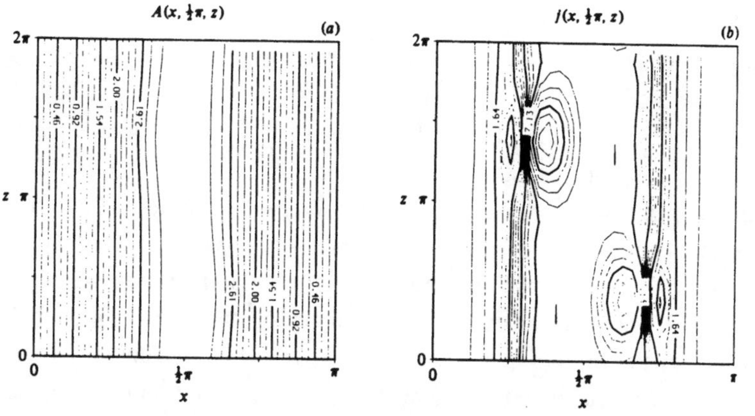

Figure 7.7: Same quantities as in Fig. 7.6, but in a plane of constant y. The helical current filament is clearly visible (from Dahlburg et al., 1985).

energy release is related to the disruption of internal magnetic-field lines. The distribution of magnetic field potential A and current density j near the time of maximum kinetic energy is shown in Fig. 7.6 (midplane of cylinder) and Fig. 7.7 (axial cut through cylinder).

Although very little perturbation is seen in the contours of A, the current density contours clearly reveal a helical $m = 1, n = 1$ current filament which wraps itself around the axis of the cylinder. (Contour plots tend to emphasize structure at the small scales of any function after ∇^2 has been applied.)

The evolution of the largest spatial scales alone in the decay runs were examined by means of a three-mode (Lorentz-like) truncation of the Strauss equations. Some time-dependent solutions of the low-order truncation system did suggest a qualitative agreement with the fully resolved solutions of the Strauss equations, while other solutions exhibited some interesting dynamical-systems behavior which were unparalleled in the fully resolved simulation results. This qualitative change in the behavior as the number of modes is increased strongly parallels the change in behavior of the Lorentz model in the same situation.

A single example of the forced Strauss equation with $E_{ext} \neq 0$ where an external electric field E_z is imposed is shown in Fig. 7.8, which shows the temporal evolution of: a) the total kinetic energy, and b) the poloidal magnetic energy.

We recognize a periodic behavior which suggests a sawtooth-like instability. Some important features are missing, however, when comparing Fig. 7.8 with Fig. 7.2: there is no clear distinction between rise time and decay time in Fig. 7.8, and the numerical simulations do not seem to relax completely to their initial states, so that a total reproduction of the internal disruption does not occur. These discrepancies indicate the importance of relaxing the incompressibility assumption (amongst other things) when trying to model the behavior of real-life disruptions on a computer. The subject of driven, nonlinear behavior according to the Strauss equations continues to evolve, including the addition of a temperature equation and temperature-dependent transport coefficients. See, e.g., Theobald et al. (1988).

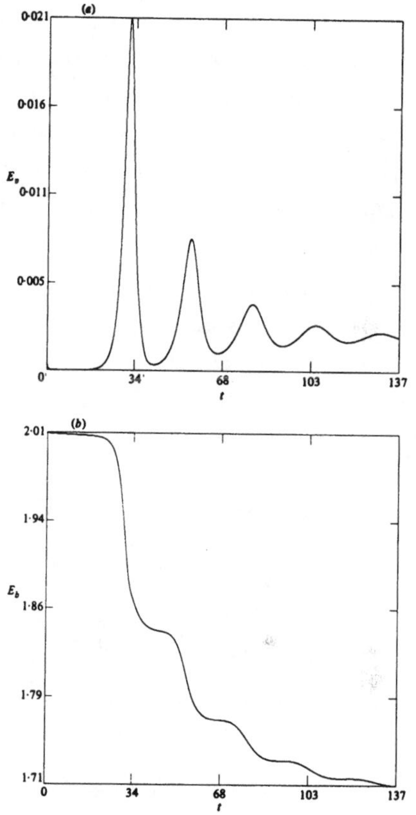

Figure 7.8: Total kinetic energy *E*v and poloidal magnetic energy *Eb* for driven Strauss equation computation (from Dahlburg et al., 1986).

Dahlburg, D., D. Montgomery, and W. H. Matthaeus, 1985: *J. Plasma Phys.*, **34**, 1.

Dahlburg, D., G. D. Doolen, and W. H. Matthaeus, 1986: *J. Plasma Phys.*, **35**, 1.

Montgomery, D., 1982: *Phys. Scripta*, **T2/1**, 83.

Shebalin, J. V., W. H. Matthaeus, and D. Montgomery, 1983: *J. Plasma Phys.*, **29**, 525.

Strauss, H. R., 1976: *Phys. Fluids*, **19**, 134.

Theobald, M., et al., 1988: "Sawtooth Oscillations about Helical Current Channels," Los Alamos Report LA-UR-88-939 (submitted to *Physics of Fluids*).

Chapter 8

COMPRESSIBLE MHD TURBULENT RELAXATION

In the past chapters, we have concentrated on incompressible fluids and magnetofluids. We shall now briefly address compressible magnetofluids, in particular, their relaxation via "selective decay" processes. Here, we describe a numerical test of the extent to which relaxation to a "minimum energy" state may occur, for an isolated magnetofluid, in this case a compressible one with very high thermal conductivity.

In the numerical work of Riyopoulos et al. (1982), they consider a compressible magnetofluid in a compact torus. A compact torus is a doughnut without the central hole, a simply-connected device. The confining wall is like a pill box with conducting end plates at $z = -L/2$ and $z = L/2$, and a cylindrical boundary at $r = (x^2 + y^2)^{1/2} = R_0$ (see Fig. 8.1). The length scale and time scale can be expressed in units of R_0 and R_0/V_A, respectively, where V_A is the Alfvén speed determined by the typical magnetic field strength and plasma density.

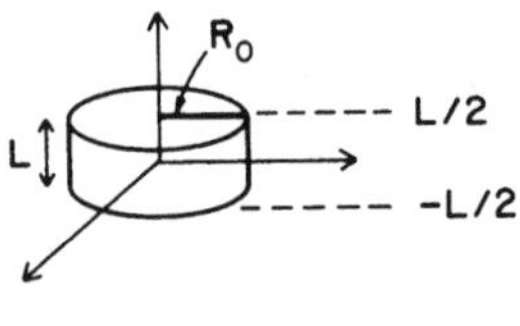

Figure 8.1: Basic geometry of the compact torus.

The compressible MHD equations are:

$$\frac{\partial \rho}{\partial t} + \nabla \cdot (\rho \mathbf{v}) = 0 , \tag{8.1}$$

$$\rho \left(\frac{\partial \mathbf{v}}{\partial t} + \mathbf{v} \cdot \nabla \mathbf{v} \right) = (\nabla \times \boldsymbol{B}) \times \boldsymbol{B} - \nabla p + \frac{\rho}{R} \left[\nabla^2 \mathbf{v} + \frac{1}{3} \nabla (\nabla \cdot \mathbf{v}) \right] , \tag{8.2}$$

$$\frac{\partial \boldsymbol{B}}{\partial t} = \nabla \times (\mathbf{v} \times \boldsymbol{B}) + \frac{1}{R_m} \nabla^2 \boldsymbol{B}, \tag{8.3}$$

where $\mathbf{v} = (v_r, v_\phi, v_z)$, and $R = LV_A/\nu$, $R_m = LV_A/\eta$, where ν and η are the kinematic viscosity and magnetic diffusivity, respectively.

Riyopoulos et al. used the Braginskii coefficients of unmagnetized plasmas for the viscosity ν and the resistivity η. Other parameters matched the PS-1 spheromak experiment of the University of Maryland in 1981, namely,

$$\begin{aligned} R_0 &= 10 \text{ cm} , \\ B_0 &= 4 \text{ kilogauss} , \\ T &= 30\ eV , \\ n &= 5 \times 10^{15} \text{ protons/cm}^3. \end{aligned} \tag{8.4}$$

They also assume infinite thermal conductivity, T = const. It follows that $p = 0.284\rho$ is the equation of state. The boundary conditions are $\mathbf{v} = 0$ and $\hat{n} \cdot \boldsymbol{B} = 0$, $\hat{n} \cdot (\nabla \times \boldsymbol{B}) = 0$ at the walls.

Axisymmetry is imposed throughout the computer simulation. Hence the magnetic field can be conveniently expressed in terms of two scalars,

$$\boldsymbol{B} = \nabla \times \left(\frac{\psi}{r} \hat{e}_\phi \right) + \frac{I}{r} \hat{e}_\phi. \tag{8.5}$$

ψ is the poloidal flux and I is the poloidal current function,

$$B_r = -\frac{1}{r}\frac{\partial \psi}{\partial z} , \; B_\psi = \frac{I}{r} , \; B_z = \frac{1}{r}\frac{\partial \psi}{\partial r} , \tag{8.5}$$

and

$$J_r = \frac{-1}{r}\frac{\partial I}{\partial z} , \tag{8.6}$$

$$J_\phi = -\frac{1}{r}\frac{\partial^2 \psi}{\partial z^2} - \frac{\partial}{\partial r}\frac{1}{r}\frac{\partial}{\partial r}\psi \equiv -\frac{1}{r}\Delta^* \psi , \tag{8.7}$$

$$J_z = \frac{1}{r}\frac{\partial I}{\partial r}. \tag{8.8}$$

With this representation of magnetic field, the induction equation is simplified and becomes

$$\frac{\partial \psi}{\partial t} + \mathbf{v} \cdot \nabla \psi = -\frac{1}{R_m}\Delta^* \psi , \tag{8.9}$$

$$\frac{\partial I}{\partial t} = -r\frac{\partial}{\partial r}\left(\frac{\mathrm{v}_r}{r} I + \frac{\mathrm{v}_\phi}{r}\frac{\partial \psi}{\partial z} \right) + \frac{\partial}{\partial z}\left(\mathrm{v}_\phi \frac{\partial \psi}{\partial r} - I \mathrm{v}_z \right) + \frac{1}{R_m}\Delta^* I. \tag{8.10}$$

All field variables are advanced by finite differences. Symmetry is assumed about the $z = 0$ midplane. The resolution is 44 grid points in the r-direction and 30 grid points in the z-direction in the computational domain $0 \leq r < 1$,

$0 \leq z < L_0/2R_0 = 0.72$. (The fields can be inferred elsewhere in the cylinder by symmetry.) Initial contours of relevant fields are shown in Fig. 8.2.

Initially, the run starts with zero velocity, and ψ and I are a superposition of force-free states but the composite state is not itself force free:

$$\psi = \sum_{i,j} A_{i,j}^{\pm} r J_1 (j_{1,i} r) \cos k_j z \,, \tag{8.11}$$

$$I = \sum_{i,j} A_{i,j}^{\pm} \left(j_{1,i}^2 + k_j^2\right)^{1/2} r J_1 (j_{1,i} r) \cos k_j z, \tag{8.12}$$

where $j_{1,i}$ are the i-th zeros of the Bessel function J_1, and + and - refer to the sign of the eigenvalue λ of the force-free equation

$$\nabla \times \boldsymbol{B} = \lambda \boldsymbol{B}. \tag{8.13}$$

The coefficients $A_{i,j}^{\pm}$ are chosen such that

$$A_{2,2}^{+} = 0.6,\ A_{1,3}^{+} = 0.15,\ A_{3,2}^{+} = -0.05,\ A_{4,3}^{-} = -0.05,\ A_{5,3}^{-} = 0.05 \tag{8.14}$$

and all others are zero at $t = 0$.

At this point, it is useful to introduce two global ideal invariants. The total energy E is

$$E = W + \int \boldsymbol{p} \ln \boldsymbol{p} \, d^3x \tag{8.15}$$

where W is the usual kinetic energy and magnetic energy;

$$W = \frac{1}{2} \int \left(\rho v^2 + B^2\right) d^3x \,. \tag{8.16}$$

It is expected that W, not E, is the appropriate decay quantity for compressible flow.

The expression of the internal energy has incorporated the isothermal assumption. The other quantity is the magnetic helicity

$$H_m = \frac{1}{2} \int \boldsymbol{A} \cdot \boldsymbol{B} d^3x \rightarrow \int \frac{I\psi}{r^2} d^3x \,. \tag{8.17}$$

Throughout the run, the plasma β rises from an initial 0.16 to a value about 2. That is, the system has converted 90 percent of its magnetic energy to particle thermal energy. The maximum Reynolds numbers, using the length scale of the dominant fluctuation, are $(R_m)_\lambda \cong 300$ and $(R)_\lambda \cong 10$. The ratio W/H_m decreases monotonically in time, consistent with the scenario of selective decay for incompressible MHD. The energy decays simultaneously with bursts of magnetic energy conversion into kinetic energy, which is then rapidly dissipated into heat. These dissipation processes result primarily at the x-point magnetic reconnection.

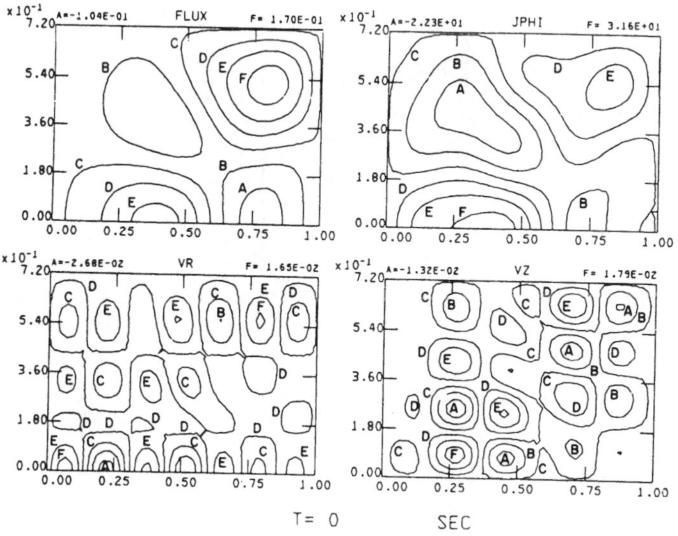

Figure 8.2: Contours of constant ψ, J_θ, v_r, and v_z at 10^{-9} sec. These contours in effect represent the initial conditions. Only the upper right quadrant, $0 < r < 1$, $0 < z < 0.72$, of the cylinder is shown. Symmetry about the midplane is assumed.

Toward the end of the run, the poloidal magnetic field appears to be a large island (see Fig. 8.3), filling the box of the computation, which reveals the final phase of turbulent relaxation. This work indeed suggests that in a compressible magnetofluid with initially small kinetic energy and no external driving:

(1) W/H_m decreases toward its lower bound.

(2) The way in which the relaxation is achieved is primarily due to intermittent, yet intense, magnetic reconnection. The reconnection and energy dissipation are all one process and it is misleading to consider either as prior to the other.

The relaxation that occurs takes the magnetofluid close to the minimum energy state, but the differing boundary conditions on $\boldsymbol{j}$ and $\boldsymbol{B}$ lead to a persistent boundary layer near the conductor. This effect has been observed repeatedly in other computations since.

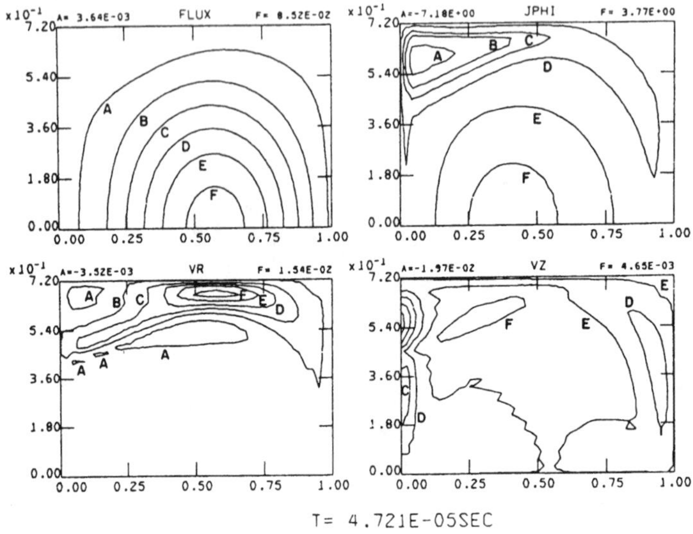

Figure 8.3: Contours of ϕ, J_θ, v_r, and v_z at time 47.21 μ sec. The magnetic reconnection is essentially complete.

A second relaxation process that plays a sometimes significant role in decaying MHD turbulence may be called "dynamic alignment"; it has already been discussed to some extent in Chapter 6. It can result from the more rapid decay of energy relative to cross helicity, instead of energy relative to the magnetic ideal invariant (magnetic helicity in 3D, mean-square-vector potential in 2D). Clearly, in situations involving decaying turbulence and finite amounts of all three quantities initially, there is a real question as to the nature of the "relaxed" state: Is it "minimum energy" ($\mathbf{v} = 0$), is it "dynamically aligned" ($\mathbf{v} \approx \pm \boldsymbol{B}$), or is it something else?

Ting et al. (1986) have attempted to identify different regimes, in the phase space of initial conditions, in which different kinds of relaxations would be observed for the decaying initial value problem. A combination of low-order and high-order truncations of the Fourier-expanded MHD equations was used. Roughly speaking, four regimes were observed in the phase space, though no sharp boundaries between them were suggested. The case of small initial $H_c = \int \mathbf{v} \cdot \boldsymbol{B} d^2x$ was characterized by the "selective decay" of energy relative to the magnetic invariant and the virtual disappearance of the kinetic energy. If the inital kinetic energy was much larger than the initial magnetic energy, the behavior observed was essentially that of a Navier-Stokes fluid, with whatever magnetic evolution there was being essentially passive. If there was considerable cross helicity and comparable magnetic and kinetic energies initially, the "dynamic alignment" process prevailed. Finally, a fourth erratic "transition regime" was contiguous with the other three, in which the magnetofluid seemed initially to have a hard time deciding which limit it wanted to head for.

The astrophysical implications of the wide range of relaxed behavior are enormous, since an initially turbulent cloud of conducting fluid seems to have in it the capability for any ratio of kinetic to magnetic energy in the evolved (relaxed) state. However, a universal result seemed to be the approach of the pointwise alignment cosine, $\boldsymbol{B}\cdot\mathbf{v}/Bv$, to ± 1, regardless of the ratio.

Riyopoulos, S., A. A. Bondeson, and D. Montgomery, 1982: *Phys. Fluids*, **25**, 107.
Ting, A. C., W. H. Matthaeus, and D. Montgomery, 1986: *Phys. Fluids*, **29**, 3261.

Chapter 9

PART I: MHD TURBULENCE IN REVERSED-FIELD PINCH PLASMAS

Before we begin our discussion of turbulence in RFP plasmas, I would like to add a footnote to the comments on solar wind turbulence discussed in the previous chapter. In Figure (9.1) (from a paper by Belcher and Davis, 1971), we show how the magnetic and velocity fields in the solar wind can be both turbulent and co-aligned. This is a plot of the three components of the magnetic and velocity fields in the solar wind over a 24-hour period. The observations are from a spacecraft that passed through the solar wind, and so arise from a steadily varying distance from the sun. Note that while the magnetic- and velocity-field components are very variable, the total magnetic-field magnitude and particle-number density are nearly constant. This type of turbulent system cannot readily be reproduced in a laboratory.

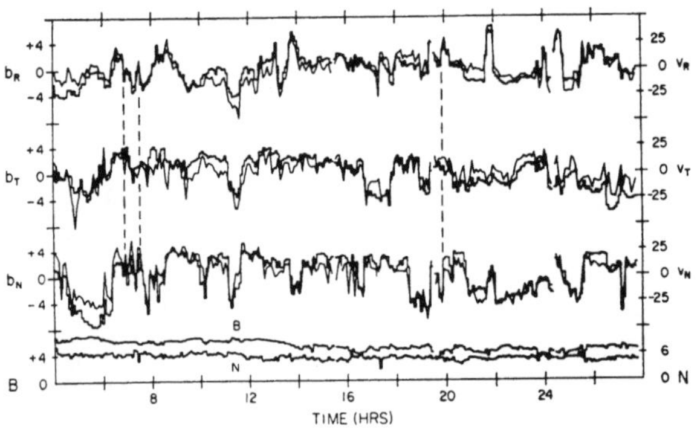

Figure 9.1: Twenty-four hours of magnetic-field and plasma data demonstrating the presence of nearly pure Alfvénic states. The upper six curves are 5.04-min bulk velocity components in km/sec (diagonal lines) and magnetic field components, in gammas (one gamma = 10^{-5}, averaged over the plasma probe sampling period (horizontal and vertical lines). The lower two curves are magnetic field strength and proton density (from Belcher and Davis, 1971).

This is an additional and compelling piece of evidence concerning the dynamic alignment process. We now return to the main subject of this chapter.

We shall see that a fundamental property of an electrically-conducting fluid is its tendency to form helical kinks when it carries a large electric current down a dc magnetic field. We will examine this property in the toroidal Z-pinch geometry. Consider a torus, that we will unwrap into a cylinder, with a toroidal applied magnetic field B_ϕ or B_z in the cylinder as shown in Fig. 9.2. If we force an electric current to flow in the toroidal direction by applying a toroidal electric field (usually inductively), a poloidal (azimuthal) magnetic field will be added to the toroidal one. In the cylindrical approximation, we will call the poloidal magnetic field $B_\theta(r)$, hoping that this switch of coordinates will not generate confusion.

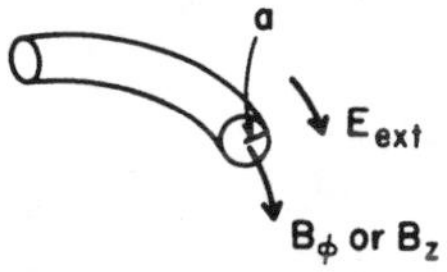

Figure 9.2: If a toroidal current is forced to flow along an externally supported toroidal magnetic field, the resulting field lines are nested helices.

If the toroidal field changes sign near the wall, at $r = a$, say, then we have a reversed field pinch (RFP). In the quiescent state (v = 0), the magnetic field configuration is, to a first approximation,

$$\boldsymbol{B} = B_\theta(r)\,\hat{e}_\theta + B_z(r)\,\hat{e}_z\ , \tag{9.1}$$

with $B_\theta \sim B_z$, unlike the case for the tokamaks, which have $B_z \gg B_\theta$. The poloidal field must be maintained by a toroidal current in Z-pinch machines. These devices tend to be noisy ($\delta B/B \sim$ 1 to 5 percent as measured *externally*); this is an underestimate of the internal noise. However, RFPs are in many ways easier to operate than tokamaks. They apparently do not experience major disruptions and are not as sensitive to minor variations in operating conditions.

The typical mode of operation is as follows: the plasma is pre-ionized, a large toroidal voltage is applied, the toroidal current surges up to hundreds of *kA*, the plasma becomes highly turbulent, and then "relaxation" occurs in 1 or 2 microseconds (roughly the Alfvén transit time of the minor radius a); the current falls, the noise level drops dramatically. Operation continues over a wide range of currents for as long as any "volt-seconds" remain to support the current. The two most common variables used to classify the operating regimes are $F \equiv \langle B_z(r)\rangle^{-1}B_z(a)$, which measures the depth of the field reversal when

$F < 0$, and $\Theta \equiv \langle B_z(r)\rangle^{-1} B_\theta(a)$, which measures the strength of the toroidal current. The fields $\langle B_z(r)\rangle$ are averaged over the cross section of the cylinder. If we have a constant dc toroidal field, $F = 1$. Figure 9.3 shows a typical time history of the current. We can see the rise as the voltage is applied, followed by a rapid relaxation to a roughly constant current level. Figure 9.4 shows $F - \Theta$ curves for three experiments. The field at the wall falls quickly as the current increases, crossing the $F = 0$ axis near $\Theta = 1.6$, relaxation occurs, and the system remains near $\Theta = 1.6$ with $F < 0$.

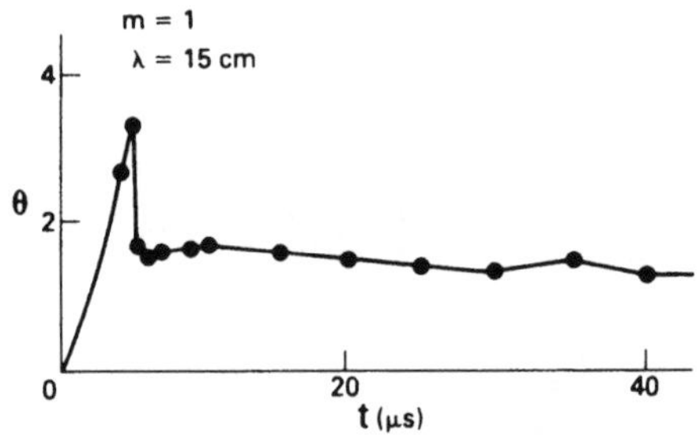

Figure 9.3: Limitation of θ, for *HBTX*1, (from Bodin and Newton, 1980).

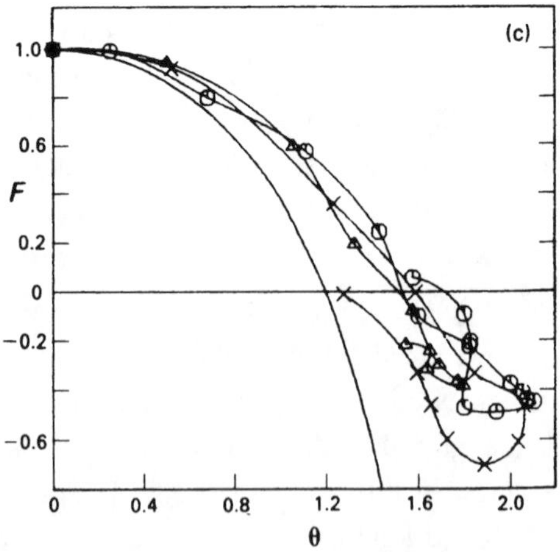

Figure 9.4: Time-dependent $F - \theta$ curve. REPUTE experiment (From Toyama et al., 1985).

The question of how this relaxation might occur was initially addressed by Taylor (1974; 1976). He hypothesized that when the energy (essentially all magnetic) of the system decays to the minimum value, it can have consistency with magnetic helicity conservation. That is $\int \boldsymbol{B}^2 d^3x$ becomes a minimal subject to the constraint $\int \boldsymbol{A} \cdot \boldsymbol{B} d^3x$ = const. The dynamical basis of this argument was vague, but it explained with surprising accuracy many features of the operation of Z-pinch machines. We will examine this question in some detail.

We will start by showing that both the magnetic energy and the magnetic helicity are ideal invariants for certain boundary conditions. We will then formulate a variational principle to determine the form of the magnetic field consistent with a "selective decay" hypothesis.

The induction equation for the magnetic field and the vector potential can be written as

$$\frac{\partial \boldsymbol{B}}{\partial t} = \nabla \times (\mathbf{v} \times \boldsymbol{B}) - \eta \nabla \times (\nabla \times \boldsymbol{B}), \tag{9.2}$$

and

$$\frac{\partial \boldsymbol{A}}{\partial t} = \mathbf{v} \times \boldsymbol{B} - \nabla \phi - \eta \nabla \times \boldsymbol{B}\,, \tag{9.3}$$

where ϕ is a scalar potential.

Taking $\boldsymbol{A}\cdot$ into the former and $\boldsymbol{B}\cdot$ into the latter yields

$$\begin{aligned} \frac{\partial (\boldsymbol{A} \cdot \boldsymbol{B})}{\partial t} &= \nabla \cdot [(\mathbf{v} \times \boldsymbol{B}) \times \boldsymbol{A}] + \nabla \times \boldsymbol{A} \cdot (\mathbf{v} \times \boldsymbol{B}) \\ &+ \boldsymbol{B} \cdot (\mathbf{v} \times \boldsymbol{B}) - \nabla \cdot (\boldsymbol{B}\phi) \\ &- \eta \nabla \cdot [(\nabla \times \boldsymbol{B}) \times \boldsymbol{A}] - \eta (\nabla \times \boldsymbol{A}) \cdot (\nabla \times \boldsymbol{B}) \\ &- \eta \nabla \cdot [\boldsymbol{B} \times \boldsymbol{B}] - \eta (\nabla \times \boldsymbol{B}) \cdot \boldsymbol{B}. \end{aligned} \tag{9.4}$$

We integrate this result over the plasma volume and use the divergence theorem. The surface integrals will vanish if $\hat{n} \cdot \mathbf{v} = 0$, $\hat{n} \times \boldsymbol{j} = 0$, and $\hat{n} \cdot \boldsymbol{B} = 0$; or if periodic boundaries are used. Then the helicity, $H_m \equiv \int \boldsymbol{A} \cdot \boldsymbol{B} d^3x$, is an ideal invariant, which in the presence of finite resistivity decays according to

$$\frac{\partial}{\partial t} \int \boldsymbol{A} \cdot \boldsymbol{B} d^3x = -2 \int \eta \boldsymbol{j} \cdot \boldsymbol{B} d^3x. \tag{9.5}$$

The energy, $E \equiv \int (1/2)(\mathrm{v}^2 + B^2) d^3x$ is also an ideal invariant with the above boundary conditions. (Note: for non-simply-connected geometries, $H_m \equiv \int \boldsymbol{A} \cdot$

$Bd^3x - \oint A \cdot dl \oint A \cdot ds$, is the gauge-invariant magnetic helicity, which becomes relevant in the presence of applied voltages.) The time rate of change of the magnetic helicity and the energy can be written:

$$\frac{\partial}{\partial t} H_m = -2\eta \int \boldsymbol{j} \cdot \boldsymbol{B} d^3x \ , \tag{9.6}$$

$$\frac{\partial}{\partial t} E = -\eta \int j^2 d^3x - \nu \int \omega^2 d^3x \ . \tag{9.7}$$

E can decay faster than H_m if the viscosity is large compared with the magnetic diffusivity; $\eta \ll \nu$. It can also decay more quickly if there is a process similar to inverse cascade occurring in the magnetofluid known as selective decay (see Montgomery et al., 1978; Frisch et al., 1975). Suppose we now make this "Taylor hypothesis," that energy decays to the minimum value it can have, subject to a given value of H_m.

Clearly, the minimum energy occurs for $\mathbf{v} = 0$, and the variational problem leads to $\delta(H_m + \lambda' E) = 0$ where λ' is a Lagrange multiplier. The variation of the helicity is

$$\begin{aligned} \delta \int \boldsymbol{A} \cdot \boldsymbol{B} \, d^3x &= \int \delta \boldsymbol{A} \cdot \nabla \times \boldsymbol{A} + \int \boldsymbol{A} \cdot \nabla \times \delta \boldsymbol{A} \\ &= \int \delta \boldsymbol{A} \cdot \nabla \times \boldsymbol{A} + \int \nabla \cdot (\delta \boldsymbol{A} \times \boldsymbol{A}) + \int \nabla \times \boldsymbol{A} \cdot \delta \boldsymbol{A} \\ &= 2 \int \delta \boldsymbol{A} \cdot \nabla \times \boldsymbol{A} \ , \end{aligned} \tag{9.8}$$

where we have assumed that $\int \nabla \cdot (\delta \boldsymbol{A}) \times \boldsymbol{A} = \int dS \, \hat{n} \cdot (\delta \boldsymbol{A} \times \boldsymbol{A}) = 0$ if $\hat{n} \times \delta \boldsymbol{A} = 0$. The variation of the energy can be expressed in a similar manner

$$\delta \int B^2 \, d^3x = 2 \int (\nabla \times \delta \boldsymbol{A}) \cdot (\nabla \times A) \ , \tag{9.9}$$

$$= 2 \int \nabla \cdot [\delta \boldsymbol{A} \times (\nabla \times \boldsymbol{A})] + 2 \int [\nabla \times (\nabla \times \boldsymbol{A})] \cdot \delta \boldsymbol{A} \ , \tag{9.10}$$

$$= 2 \int \delta \boldsymbol{A} \cdot \nabla \times (\nabla \times \boldsymbol{A}) \ . \tag{9.11}$$

The statement that $\delta(H_m + \lambda' E) = 0$ implies that the Euler equation is

$$\nabla \times (\nabla \times \boldsymbol{A}) = \lambda \nabla \times \boldsymbol{A}, \tag{9.12}$$

or

$$\nabla \times \boldsymbol{B} = \lambda \boldsymbol{B} = \boldsymbol{j} \ , \tag{9.13}$$

where $\lambda = -(\lambda')^{-1}$. A *force-free state* results from the variational analysis.

The force-free magnetic field can be expressed in terms of Chandrasekhar-Kendall (1957). Define a function ψ_λ such that

$$\left(\nabla^2 + \lambda^2\right)\psi_\lambda = 0, \tag{9.14}$$

then let

$$\boldsymbol{a}_\lambda \equiv \nabla \times \hat{\boldsymbol{e}}\psi + \frac{\nabla \times (\nabla \times \hat{\boldsymbol{e}}\psi)}{\lambda}, \tag{9.15}$$

where $\hat{\boldsymbol{e}}$ is any unit vector. It follows that

$$\begin{aligned} \nabla \times \boldsymbol{a}_\lambda &= \nabla \times (\nabla \times \hat{\boldsymbol{e}}\psi) + \frac{\nabla \times (\nabla \times (\nabla \times \hat{\boldsymbol{e}}\psi))}{\lambda} \\ &= \nabla \times (\nabla \times \hat{\boldsymbol{e}}\psi) + \frac{\nabla(\nabla \cdot \nabla \times \hat{\boldsymbol{e}}\psi) - \nabla^2(\nabla \times \psi)}{\lambda} \\ &= \lambda\left[\nabla \times \hat{\boldsymbol{e}}\psi + \frac{1}{\lambda}\nabla \times (\nabla \times \hat{\boldsymbol{e}}\psi)\right] \end{aligned} \tag{9.16}$$

or

$$\nabla \times \boldsymbol{a}_\lambda = \lambda \boldsymbol{a}_\lambda\ ; \tag{9.17}$$

thus any $\boldsymbol{a}_\lambda$ defined by (9.15) is a force-free solenoidal field. For a cylinder, the ψs can be expressed in terms of Bessel functions

$$\psi_\lambda \Rightarrow J_m(\gamma_{mn}r)e^{(im\theta + ik_n z)}, \tag{9.18}$$

where $k_n = 2\pi n/L$. Using the defining relation for ψ_λ yields Bessel's equation (if we are in cylindrical coordinates), and the unit vector is in the z-direction

$$\frac{1}{r}\frac{d}{dr}r\frac{dJ_m(\gamma_{mn}r)}{dr} - \frac{m^2}{r^2}J_m(\gamma_{mn}r) - k_n^2 J_m(\gamma_{mn}r) = -\lambda^2 J_m(\gamma_{mn}r), \tag{9.19}$$

if $\gamma_{mn}^2 + k_n^2 = \lambda^2$. For low enough currents (or low helicity), the minimum energy state is

$$\boldsymbol{B} = \alpha\left[J_0(\lambda r)\,\hat{\boldsymbol{e}}_z + J_1(\lambda r)\,\hat{\boldsymbol{e}}_\theta\right], \tag{9.20}$$

which leads to a pinch ratio

$$\Theta = \frac{\lambda a}{2}, \tag{9.21}$$

and

$$F = \frac{\alpha J_0(\lambda a)}{\langle B_z \rangle}. \tag{9.22}$$

The $F - \Theta$ curve is shown in Fig. 9.5.

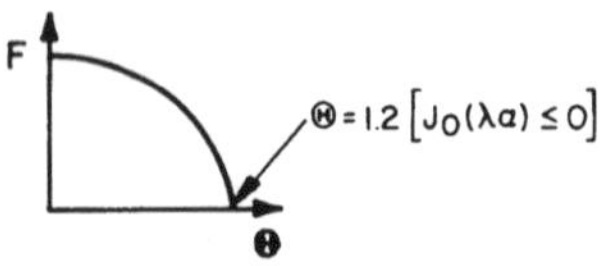

Figure 9.5: $F - \Theta$ plot resulting from the state (9.20).

As H_m/E increases (I_z increases) a point is reached at which the (0,0,1) state (9.20) is no longer the minimum energy state. At $\Theta \cong 1.6$ ($\lambda a = 3.11$, or $k_n a \cong 1.25$), a (1,1,n) helical state begins to contribute to the minimum energy state. The plasma becomes more turbulent above $\Theta \cong 1.6$. This is also the threshold at which *ideal* MHD instability sets in.

This is an extremely abbreviated and hasty sketch of the first influential theory of RFP operation. Turbulence is important primarily for its role in the achievement of the state from something which looks initially very unlike it. The steady-state level of turbulence that is achieved is much milder than during the formation phase. For a detailed introduction and a more extensive bibliography, see the review article of Taylor (1986). For our purposes here, the most significant thing about the RFP is that it is by far the most important example of real MHD turbulence in captivity: an existing MHD device that produces copious amounts of MHD turbulence of different kinds that is available for study.

Chapter 9

PART II: 3D REVERSED-FIELD PINCH SIMULATION

The turbulent relaxation observed in the reversed-field pinch (RFP) experiments, and postulated in the selective decay process just described, has been numerically simulated by Dahlburg et al. (1986; 1987) with an incompressible 3D code. Though most RFP devices have circular cross sections, a square cross section is used in the simulation in order to be able to use Fast Fourier Transforms (FFTs) to facilitate the computations, and to avoid all the unphysical complications perennially associated with the point $r = 0$. So the magnetofluid is supposed to be confined by rigid, free-slip, perfectly conducting square boundaries in the x, y directions, and periodic boundary conditions are used in the z-direction which corresponds to the toroidal direction of the RFP device. The corners of the square are taken at (0,0), $(0, \pi)$, $(\pi, 0)$ and (π, π) in the (x,y) plane, and boundaries in the (periodic) z-direction are taken at a distance of 4π apart. A constant, uniform dc magnetic field B_0 is assumed in the z-direction, so that the total magnetic field can be written as

$$\boldsymbol{B}_{tot} = B_0 \hat{\boldsymbol{e}}_z + \boldsymbol{B}. \tag{9.23}$$

To start the run, a non-uniform current in the z-direction is set up at time $t = 0$. Since this current would give rise to unbalanced Lorentz forces, the system is not in equilibrium and evolves according to the following equations for the velocity field $\mathbf{v}$ and the vector potential $\boldsymbol{A}(\nabla \times \boldsymbol{A} = \boldsymbol{B})$:

$$\frac{\partial \mathbf{v}}{\partial t} = \mathbf{v} \times \omega - \nabla \pi + \boldsymbol{j} \times \boldsymbol{B}_{tot} \,, \tag{9.24}$$

$$\frac{\partial \boldsymbol{A}}{\partial t} = \nabla \phi + \mathbf{v} \times \boldsymbol{B} + \frac{1}{S} \nabla^2 \boldsymbol{A}. \tag{9.25}$$

Here $\pi = (1/2)|\mathbf{v}|^2 +$ mechanical pressure, ϕ is a scalar potential and S is the Lundquist number, taken to be 250 for the run. We choose the Coulomb gauge and have

$$\nabla \cdot \boldsymbol{A} = 0 \,, \qquad \nabla \cdot \mathbf{v} = 0. \tag{9.26}$$

The boundary conditions on the sides of the square are

$$\mathbf{v} \cdot \hat{n} = \boldsymbol{B} \cdot \hat{n} = \nabla\pi \cdot \hat{n} = 0 \ , \tag{9.27}$$

$$\boldsymbol{j} \times \hat{n} = 0, \tag{9.28}$$

where everything is periodic in z with a period of 4π.

The code is spectral in character, using the Turner-Christiansen functions as the basis of expansion for all the variables (see Appendix A of Dahlburg et al., 1987 paper for details). The typical spatial resolution is $16 \times 16 \times 128$. The initial current profile chosen is

$$j_0(x,y) = \sin x \sin y \left[10 - 9\exp\left(-3\left(x - \frac{1}{2}\pi\right)^2 - 3\left(y - \frac{1}{2}\pi\right)^2\right)\right] . \tag{9.29}$$

A three-dimensional perspective plot of the current is seen in Fig. 9.6.

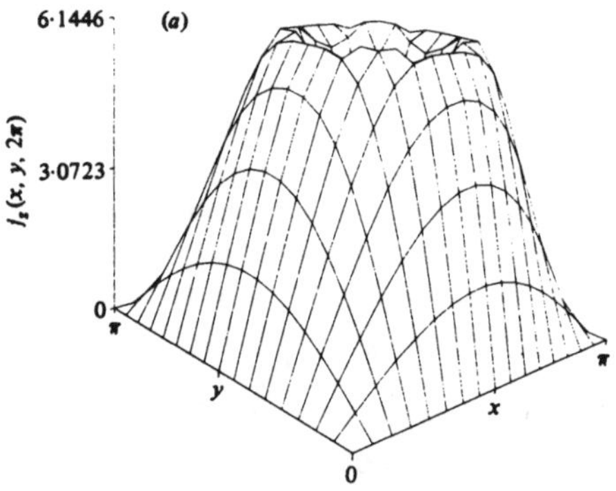

Figure 9.6: Perspective plots of the initial j_z current distribution (9.29) as a function of the poloidal coordinates x and y (from Dahlburg et al., 1987).

Measuring time in units of Alfvén transit time for unit distance, it is found that the system goes through a turbulent phase lasting until about $t = 25$, after which a relaxed state is reached that decays slowly due to Ohmic dissipation. Indications of this relaxation process can be seen in Figs. 9.7 and 9.8, showing the time-evolutions of different global quantities.

One of the most striking features of the RFP experiments is certainly the appearance of a reversed toroidal field at the wall. This can be seen clearly by plotting the evolution of average $\langle B_{z_{tot}} \rangle$ at the wall.

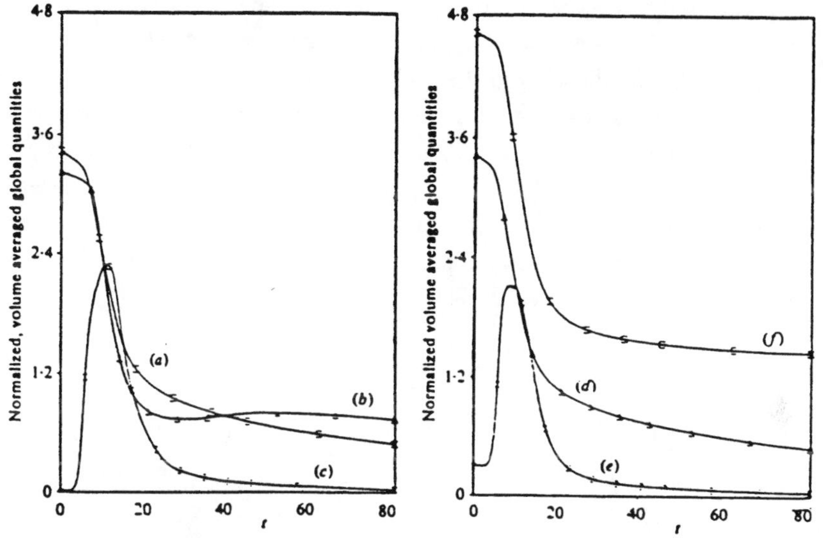

Figure 9.7: Decay of global integral quantities: (a) total energy; (b) net toroidal current; (c) kinetic energy × 10; (d) total magnetic energy; (e) total Ohmic dissipation × 10; (f) ratio of total energy to magnetic helicity. Time is in units of Alfvén transit time of unit distance. The lowest possible value the magnetic energy can assume is 0.25 (from Dahlburg et al., 1987).

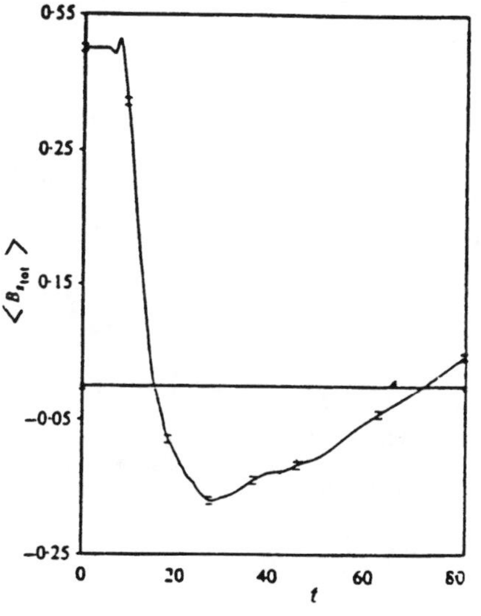

Figure 9.8: The average total toroidal magnetic field $\langle B_{z_{tot}} \rangle$ at the wall, plotted vs. time (from Dahlburg et al., 1987).

A reversed field appears around $t = 20$ when the magnetofluid is sufficiently relaxed. If a force-free state is reached during the relaxed phase, then the current should be aligned with the field, making the alignment cosine $\cos\theta = \boldsymbol{j}\cdot\boldsymbol{B}_{tot}/jB_{tot}$ equal to 1 everywhere. In addition, if Taylor's hypothesis is correct, then the quantity $\lambda = \boldsymbol{j}\cdot\boldsymbol{B}_{tot}/B_{tot}^2$ should be a uniform constant across a cross section. Plots of $\cos\theta$ and λ at different instants of time are shown in Figs. 9.9 and 9.10. It is seen that $\cos\theta$ relaxes to a value of 1 everywhere, but λ does *not* relax to a constant value. This implies that the final relaxed field is force free, but only approximately a Taylor field.

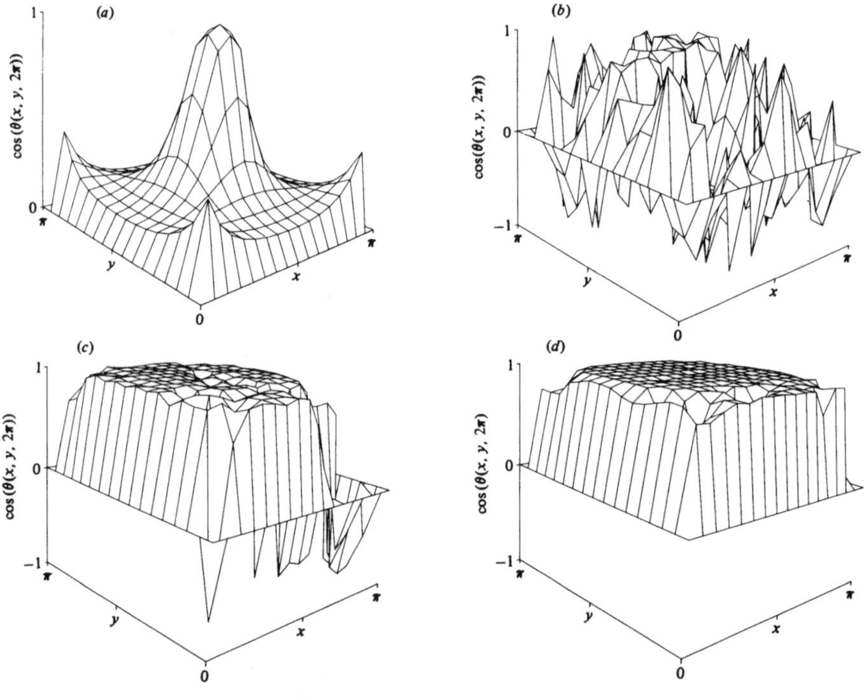

Figure 9.9: Three-dimensional perspective plot in the (x,y) plane at $z = 2\pi$ of the alignment cosine $\boldsymbol{j}\cdot\boldsymbol{B}_{tot}/(jB_{tot})$: a) $t = 0$; b) 12.0; c) 39.3; d) 74.4; relaxation toward a force-free state is apparent (from Dahlburg et al., 1987).

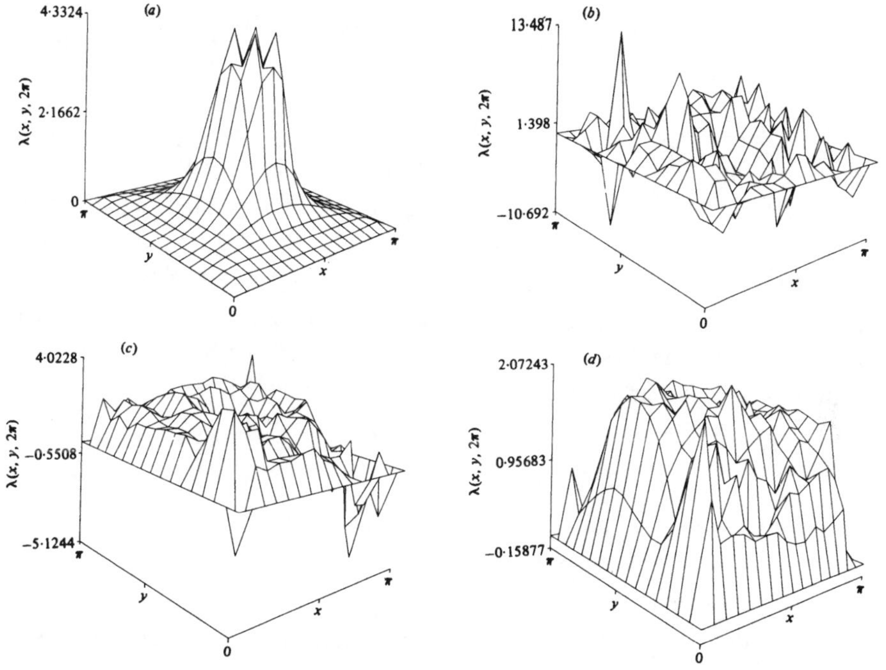

Figure 9.10: Three-dimensional perspective plot in the (x,y) plane at $z = 2\pi$ of $\lambda \equiv \boldsymbol{j} \cdot \boldsymbol{B}_{tot}/B_{tot}^2$. a) $t = 0$; b) 12.0; c) 39.3; d) 74.4 (from Dahlburg et al., 1987).

Belcher, J. and L. Davis, 1971: *J. Geophys. Res.*, **76**, 3534.

Bodin, H. A. B., and A. A. Newton, 1980: *Nuclear Fusion*, **20**, 1255.

Chandrasekhar-Kendall, 1957: *Ap. J.*, **126**, 457.

Dahlburg, J. P., D. Montgomery, G. D. Doolen, and L. Turner, 1986: *Phys. Rev. Lett.*, **57**, 428.

Dahlburg, J. P., D. Montgomery, G. D. Doolen, and L. Turner, 1987: *J. Plasma Physics*, **57**, 299.

Frisch, U., A. Ppuquet, J. Léovat, and M. Maure 1975: *J. Fluid Mech.*, **68**, 769.

Montgomery, D., L. Turner, and G. Vahala, 1978: *Phys. Fluids*, **21**, 757.

Taylor, J. B., 1974: *Phys. Rev. Lett.*, **33**, 1139.

Taylor, J. B., 1976: "Pulsed, High-Beta Plasmas", ed. D. E. Evans, Pergamon Press, N. Y.

Taylor, J. B., 1986: *Revs. Mod. Phys.*, **58**, 741.

Toyama, H., et al., 1985: *Proc. 12th European Conf. on Controlled Fusion and Plasma Physics*, Budapest: ed. L. Pacs and A. Montrai, Europ. Phys. Soc., Geneva, Vol. 1, 602.

Chapter 10

MHD (PLASMA) TURBULENCE TO THE YEAR 2000

A. Introduction

In this final chapter we shall list and discuss among ourselves possible directions for MHD research up to the end of the century. The subjects to be pursued fall naturally into two categories, which we will denote as "aesthetically driven" and "phenomenon driven." The former are pursued basically for their intrinsic intellectual interest, or appeal to a sense of symmetry, while the latter subjects arise from observations or experiments which seem to require explanations, or technological needs which involve turbulence.

B. Aesthetically Driven Problems

1. Elaboration of Kolmogorov ideas in two- and three-dimensional MHD; calculation of "universal constants"; description of intermittency; departures from Gaussianity.
2. Generalization of cascades to anisotropic and inhomogeneous cases (particularly with non-vanishing mean magnetic field and conduction current). How do selective decays change with Reynolds-like numbers and magnetic Prandtl numbers? How do selective decay and dynamic alignment processes divide up the initial phase space?
3. Identification of small-scale dissipative structures and their large-scale byproducts, especially in 3D MHD; 3D analogues of "X-points" and current filaments.
4. Demonstration of the rigorous properties of the MHD equations, e.g., singularities in ideal and non-ideal equations in the real and complex domain of independent variables. Does dissipation persist in the limit of an infinite Reynolds number? If so, dissipation of which quantities?
5. Basic questions associated with "first principles" calculations of the dynamics of symmetric turbulence, e.g., closures and renormalization group. The latter may be more useful in MHD than in Navier-Stokes turbulence.

6. Saying what is important about dynamical systems theory for turbulence or else confining it once and for all to the stability-instability transition region. Do low-order truncations imply anything at all about fully developed turbulence?

7. Vlasov turbulence phenomena: Are Zakharov equations and Langmuir solutions enough? Are they even right? That is, how adequately do their results predict the behavior of the full two-species behavior of the electrostatic Vlasov plasma?

8. Starting to think seriously about compressible MHD turbulence. For example, a calculation of the density fluctuations in the solar wind and local interstellar medium (Montgomery et al., 1987), in which weak compressibility is assumed, yields a $k^{-5/3}$ spectrum, thus suggesting a resolution to a previous discrepancy between theory and observation (Armstrong et al., 1981). These results remain to be generalized to cases with fewer symmetries.

C. Phenomenon-Driven Problems

1. MHD flow around obstacles: theory, experiment, and simulation. Modelling of magnetospheres, and the flows around them and in them.

2. Shock-wave-generated MHD turbulence (earth's bow shock, interplanetary shock).

3. Solar prominences ("loops," "ropes," "arcades," etc., all highly asymmetric configurations).

4. Acquisition of a body of knowledge of MHD turbulence in laboratory-confined plasmas based on measurements with internal probes. Development of techniques for manipulating MHD turbulence. What are "good turbulence" and "bad turbulence," vis-à-vis heating and confinement? For example, in confined plasmas, distinctions may be drawn between turbulence which transports material or heat out of the machine effectively and that which does not.

5. Turbulence of interfaces between electrically conducting and non-conducting media (e.g., the ionosphere and the upper atmosphere).

6. Variational principles which predict the balance between driving agencies (electric fields, pressure gradients) and dissipation in steady-state MHD configurations (selective decay and dynamic alignment, as so far developed, are inadequate to derive steady states).
7. Electrostatic particle diffusion across magnetic fields (requires closer collaboration between theory and experiment; see Montgomery [1975]). How are these Bohm-like diffusion behaviors modified for the case of variable magnetic fields?
8. Development of a more geometrically realistic formulation of dynamo theory for solar, geophysical, and astrophysical applications.

D. Conclusions

We are at an early stage in deciding what the central questions in MHD turbulence theory ought to be. The level of advancement of the subject is, just possibly, about where fluid turbulence was fifty or sixty years ago. The going is slow for two reasons, which are pretty good ones, and a third which is not good at all. The good reasons are: (1) In the two applications of MHD which are thought to be of the most importance, controlled fusion confinement and space-physics measurements, experimental diagnostics are, and are likely to remain, far less complete than corresponding diagnostics are for, say, hydrodynamics and aerodynamics; (2) Mostly because of the long-collision mean-free paths, the MHD description has less persuasive mathematical underpinnings for parameter ranges of interest, than does the Navier-Stokes description for fluids. The not-so-good reason is: (3) The plasma physics community, which guards the gate to most of the possibilities for serious MHD turbulence investigations, is enamoured of the equilibria-plus-normal-modes paradigm for plasma dynamics, and resolutely resists delving into the kinds of investigations that forced hydrodynamicists to relegate the paradigm to a place of minor ornamentation in their subject. How long this state of affairs can persist is anyone's guess, but it has gone on longer than I imagined it could, in spite of nearly total failure at the accomplishments that plasma physics was supposed to have delivered by now.

Nobody important or influential appears to be upset by this, which seems in the last analysis miraculous. Nevertheless, MHD turbulence remains an open and intriguing arena, in which most of the great experiments, calculations, and computations remain to be performed. Maybe the Twenty-First Century will have more luck with it than the Twentieth.

Armstrong, J. W., J. M. Cordes, and B. J. Rickett, 1981: *Nature*, **291**, 561.

Montgomery, D., 1975: In *Plasma Physics*, ed. C. DeWitt, 1972 (New York: Gordon and Breach).

Montgomery, D., M. R. Brown, and W. H. Matthaeus, 1987: *J. Geophys. Res.*, **92**, 282.

LECTURE NOTES

The following chapters represent the collection of my lectures which were presented at the Geophysical Turbulence Workshop held at the National Center for Atmospheric Research, Boulder, Colorado, in the summer of 1987.

Douglas K. Lilly
University of Oklahoma
2 June 1988

Chapter 1

HELICITY

The purpose of this chapter is to give an introduction to the quantity known as helicity and to show that it is a partially conserved flow variable, much like energy.

Helicity is formally defined as the volume integral of the dot product of vorticity and velocity:

$$\text{Helicity} = \iiint \boldsymbol{V}\cdot\boldsymbol{\omega} d\boldsymbol{x}. \tag{1.1}$$

We will usually find it convenient to work with the helicity density (also loosely referred to as helicity) defined as the above integrand:

$$\text{Helicity Density} = H \equiv \boldsymbol{V}\cdot\boldsymbol{\omega}. \tag{1.2}$$

Another useful quantity is the relative helicity:

$$\text{Relative Helicity} = \frac{H}{\omega V} = \cos(\boldsymbol{\omega},\boldsymbol{V}), \tag{1.3}$$

where ω and V are the magnitudes of the helicity density and velocity vectors, respectively.

Surprisingly, helicity was not mentioned in the literature until about 25 years ago. Early work was done by Betchov (1961).

An equation for the conservation of helicity (density) can be derived from the compressible Navier-Stokes equations,

$$\frac{\partial \boldsymbol{V}}{\partial t} + \boldsymbol{V}\cdot\nabla\boldsymbol{V} + \frac{1}{\rho}\nabla p + \boldsymbol{k}g = \nu\nabla^2\boldsymbol{V}, \tag{1.4}$$

and the vorticity transport equation,

$$\frac{\partial \boldsymbol{\omega}}{\partial t} + \boldsymbol{V}\cdot\nabla\boldsymbol{\omega} + (\nabla\cdot\boldsymbol{V})\boldsymbol{\omega} - \boldsymbol{\omega}\cdot\nabla\boldsymbol{V} + \nabla\times\left(\frac{\nabla p}{\rho}\right) = \nu\nabla^2\boldsymbol{\omega}, \tag{1.5}$$

as follows: the dot product of vorticity with (1.4) is added to the dot product of velocity with (1.5) to yield

$$\frac{\partial H}{\partial t} + \nabla\cdot(\boldsymbol{V}H) + \boldsymbol{\omega}\cdot\left[\frac{\nabla p}{\rho} - \nabla\left(\frac{V^2}{2}\right)\right] + \boldsymbol{V}\cdot\nabla\times\left(\frac{\nabla p}{\rho}\right) + g\zeta = \nu\left[\nabla^2 H - 2\frac{\partial V_i}{\partial x_j}\frac{\partial \omega_i}{\partial x_j}\right]. \tag{1.6}$$

In the above, ζ is the vertical component of vorticity. Notice that the last term on the right-hand side (RHS) of (1.6) implies that helicity is not necessarily dissipated by viscosity, since it can be of either sign.

The helicity conservation equation can also be written in the form

$$\frac{\partial H}{\partial t} + \nabla\cdot(\boldsymbol{V}H - \boldsymbol{\omega}\frac{V^2}{2}) + \boldsymbol{\omega}\cdot\frac{\nabla p}{\rho} + \nabla\cdot\left[(\frac{\nabla p}{\rho} - g\boldsymbol{k})\times\boldsymbol{V}\right] = \nu\left[\nabla^2 H - 2\frac{\partial V_i}{\partial x_j}\frac{\partial \omega_i}{\partial x_j}\right]. \tag{1.7}$$

If the flow is inviscid and barotropic ($\rho = \rho(p)$), helicity is conserved in infinite or periodic domains, but can escape to or from the boundaries. For example, consider an inviscid and barotropic flow bounded at $z = 0$ and z_{max}. The contribution to change of total helicity due to flux across the lower boundary associated with the vertical vorticity component is given by

$$\int_x\int_y\int_z \frac{\partial H}{\partial t} dxdydz = \int_x\int_y \zeta\left[-\int\frac{dp}{\rho} + \frac{V^2}{2}\right]\Big|_{z=0}^{z_{max}} dxdy.$$

In the case of viscous no-slip boundaries, the RHS vanishes, as well as the normal components of ∇H. Thus no helicity can be introduced through the boundaries. The "dissipation" terms, those last in the brackets in the RHS of (1.7), may be large, however, in a viscous boundary layer. Thus helicity can be strongly emitted or absorbed in such a layer.

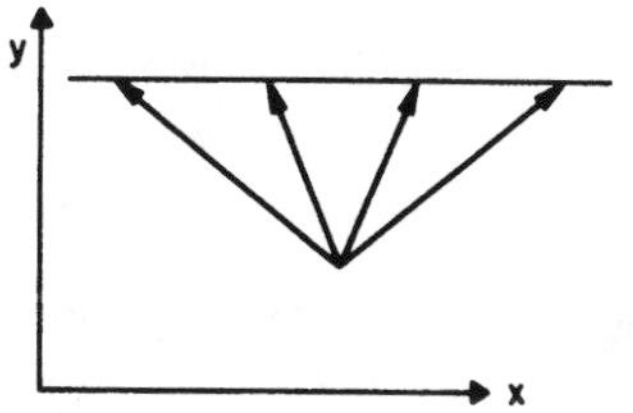

Figure 1.1: Hodograph of a unidirectional shear flow.

Helicity, like kinetic energy, is not in general a Galilean invariant. This can be seen by considering a unidirectional shear, with $\partial u/\partial z$ and v both constant, as in the hodograph in Fig. 1.1. In the coordinate system shown, the vorticity is $\boldsymbol{j}\partial u/\partial z$, so that the helicity is $v\partial u/\partial z$, a constant positive quantity. When the helicity is computed in a coordinate system which is translating in the y-direction with a constant velocity, however, the magnitude changes. In fact, if the coordinate advection speed is equal to twice v, the new helicity has the same magnitude as in the previous case, but the opposite sign!

Buoyancy and boundary effects can be better illustrated using the Boussinesq equations, neglecting viscous terms:

$$\frac{\partial \boldsymbol{V}}{\partial t} + \boldsymbol{V} \cdot \nabla \boldsymbol{V} + \nabla \Pi - \boldsymbol{k} b = \nu \nabla^2 \boldsymbol{V}, \tag{1.8}$$

$$\nabla \cdot \boldsymbol{V} = 0, \tag{1.9}$$

$$\frac{\partial \boldsymbol{\omega}}{\partial t} + \boldsymbol{V} \cdot \nabla \boldsymbol{\omega} - \boldsymbol{\omega} \cdot \nabla \boldsymbol{V} + \boldsymbol{k} \times \nabla b = \nu \nabla^2 \boldsymbol{\omega}, \tag{1.10}$$

where $\Pi = p'/\rho_0, b = -g\rho'/\rho_0$, with primes denoting deviations from mean hydrostatic values, and ρ_0 a constant reference density. Following the previous procedure, the helicity equation is now

$$\frac{\partial H}{\partial t} + \boldsymbol{V} \cdot \nabla H + \boldsymbol{\omega} \cdot \nabla (\Pi - \frac{V^2}{2}) = \zeta b + \boldsymbol{k} \cdot (\boldsymbol{V} \times \nabla b). \tag{1.11}$$

Rewriting the RHS of (1.11) as

$$\boldsymbol{k} \cdot (\boldsymbol{V} \times \nabla b) = \nabla \cdot (\boldsymbol{k} \times \boldsymbol{V} b) + \zeta b, \tag{1.12}$$

we see that both buoyancy terms contribute equally to the integral of helicity change, but can be interpreted differently locally.

The significance of helicity can be explored by considering Beltrami flow (purely helical), defined as

$$\boldsymbol{\omega} = \kappa \boldsymbol{V}, \tag{1.13}$$

where κ is constant along streamlines. The equations of motion (1.8) may be rewritten, neglecting buoyancy, as

$$\frac{\partial \boldsymbol{V}}{\partial t} + \nabla (\frac{V^2}{2} + \Pi) + \boldsymbol{\omega} \times \boldsymbol{V} = \boldsymbol{F}, \tag{1.14}$$

where $\boldsymbol{F}$ represents the friction terms. The last term on the left-hand side (LHS) is zero in helical flow, and the resulting vorticity equation is then essentially trivial,

$$\frac{\partial \boldsymbol{\omega}}{\partial t} = \nabla \times \boldsymbol{F}, \tag{1.15}$$

because the advection, tilting, and stretching terms ($\boldsymbol{V}\cdot\nabla\boldsymbol{\omega}-\boldsymbol{\omega}\cdot\nabla\boldsymbol{V}$) cancel each other. This restriction precludes development of an inertial cascade. Another aspect of this simple flow is that if κ is a constant, then

$$\nabla\times\boldsymbol{\omega} = -\nabla^2\boldsymbol{V} = \kappa^2\boldsymbol{V}, \tag{1.16}$$

so that the Beltrami flow is periodic and κ can be interpreted as a wave number.

Two Beltrami flows added together remain Beltrami only if κ is the same for both fields. Thus a multiscale flow cannot be purely helical. However,

$$(\boldsymbol{V}\cdot\boldsymbol{\omega})^2 + (\boldsymbol{V}\times\boldsymbol{\omega})\cdot(\boldsymbol{V}\times\boldsymbol{\omega}) = V^2\omega^2, \tag{1.17}$$

so that the larger the helicity, the smaller the cross product ("lift" or "Lamb force") term.

We need at this point to recognize that a Beltrami flow is not necessarily stable. For example, we can consider a rotating hodograph with a constant speed but periodically changing direction (Fig. 1.2a):

$$U = -M\cos(\kappa z), V = M\sin(\kappa z). \tag{1.18}$$

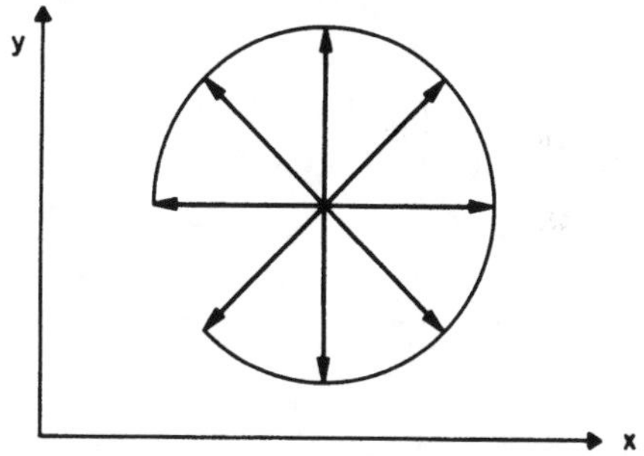

Figure 1.2a: Rotating hodograph for a constant speed flow with periodically varying direction.

This flow has a constant unit relative helicity, since the vorticity and velocity are aligned. If we look at a height profile of one of the velocity components (Fig. 1.2b), we see that there are points of inflection every half wavelength. This periodic structure apparently ensures that every level has an inviscid instability in some direction.

This instability would seem to cast doubt on the usefulness of helicity; if this simple flow is unstable everywhere, its large value of helicity is irrelevant. Why

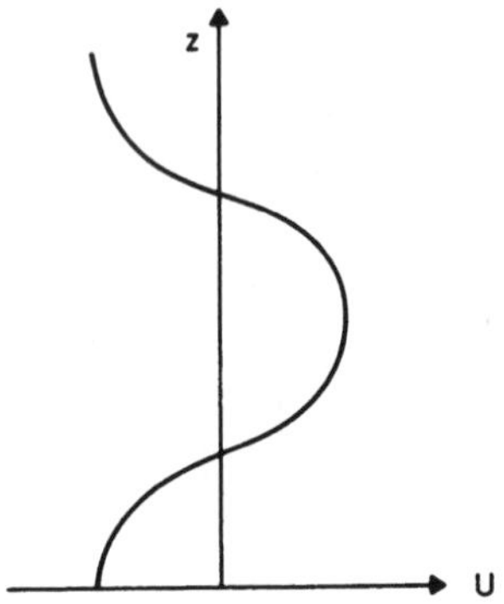

Figure 1.2b: Velocity profile.

then is helicity still a subject of interest? Part of the answer can be seen in the results of a numerical experiment done by André and Lesieur (1977). Using an eddy-damped quasi-normal Markovian closure, the experiment began with one of two possible initial states: isotropic turbulence, or turbulence with the maximum possible helicity of one sign. After four "turn-over times," the energy spectra from the isotropic case showed the characteristic inertial range energy cascade (Fig. 1.3). By contrast, the spectra for the maximum helicity case (Fig. 1.4) showed an increased concentration of energy at small wavenumbers. These results seem to indicate that helicity can block the transfer of energy from large to small scales and effectively reduce the dissipation in a turbulent flow.

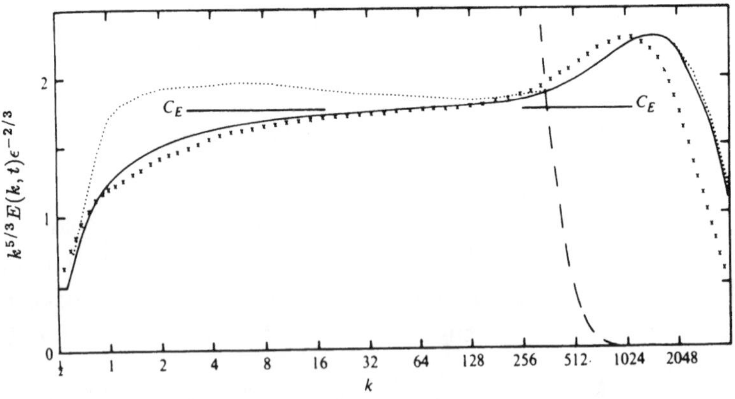

Figure 1.3: Energy compensated spectrum $\epsilon^{-2/3}k^{5/3}E(k,t)$ as a function of k: —, t = 5; …, t = 6: —, t = 8; xxx, t = 15. From results of a turbulence closure model for decay of non-helical turbulence (André and Lesieur, 1977).

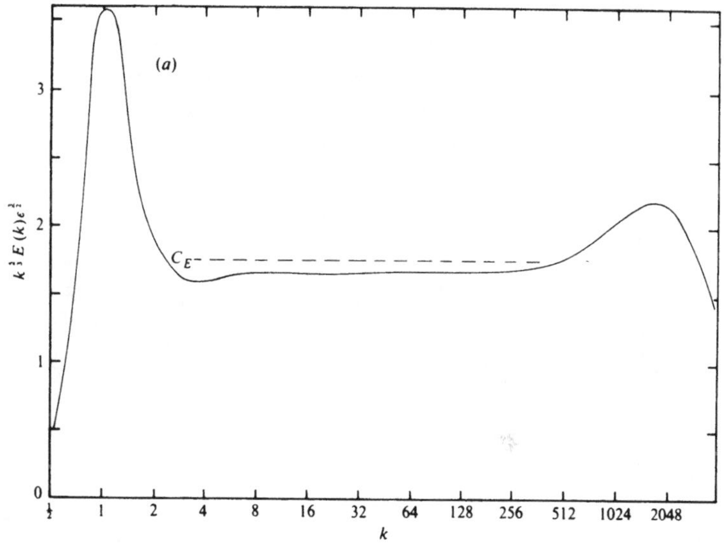

Figure 1.4: Same as Fig. 1.3 for a late time, but with the initial state highly helical.

Levich and Tsinober (1983) have hypothesized that regions of accidentally helical flow are protected from inertial cascading and become quasi-steady "islands" of low dissipation. For sufficiently high Reynolds number, these islands are thought to become closely packed helical "continents," with most of the flow contained within them. Most of the cascading and dissipation occurs in the boundaries between the continents. This hypothesis might provide a phenomenological explanation of the intermittent nature of dissipation at high wave-numbers. A problem with the hypothesis is the above mentioned instability of helical regions, especially when in contact with other regions of opposite helicity.

The Pelz et al. (1985) direct numerical simulation of turbulent plane channel and Taylor-Green vortex flows produced results which are consistent with the low-dissipation helicity island hypothesis: regions of lower than average dissipation tended to have high relative helicity and vice versa (Figs. 1.5 and 1.6). However, the simulation of plane channel flow results of Rogers and Moin (1986), showed a much weaker correlation between dissipation and helicity (Figs. 1.7 and 1.8). The reason for this great discrepancy is not clear from the information presented. The Rogers and Moin results appear, however, to be the result of a much more complete investigation. Their explanation is that the Pelz et al. calculations were presented for a time before a statistical steady state was attained, and also that those calculations were done with inadequate resolution.

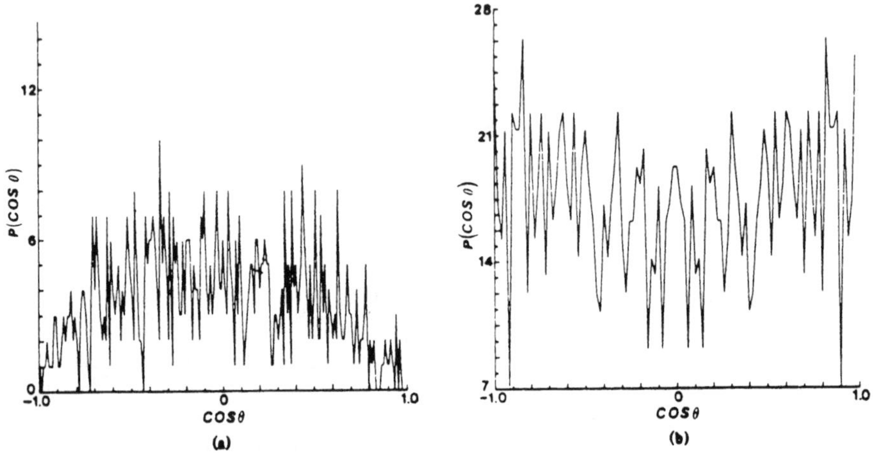

Figure 1.5: The probability density for the distribution of the angle between velocity u and vorticity ω conditionally sampled in the region where dissipation is greater than 30% of its maximum value: (a) in the outer part of the channel ($15 < z_+ < 100$); (b) in the Taylor-Green vortex at $t = 8.6$. From Pelz et al. (1985).

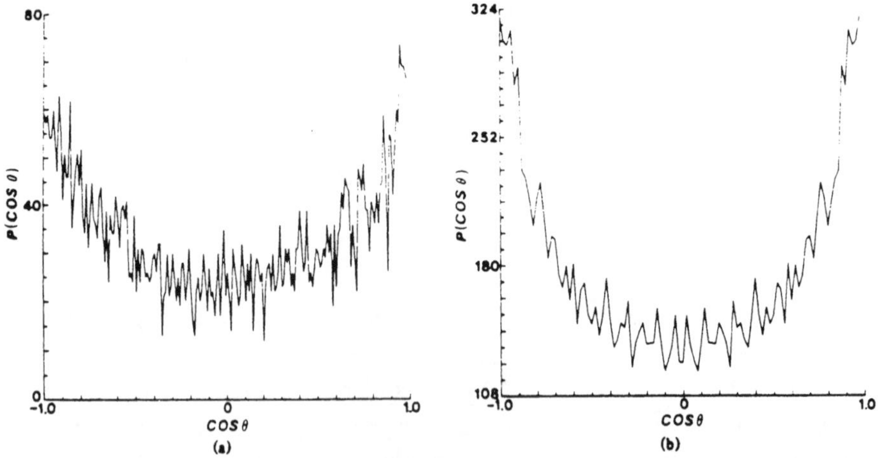

Figure 1.6: The probability density for the distribution of the angle between velocity u and vorticity ω conditionally sampled in the region where dissipation is less than 5% of its maximum value: (a) in the outer part of the channel ($15 < z_+ < 100$); (b) in the Taylor-Green vortex at $t = 8.6$. From Pelz et al. (1985).

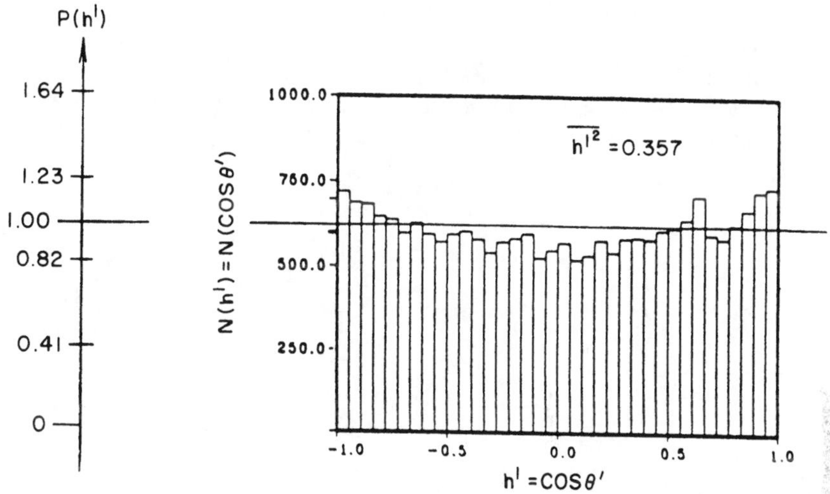

Figure 1.7: The relative fluctuating helicity pdf for isotropic flow conditioned on low dissipation. Plot contains data from 1% of the flow field. $N(h')$ represents the number of data points in each of 36 bins. From Rogers and Moin (1986).

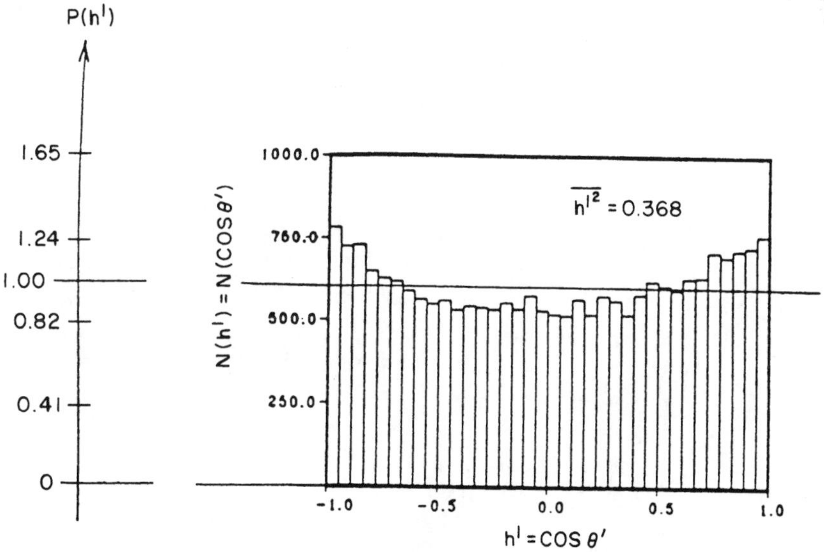

Figure 1.8: The relative fluctuating helicity pdf for isotropic flow conditioned on high dissipation. Plot contains data from 1% of the flow field. $N(h')$ represents the number of data points in each of 36 bins. From Rogers and Moin (1986).

André, J., and Lesieur, M., 1977: *J. Fluid Mech.*, **81(1)**, 187.

Betchov, R., 1961: *Phys. Fluids*, **4**, 925.

Levich, E., and Tsinober, A., 1983: *Physics Letters*, **93A**, 293.

Pelz, P., Yakhot, V., Orszag, S., Shtilman, L., and Levich, E., 1985: *Phys. Rev. Letters*, **54**, 2505.

Rogers, M., and Moin, P., 1986: *Physics of Fluids*, **30**, 2662.

Chapter 2

HELICAL CONVECTION

2.1 Regimes of Atmospheric Convection

Before considering the effects of helicity, I will first review the circumstances of buoyant atmospheric convection The simplest type is similar to the Benard cell, but with only the lower boundary defined by external conditions. Such convection is produced by heating of the ground during the diurnal cycle, or by the flow of air over a warmer surface. The convective layer is typically bounded at the top by a temperature inversion. Convective updrafts frequently overshoot this inversion and draw warmer air down into the convective layer, thus producing a downward heat flux. The classical problem of the convective boundary layer involves the prediction of the overshoot and negative heat flux along with the evolving height of the capping inversion. We shall consider this problem in a later chapter.

The convective boundary layer is complicated when latent heating, due to the condensation and evaporation of water, must be considered. A stratiform cloud under a clear sky will be cooled radiatively at its top and will release latent heat in its interior, with both processes strongly affecting the heat and buoyancy fluxes. If scattered cumuliform clouds are present, the turbulence exhibits strong horizontal intermittency. The compensating subsidence in the clear air around the clouds produces stable stratification, so that turbulence is confined mainly to the clouds. There are thus three regimes of convection: dry heated boundary-layer convection; cloud-filled or cloud-topped boundary layers; and dry boundary layer convection with intermittent cumuliform clouds above. The primary interest is often in the effects of these regimes on the larger scale, in which case they are treated statistically.

2.2 Convective Storm Evolution

The dynamics of individual convective elements become important when latent heating produces intense updrafts, downdrafts, and thunderstorms. These storms also have important interactions with the larger scales. Let us first look at the classical life cycle of a convective storm.

The early stage of the cumulus-cloud life cycle is characterized by a rising motion throughout the cloud, with a rapidly ascending cloud top and a rather flat and vertically stationary cloud base, at the level at which low-level air becomes saturated by adiabatic cooling. Rain production begins to become efficient in the middle and upper part of the cloud due to coalescence of small droplets into larger ones. In the mature stage, a large quantity of rain has formed. The weight of the raindrops, together with evaporative cooling of subsaturated air as the rain falls through it, acts to force downward motion through a significant part of the storm and below it. The dissipating stage results when the cooled air has spread horizontally beneath the storm, cutting off the supply of warm, moist air. The life cycle generally takes an hour or less. Such storms frequently occur in Florida and the Rockies.

Intense thunderstorms can be produced when the ambient wind is sheared in the vertical ($|\partial u/\partial z| > 0$). This is somewhat paradoxical, in that laboratory experiments indicate that plane shear suppresses convection by tilting it downshear. The downshear-tilted updraft has $-\overline{u'w'}(\partial u/\partial z) < 0$, so that the convection gives up energy to the mean flow. Linear theory and two-dimensional numerical simulation results are consistent with those from the laboratory.

Three-dimensional simulations have helped to explain how vertical shear can make thunderstorms more intense and long-lived (Davies-Jones, 1984; Lilly, 1986a, b). In the first stage, an updraft bends the initially horizontal vortex lines partially into the vertical (Fig. 2.1), producing regions with positive and negative vertical vorticity (denoted by ζ^+ and ζ^-). This is explained by the tilting term in the vertical vorticity equation linearized about a mean sheared background flow $U(z)$, i.e.,

$$\frac{\partial \zeta}{\partial t} = -U\frac{\partial \zeta}{\partial x} + \frac{\partial w}{\partial y}\frac{\partial U}{\partial z}. \tag{2.1}$$

(The final term is the tilting term.) Here w is the perturbation vertical motion.

Through centrifugal action, low pressure tends to develop in the center of each region of vertical vorticity, in approximate satisfaction of cyclostrophic balance. This is represented as

$$\frac{1}{\rho}\frac{\partial p}{\partial r} = \frac{v_\phi^2}{r}, \tag{2.2}$$

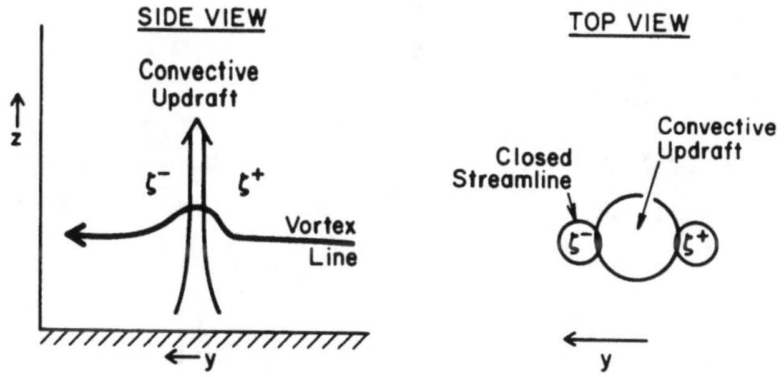

Figure 2.1: (left) Tilting of a horizontal vortex line by a convective updraft. (right) The two resulting centers of vertical vorticity.

where r is the radial distance from the vortex center and v_ϕ is the azimuthal velocity about the center. The low pressure exists only above the surface, because tilting cannot produce vertical vorticity at the ground (unless there is a horizontal velocity discontinuity), so upward motion is forced into the bottom of each vortex. The original single updraft thus becomes two separate updrafts, one in each vertical vortex. The two counter-rotating updrafts tend to move apart, each propagating relative to the ambient flow toward the ambient horizontal vortex lines, which enter the flow at low levels; each updraft thus becomes helical. The ζb term in the helicity equation and the bw term in the kinetic energy equation predict this tendency for maximum buoyancy, vertical velocity, and vertical vorticity to occur in coincidence.

2.3 Motion of Rotating Updrafts

The transfer of horizontal environmental vorticity to vertical updraft vorticity can be more closely examined by looking at the vertical equation of motion in the updraft's frame of motion. With the updraft observed to have a y-component of propagation, say C_y, with respect to the mean flow, we can write

$$\frac{\partial \zeta}{\partial t} = -C_y \frac{\partial \zeta}{\partial y}. \tag{2.3}$$

The vorticity equation then becomes

$$-C_y \frac{\partial \zeta}{\partial y} \approx \frac{\partial w}{\partial y} \frac{\partial U}{\partial z}. \tag{2.4}$$

Integrating over y,

$$\zeta \approx \frac{\partial U/\partial z}{C_y} w. \tag{2.5}$$

This gives the updraft vertical vorticity in terms of the updraft vertical velocity, the mean wind shear, and the updraft's horizontal propagation. The problem is closed when C_y can be found.

The helicity of the mean flow in the frame of reference of the updraft is given by the product of the propagation speed and the vorticity component in that direction, i.e., $H = C_y \partial U/\partial z$. An expression for C_y may be found by equating the helicity wavenumbers (vorticity/velocity) of the mean flow and the disturbance. This equation is based upon the assumption that helical flows are favored to persist, and the knowledge that if two flows are individually Beltrami flows, the sum of the flow is a Beltrami flow only if the wavenumbers of both are the same. The wavenumber of the mean flow, say, κ, is

$$\kappa = \frac{\partial U/\partial z}{C_y}, \tag{2.6}$$

A characteristic wavenumber of the disturbance, say, k, is

$$k = \left[\left(\frac{2\pi}{L_x}\right)^2 + \left(\frac{2\pi}{L_y}\right)^2 + \left(\frac{\pi}{h}\right)^2\right]^{\frac{1}{2}}, \tag{2.7}$$

where L_x and L_y are the horizontal wavelengths of the actively rotating part of the disturbance and h is its height.

Generally $L_x \approx L_y \approx 2h$, so that after equating (2.6) and (2.7),

$$C_y \approx 0.2h \frac{\partial U}{\partial z}, \tag{2.8}$$

yielding a typical value of C_y in the range 5 to 10 m s^{-1}.

A schematic representation of the motion of both the ζ^+ and ζ^- updrafts in a mean flow with unidirectional shear is given by the wind hodograph below, in which L, M, and H denote the low, middle, and high mean-wind vectors relative to the ground:

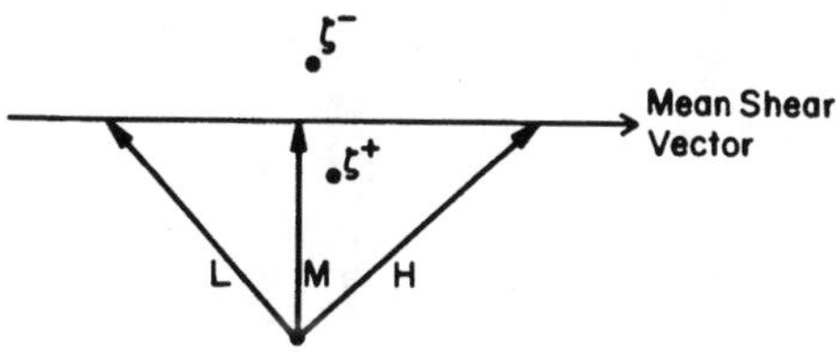

Figure 2.2: Schematic mean flow hodograph in convective storm updrafts with environment. Points marked ζ^+ and ζ^- indicate motions of positive and negative vorticity.

The ζ^+ updraft travels slower than the mean wind, while the ζ^- updraft travels faster than the mean wind.

Numerical simulations and radar observations of severe storms show that w and ζ are well correlated. The horizontally averaged relative helicity from radar data is noisy but tends to be strongly positive. This may account for the long life and high energy density of these storms.

Davies-Jones, R. P., 1984: *J. Atmos. Sci.*, **41**, 2991.
Lilly, D. K., 1986a: *J. Atmos. Sci.*, **43**, 113.
Lilly, D. K., 1986b: *J. Atmos. Sci.*, **43**, 126.

Chapter 3

FLOW FIELDS IN HELICAL CONVECTION

3.1 An Idealized Rotating Thunderstorm

In the previous chapter, I described how helicity develops in a sheared environment. Here, I will show how a rotating thunderstorm within its sheared environment may be idealized as a Beltrami (purely helical) flow. This model cannot be an exact expression for the motion field, because it neglects buoyancy effects. The trajectories from it show a striking resemblance, however, to those observed from both dual-Doppler observations and dynamic simulations. Also the mean flow in this model, which varies in direction only, differs from the unidirectionally sheared environment of the previous chapter. Rotating updrafts can develop in both environments, and a typical pre-storm environment is often a compromise between the two.

Assume a velocity field with horizontal components given by

$$\boldsymbol{V}_H = \nabla_H \phi + \boldsymbol{k} \times \nabla_H \psi, \tag{3.1}$$

so that the horizontal divergence and vertical vorticity are

$$\nabla_H \cdot \boldsymbol{V}_H = \nabla_H^2 \phi \tag{3.2}$$

and

$$\boldsymbol{k} \cdot \nabla_H \times \boldsymbol{V}_H = \nabla_H^2 \psi. \tag{3.3}$$

We choose a vertical velocity of periodic form as follows

$$w = W \cos kx \cos ly \sin mz, \tag{3.4}$$

and solve for ϕ and ψ, using incompressible continuity and the Beltrami condition (1.13), to obtain

$$\phi = \frac{m}{k^2 + l^2} W \cos kx \cos ly \cos mz, \tag{3.5}$$

$$\psi = \frac{-\kappa}{k^2 + l^2} W \cos kx \cos ly \sin mz, \tag{3.6}$$

$$\kappa^2 = k^2 + l^2 + m^2. \tag{3.7}$$

For $|kx|$ and $|ly| < \pi/2$ and $mz < \pi$, this is a simple helical updraft with inflow at $z = 0$ and outflow at $z = \pi/m$ (Fig. 3.1).

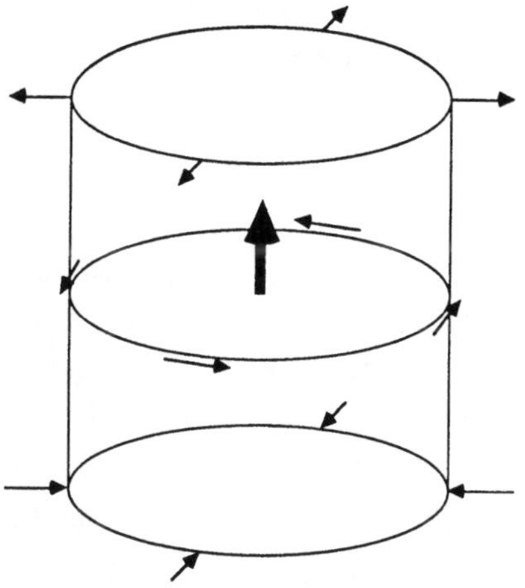

Figure 3.1: Beltrami rotating updraft described by equations (3.5) to (3.7).

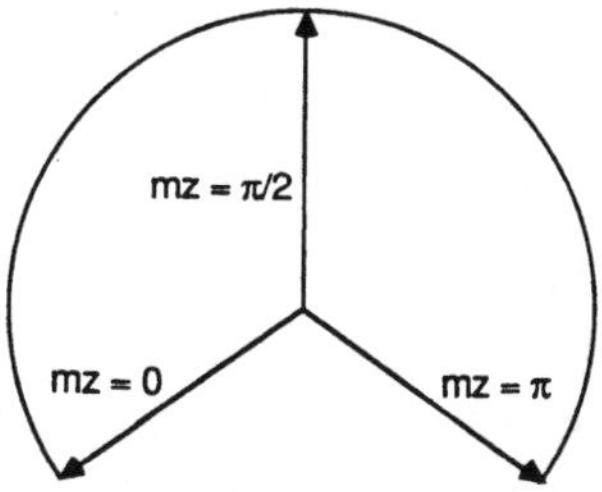

Figure 3.2: Beltrami mean flow hodograph described by equations (3.8) and (3.9).

Now we add this Beltrami flow to a helical mean flow which has a rotating hodograph (Fig. 3.2):

$$\bar{u} = M \sin \kappa \left(z - \frac{\pi}{2m} \right) , \tag{3.8}$$

$$\bar{v} = M \cos \kappa \left(z - \frac{\pi}{2m} \right) . \tag{3.9}$$

The net flow is now a simple kinematic model for a rotating thunderstorm (i.e., a mesocyclone). One can trace trajectories that enter at low levels, twist around each other and exit at high levels (see Fig. 3 in Lilly, 1986). The direction of rotation of the horizontal projections of these trajectories is in the same sense as the rotation of the hodograph, i.e., clockwise. This is opposite to that which one might naively expect from the sign of the vertical vorticity in the updraft.

3.2 Tornadic Flow Fields

The study of tornadoes may be the "last frontier" of meteorology, since the flow fields within tornadoes have still not been measured accurately, either directly or indirectly. There is no doubt that most major tornadoes arise from rotating thunderstorms, however, and the processes involved are becoming clearer. Davies-Jones (1986) and Snow (1987) have recently reviewed the subject.

Nested-mesh numerical simulations by Rotunno and Klemp (1985) have windowed down to physical scales nearly capable of resolving tornadoes, using their thunderstorm simulations to provide boundary conditions. It appears that the discontinuity in flow fields along the outflow gust front is instrumental in producing strong vortical flow. Other simulations have been done in axisymmetric environments, but must be interpreted with caution, as thunderstorms are not axisymmetric.

Laboratory simulations of helically influenced vortex flows have led to a good understanding of the conditions required for the formation of strong vortices. Early work was begun by Ward (1972), and continues at Purdue University (Snow et al., 1980), and at the University of Oklahoma (Wilkins and Diamond, 1987). One of the more striking recent results (Rothfusz, 1986) is the production of an intense vortex from a sheared mean flow across the test chamber. Evidently, horizontal vorticity is tilted into the vertical, resulting in a rotating updraft at middle and higher levels. This rotating updraft also includes a strong vortex linked to the surface, quite analogous to the tornado and its environment.

3.3 Other Candidates for Helical Flow in the Atmosphere

Hurricanes are obviously rotating storm systems, but are they helical? A simple model of a hurricane suggests not. Assume radial symmetry of the flow field, so that the radial, tangential, and vertical components, respectively, of the vorticity vector are given by:

$$\omega_r = -\frac{\partial v_\phi}{\partial z} \,, \tag{3.10}$$

$$\omega_\phi = \frac{\partial v_r}{\partial z} - \frac{\partial w}{\partial r}, \tag{3.11}$$

$$\zeta = \frac{1}{r}\frac{\partial}{\partial r}(r v_\phi). \tag{3.12}$$

If the flow is helical, then,

$$\frac{\zeta + f}{w} = \frac{\omega_r}{v_r} = \frac{\omega_\phi}{v_\phi}, \tag{3.13}$$

where f is the vertical component of the planetary vorticity. If angular momentum is conserved, then

$$v_r\left(\frac{\partial v_\phi}{\partial r} + \frac{v_\phi}{r} + f\right) + w\frac{\partial v_\phi}{\partial z} = 0 \tag{3.14}$$

and the first equality in (3.13) is true. Next we examine whether

$$\frac{\omega_\phi}{v_\phi} = \frac{\omega_r}{v_r}.$$

For flows that are hydrostatic and roughly cyclostrophic,

$$\begin{aligned} \frac{\omega_\phi}{v_\phi} &\simeq \frac{\partial v_r/\partial z}{v_\phi}, \\ \frac{\omega_r}{v_r} &\simeq \frac{\partial v_\phi/\partial z}{v_r}, \end{aligned} \tag{3.15}$$

but since v_ϕ and its vertical gradients are about five times as large as v_r and its vertical gradients, the second equality in (3.13) is not well approximated. Since the horizontal components of velocity and vorticity are generally larger than the vertical components, it is evident that the vorticity and velocity vectors are not closely aligned.

Another possible helical flow is the longitudinal convective roll in the planetary boundary layer. The large eddies are weak vortices roughly aligned with the mean flow, with diameters comparable to the depth of the boundary layer. To be helical they would need to have jets down the center of each vortex. Observational data is inadequate for a definite conclusion on whether such jets exist.

Davies-Jones, R. P., 1986: Tornado dynamics, Chap. 10 (pp. 297-361) in *Thunderstorms: A Social, Scientific and Technological Documentary, Vol. 2: Thunderstorm Morphology and Dynamics* (E. Kessler, ed.), Univ. of Oklahoma Press, Norman, Okla., 432 p.

Lilly, D. K., 1986: *J. Atmos. Sci.,* **42,** 126.

Rothfusz, L. P., 1986: *J. Atmos. Sci.* **43,** 2677.

Rotunno, R. and J. B. Klemp, 1985: *J. Atmos. Sci.,* **42,** 271.

Snow, J. T., 1987: *Rev. Geophys.* **25,** 371.

Snow, J. T., C. R. Church, and B. J. Barnhart, 1980: *J. Atmos. Sci.,* **37,** 1013.

Ward, N. B., 1972: *J. Atmos. Sci.,* **29,** 1194.

Wilkins, E. M., and C. J. Diamond, 1987: *J. Atmos. Sci.,* **44,** 140.

Chapter 4

TURBULENCE IN STABLY STRATIFIED FLUIDS

4.1 The Ozmidov Length Scale and Application

Turbulence in a stably stratified fluid has to work against a negative buoyancy flux, so that potential energy is generated at the expense of kinetic energy. Despite this adverse environment, stratified turbulence is widely observed in the nocturnal boundary layer, the lower stratosphere, and the oceanic thermocline. While studies of turbulence in these environments have proceeded largely independently, there are a number of unifying concepts. A good review of this topic is that by Hopfinger (1987). Two other useful references are Hunt (1985) and Lin and Pao (1979).

Stratification introduces new length scales into the turbulence problem even in an unbounded domain. The kinetic energy required to displace a stably stratified fluid a vertical distance L is of order L^2N^2, while that associated with an inertial range eddy of length scale L is of order $(\epsilon L)^{2/3}$, where ϵ is the energy dissipation rate. The implication is that advective stirring overwhelms buoyant energetics, and an inertial range can therefore exist only for length scales smaller than that for which the above two expressions coincide, i.e., for $L < L_0$, where $L_0 = (\epsilon/N^3)^{1/2}$. L_0 is called the Ozmidov scale in oceanography (Ozmidov, 1965). In meteorology and engineering fluid dynamics the concept is commonly attributed to Lumley (1964), but was introduced earlier by Dougherty (1961). For motions at scales $L \gg L_0$, buoyancy effects dominate and gravity waves are expected, with strong anisotropy possible. When the Ozmidov length scale approaches the Kolmogorov microscale, $L_\epsilon = \left(\nu^3/\epsilon\right)^{1/4}$, no inertial range can exist and turbulence collapses.

The above analysis has been applied successfully to the experimental study of buoyant collapse of turbulence in a stratified medium. The decay of turbulence injected into a vertically stratified tank shows a collapse phenomenon when the length scale of the turbulence approaches L_0. This is illustrated in Fig. 2 of the review of Hopfinger (1987), reproduced here as Fig. 4.1.

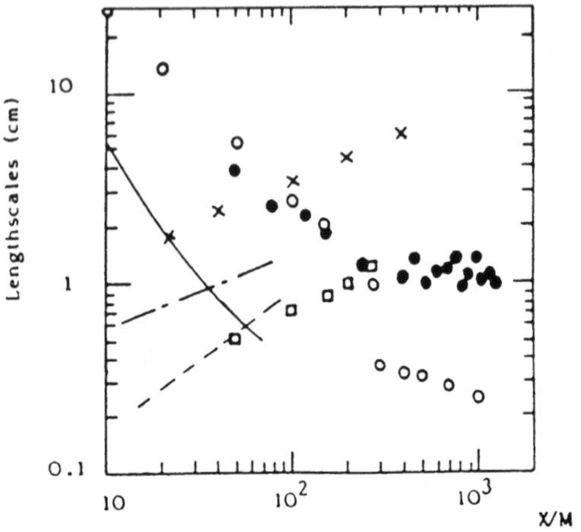

Figure 4.1: Length scale evolutions for the data of Dickey and Mellor (1980), $N = 0.37$: solid circles are L_b; open circles are 1.4 L_R; squares are 15 L_K; crosses are L. Length scale evolution is as follows for the data of Stillinger et al. (1983), $N = 0.73$: the solid curve is 1.4 L_R; the dashed curve is 10 L_K; and the long and short dashed curve is L.

4.2 Waves and Quasi-Horizontal Turbulence

A more formal analysis, using similar concepts but adding a new element, was produced by Riley et al. (1981) and extended further by Lilly (1983). Riley et al. proposed that stratified turbulence, after it starts to decay, should include a combination of waves and quasi-horizontal turbulence, since the latter can exist at any scale without interference by buoyant forces.

The equations of motion for a Boussinesq incompressible flow are

$$\frac{\partial \boldsymbol{V}_H}{\partial t} + \boldsymbol{V}_H \cdot \nabla \boldsymbol{V}_H + w\frac{\partial \boldsymbol{V}_H}{\partial z} + \nabla_H \pi = 0 \ , \tag{4.1}$$

$$\frac{\partial w}{\partial t} + \boldsymbol{V}_H \cdot \nabla w + w\frac{\partial w}{\partial z} + \frac{\partial \pi}{\partial z} - b = 0, \tag{4.2}$$

where ∇ is the horizontal gradient, $\boldsymbol{V}_H$ is the horizontal velocity, w is the vertical velocity, and π is pressure divided by reference density. The buoyancy equation is

$$\frac{\partial b}{\partial t} + \boldsymbol{V}_H \cdot \nabla b + w\frac{\partial b}{\partial z} + N^2 w = 0, \tag{4.3}$$

and the continuity equation is

$$\nabla \cdot \boldsymbol{V}_H + \frac{\partial w}{\partial z} = 0, \tag{4.4}$$

where buoyancy has been decomposed into a horizontally averaged buoyancy $\bar{b}$ and its departure b and N^2 is $\partial \bar{b}/\partial z$. Two regimes of scaling are then applied, one for turbulence and the other for waves. In the case of an initially 3D isotropic turbulence regime, the space variables are scaled by L, the scale at which turbulence is injected, and the velocity scale is V, the initial turbulent velocity amplitude. Pressure is scaled by V^2 and time by L/V. In that regime, hydrostatic scaling applies in the w momentum equation so that $b \sim \pi/L$, and b is then scaled by V^2/L. In the buoyancy equations, the balance $\partial b/\partial t \sim N^2 w$ allows us to scale w as V^3/L^2N^2.

We then introduce the Froude number $F = V/NL$ and rewrite the equations as

$$\frac{\partial \boldsymbol{V}_H}{\partial t} + \boldsymbol{V}_H \cdot \nabla \boldsymbol{V}_H + F^2 w \frac{\partial \boldsymbol{V}_H}{\partial z} + \nabla_H \pi = 0\ , \tag{4.5}$$

$$F^2 \left(\frac{\partial w}{\partial t} + \boldsymbol{V}_H \cdot \nabla w + F^2 w \frac{\partial w}{\partial z} \right) + \frac{\partial \pi}{\partial z} - b = 0\ , \tag{4.6}$$

$$\frac{\partial b}{\partial t} + \boldsymbol{V}_H \cdot \nabla b + F^2 w \frac{\partial b}{\partial z} + N^2 w = 0\ , \tag{4.7}$$

$$\nabla \cdot \boldsymbol{V}_H + F^2 \frac{\partial w}{\partial z} = 0. \tag{4.8}$$

At the lowest order in F^2, motions are hydrostatic and have negligible vertical advection and horizontal divergence. Fluid motions correspond to horizontal layers evolving independently, equivalent to 2D turbulence. Layers will interact at higher orders. As they evolve out of phase, they develop shearing instabilities and vertical pressure gradients which may recouple them.

Another scaling can be introduced to describe linear wavelike motions, with pressure gradients balancing accelerations. Using N^{-1} as a time scale, one gets the following scaled equations:

$$\frac{\partial \boldsymbol{V}_H}{\partial t} + F \left(\boldsymbol{V}_H \cdot \nabla \boldsymbol{V}_H + w \frac{\partial \boldsymbol{V}_H}{\partial z} \right) + \nabla_H \pi = 0\ , \tag{4.9}$$

$$\frac{\partial w}{\partial t} + F\left(\boldsymbol{V}_H \cdot \nabla w + w\frac{\partial w}{\partial z}\right) + \frac{\partial \pi}{\partial z} - b = 0\,, \tag{4.10}$$

$$\frac{\partial b}{\partial t} + F\left(\boldsymbol{V}_H \cdot \nabla b + w\frac{\partial b}{\partial z}\right) + w = 0\,, \tag{4.11}$$

$$\nabla \cdot \boldsymbol{V}_H + \frac{\partial w}{\partial z} = 0. \tag{4.12}$$

The nonlinear terms are scaled by F. The lowest order in F corresponds to linearized internal gravity waves. F can also be interpreted as the ratio of the time scale characterizing wave motions to the time scale characterizing turbulent motions: $F = N^{-1}/(L/V)$. Hence if $F \ll 1$, one would expect little interaction between waves and turbulence. Initially, when turbulence is injected, F is possibly greater than unity, and one can write an evolution equation for F based on the kinetic energy equation

$$\frac{d}{dt}\left(\frac{V^2}{2}\right) = -\frac{V^3}{L}, \tag{4.13}$$

or

$$\frac{dV}{dt} = -\frac{V^2}{L}. \tag{4.14}$$

If one assumes L to be constant on the variation time scale of F, one obtains

$$\frac{dF}{dt} = -NF^2, \tag{4.15}$$

whose solution is

$$\frac{1}{F} = \frac{1}{F_0} + Nt. \tag{4.16}$$

This shows that no matter how large the initial F is, within one buoyancy time scale F will be reduced to O(1) and the 3D turbulent regime ends. In its place, there will be gravity waves that propagate away from the generation region (at least partly in the vertical direction) and 2D turbulence that grows in scale.

In wavenumber space, the velocity vector lies orthogonal to the propagation vector $\boldsymbol{\kappa}$. In the plane perpendicular to $\boldsymbol{\kappa}$, one introduces unit vectors $\boldsymbol{\kappa} \times \boldsymbol{k}/\kappa$ and $\boldsymbol{\kappa} \times (\boldsymbol{\kappa} \times \boldsymbol{k})/\kappa^2$, where $\boldsymbol{k}$ is a unit vector in the vertical direction. Turbulent motions correspond to $\boldsymbol{V}$ in the direction $\boldsymbol{\kappa} \times \boldsymbol{k}/\kappa$, wavelike motions to the direction $\boldsymbol{\kappa} \times (\boldsymbol{\kappa} \times \boldsymbol{k})/\kappa^2$. The equations of motion in wavenumber space are

$$\frac{\partial \hat{\boldsymbol{V}}}{\partial t} + i\boldsymbol{\kappa}\hat{\pi} - \boldsymbol{k}\hat{b} = 0, \tag{4.17}$$

$$\frac{\partial \hat{b}}{\partial t} + N^2\hat{w} = 0, \tag{4.18}$$

$$\boldsymbol{\kappa} \cdot \hat{\boldsymbol{V}} = 0, \tag{4.19}$$

where $\hat{\boldsymbol{V}}, \hat{\pi}, \hat{b}$ are the transformed variables. An equation for pressure is obtained by taking the dot product of (4.17) with $\boldsymbol{\kappa}$ and using the continuity equation to obtain

$$\kappa^2\hat{\pi} = -i(\boldsymbol{k} \cdot \boldsymbol{\kappa})\hat{b}. \tag{4.20}$$

Substitution of (4.20) into the momentum and buoyancy equations yields

$$\frac{\partial \hat{\boldsymbol{V}}}{\partial t} + \left[\frac{\boldsymbol{\kappa}(\boldsymbol{k} \cdot \boldsymbol{\kappa})}{\kappa^2} - \boldsymbol{k}\right]\hat{b} = 0 \tag{4.21}$$

and

$$\frac{\partial^2 \hat{w}}{\partial t^2} + \left(1 - \frac{\kappa_z^2}{\kappa^2}\right)N^2\hat{w} = 0. \tag{4.22}$$

Note that the factor in brackets in the equation of horizontal notion is identical to $[\boldsymbol{\kappa} \times (\boldsymbol{\kappa} \times \boldsymbol{k})]/\kappa^2$, which supports the previous identification of wave and turbulence motion directions. The associated dispersion relation $\omega^2 - N^2 \cos^2\theta = 0$, where ω is frequency and $\cos\theta = \kappa_H/\kappa$, characterizes gravity waves. If rotation effects and an aspect ratio between vertical and horizontal scales are included, inertio-gravity waves and geostrophic turbulence play a similar role in the low-Rossby-number regime.

Whether the time scale of injection of turbulence is greater or smaller than N will determine the ratio of gravity wave energy to turbulence energy in the initial generation. If the injection time scale is $\ll N^{-1}$, gravity waves are mainly generated. A spectrum of horizontal velocities proportional to $k^{-5/3}$ has been observed in the atmosphere (Nastrom and Gage, 1985), possibly created by upscale transfer of energy by the 2D turbulence under the constraint of stratification.

Dickey, T. D., and G. L. Mellor, 1980: *J. Fluid Mech.*, **99**, 13.

Dougherty, J. P., 1961: *J. Atmos. Terr. Phys.*, **21**, 210.

Hopfinger, E., 1987: *J. Geophys. Res.*, **92**, 5287.

Hunt, J. C. R., ed., 1985: *Turbulence and Diffusion in Stable Environments.* Clarendon Press, Oxford, 319 p.

Lilly, D. K., 1983: *J. Atmos. Sci.*, **40**, 749.

Lin, J. T., and Y.-H. Pao, 1979: *Ann. Rev. Fluid Mech.*, **11**, 317.

Lumley, J. L., 1964: *J. Atmos. Sci.*, **21**, 99.

Nastrom, G. D. and K. S. Gage, 1985: *J. Atmos. Sci.*, **42**, 950.

Ozmidov, R. V., 1965: *Atm. Ocean. Phys.*, **1**, 493.

Riley, J. J., R. W. Metcalfe, and M. A. Weissman, 1981: Proc. AIP Conf. Nonlinear Properties of Internal Waves, ed. B. J. West, 79.

Stillinger, D. C., K. N. Helland, and C. W. Van Atta, 1983: *J. Fluid Mech.*, **131**, 91.

Chapter 5

MODES OF TURBULENCE IN STABLY STRATIFIED ENVIRONMENTS: PART I

In this and the next chapter we will consider four modes of turbulence which may occur in stably stratified environments. These are: 1) grid–generated or numerically initiated homogeneous decaying turbulence; 2) collapse of turbulent wakes or cloud plumes; 3) continuously forced turbulence bounded by stable interfaces; and 4) breaking gravity waves. In the first three examples the turbulence is generated by mechanical means or by convective forcing. In the fourth example the waves generate turbulence by nonlinear interactions. Related phenomena are discussed by Hopfinger (1987).

5.1. Grid-Generated or Numerically Initiated Turbulence

In the typical laboratory configuration a grid is arranged to pass through a volume of salt-stratified fluid, either horizontally or vertically, leaving a turbulent wake, which quickly becomes nearly homogeneous and then undergoes temporal decay. The results of such experiments have been described by Britter (1985) and Itsweire et al. (1986). Similar experiments have been conducted by numerical simulation, by Riley et al. (1981) and by Métais et al. (1987). The results of the two kinds of experiments seem to be in general agreement, although the analyses tend to emphasize different features.

In general the initial decay rate of velocity fluctuations is either unchanged or increased by stratification. After a few Brunt-Vaisala periods the flow becomes anisotropic, with the horizontal scales growing and the vertical scales shrinking, at least by comparison with the evolution of turbulence in an unstratified environment. Figure 5.1 shows results of such an experiment. Because the Reynolds number and/or numerical resolution is inevitably limited in these experiments, the decay process soon becomes laminar. For this reason it remains unclear to what extent the process of separation and separate evolution of the gravity wave and quasi-two-dimensional turbulence components, as predicted by Riley et al. (1981) and Lilly (1983), occurs.

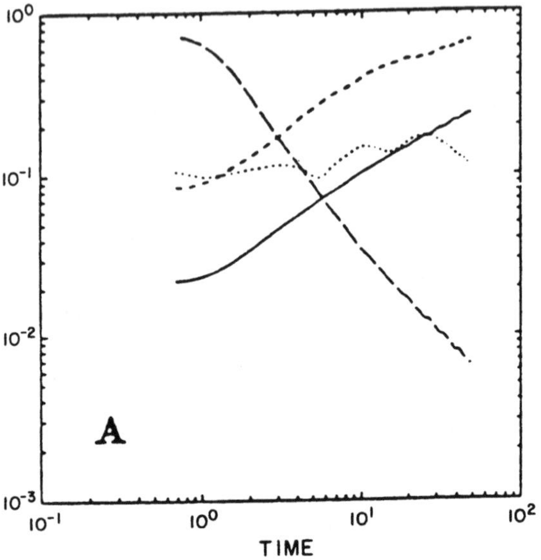

Figure 5.1: Evolution with time of the Kolmogorov dissipative scale (solid), the Ozmidov scale (long dashes), the horizontal integral scale (short dashes), and the vertical integral scale (dots) from a numerical simulation of initially isotropic, decaying turbulence in a stratified environment. The abscissa is in units of N^{-1}. From Métais et al (1987).

The Schmidt number (ratio of viscosity to salt diffusivity) is very high, so that in the lab experiments, density fluctuations tend to remain after velocity fluctuations have been damped out. The residue has often been referred to as "fossil turbulence."

5.2. Wake Collapse

In the wake collapse phenomenon, a relatively well-mixed wake, perhaps caused by a submarine or aircraft, is introduced into a stably stratified environment. If the wake is neutrally buoyant, then its upper portion is cooler than its surroundings at the same level and its lower portion is warmer. The top and bottom of the wake, therefore, tend to converge, or "collapse," because of the density differences, and the sides must spread out. In this process gravity waves are generated, but the remaining wake maintains large-scale quasi-horizontal irregularities. Thus the wake collapse process is an inhomogeneous version of the evolution exhibited in the decay of homogeneous turbulence in a stratified environment. Laboratory experiments and/or significant analyses have been presented by Schooley and Stewart (1963), Wu (1969), Bell and Dugan (1974), and Dugan et al. (1976).

The outflow from a thunderstorm can show similar behavior (Lilly, 1988). Air rising in a convective updraft becomes turbulent, well-mixed, and fairly isotropic. The well-mixed air usually overshoots its level of neutral buoyancy, but as it spreads out descends back to that level and undergoes the collapse phenomenon (Fig. 5.2). The rate at which the outflow plume collapses is estimated by equating the potential energy lost to the kinetic energy gained in the outflow, giving

$$v_{max} = \frac{ND_0}{2\sqrt{2}}, \tag{5.1}$$

where N is the Brunt-Vaisala frequency of the environment and D_0 is the initial diameter of the plume. This model ignores entrainment, any initial stratification of the plume, and transfer of energy into propagating waves, all of which reduce the outflow velocity.

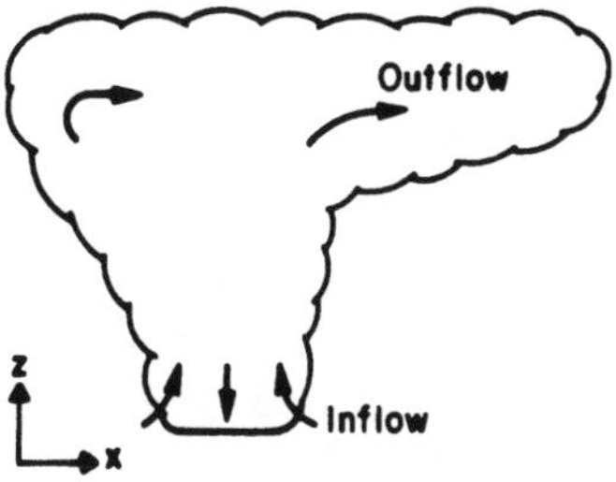

Figure 5.2: Schematic diagram of outflow of thunderstorm into a stably stratified environment.

5.3. Turbulent Layer Bounded by Stable Interfaces

A turbulent layer bounded by one stable interface is produced by either heating from below or cooling from above of a fluid which is initially stratified, or by stirring at either the top or bottom. We will consider here the typical atmospheric boundary case of surface heating. It is also possible, through artificial stirring or natural radiant heating and cooling processes of a cloud sheet, to generate turbulence in a layer bounded by stable interfaces at both the top and bottom. This has been discussed by Lilly (1988).

When turbulent convection is generated by a heated surface, plumes of heated air rise, and are carried past their equilibrium value by inertia, while air above the layer is entrained down into it by turbulence. The problem is to estimate the rate of increase of temperature by both surface heating and entrainment and the growth of the boundary layer by entrainment. This problem is simplified for analysis (Lilly, 1968) by assuming the boundary layer to be well-mixed and neutrally stable except for a thin, unstable layer at the bottom and a thin, very stable layer (temperature inversion in the atmosphere) at the top (Fig. 5.3).

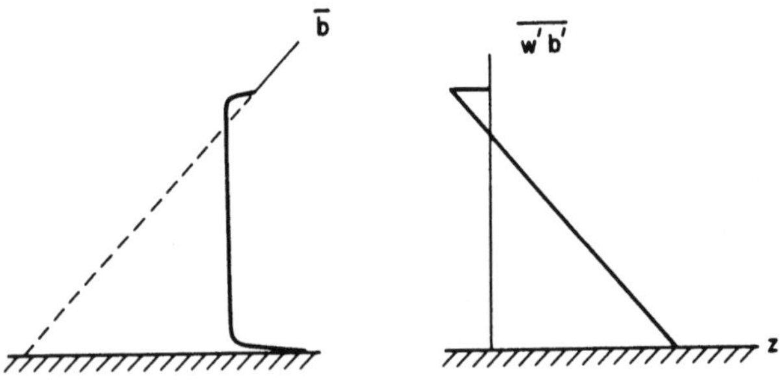

Figure 5.3: Potential temperature and heat flux profiles for boundary layer heated from below.

Horizontal gradients and advection are ignored. Thus the rate of change of mean buoyancy in this mixed layer is the negative of the gradient of the upward buoyancy flux, i.e.,

$$\frac{\partial \bar{b}}{\partial t} = -\frac{\partial(\overline{b'w'})}{\partial z} = \frac{(\overline{b'w'})_o - (\overline{b'w'})_i}{z_i}, \tag{5.2}$$

where subscripts i and o refer to the interface and the surface, respectively. The mixed layer is assumed to be growing upward into a stably stratified environment, with a buoyancy profile, b_e, assumed to be linear in z, hence

$$b_e = b_{eo} + N^2 z \qquad \text{for } z > z_i\ . \tag{5.3}$$

The buoyancy jump at the interface, say, δb, is then

$$\delta b = b_{eo} + N^2 z - \bar{b}\ . \tag{5.4}$$

We wish to relate the growth rate of the mixed layer by entrainment, dz_-/dt, to the buoyancy flux at the top of the mixed layer. This is done by integrating the left equality of (5.2) over a vanishingly small region of depth ϵ about the interface, using Leibnitz's rule, to obtain

$$\int_{z_i-\frac{\epsilon}{2}}^{z_i+\frac{\epsilon}{2}} \frac{\partial \overline{b}}{\partial t}\, dz = \frac{d}{dt}\int_{z_i-\frac{\epsilon}{2}}^{z_i+\frac{\epsilon}{2}} \overline{b}\, dz - \frac{dz_i}{dt}\delta\overline{b} = \left(\overline{w'b'}\right)_-, \tag{5.5}$$

where $(\overline{w'b'})_-$ is the negative buoyancy flux just below the interface and δb the buoyancy jump across it. Upon letting $\epsilon \to 0$, the first integral in the RHS of the first equality vanishes, and the second equality may be written

$$\frac{dz_i}{dt} = -\frac{\left(\overline{w'b'}\right)_-}{\delta\overline{b}}\ . \tag{5.6}$$

The problem is closed by determining the buoyancy fluxes at the top and bottom of the layer, which so far has always involved some empiricism.

Bell, T. H. and J. P. Dugan, 1974: *J. Enging. Math.*, **8**, 241.

Britter, R. E., 1985: Diffusion and decay in stably-stratified turbulent flows, in *Turbulence and Diffusion in Stable Environments*, J. C. R. Hunt, Ed. Clarendon Press, Oxford.

Dugan, J. P., A. C. Warn-Varnas, and S. A. Piacsek, 1976: *Computers and Fluids*, **4**, 109.

Hopfinger, E., 1987: *J. Geophys. Res.*, **92**, 5287.

Itsweire, E. C., K. N. Helland, and C. W. Van Atta, 1986: *J. Fluid Mech.*, **162**, 299.

Lilly, D. K., 1968: *Quart. J. Roy. Meteor. Soc.*, **94**, 292.

Lilly, D. K., 1983: *J. Atmos. Sci.*, **40**, 749.

Lilly, D. K., 1988: *J. Atmos. Sci.*, **45**,

Métais, O., J. R. Herring, M. Lesieur, and J. P. Chollet, 1987: Turbulence in stably stratified fluids: statistical theory and direct numerical simulations. Preprints, 3rd International Symposium on Stratified Flows, Pasadena, Calif., Feb. 3–5, 1987.

Riley, J. J., R. W. Metcalfe, and M. A. Weissman, 1981: Direct numerical simulations of homogeneous turbulence in density–stratified fluids, in *Proceedings AIP Conference on Nonlinear Properties of Internal Waves*, B. J. West, ed. 79.

Schooley, A. H. and R. W. Stewart, 1963: *J. Fluid Mech*, **15**, 83.

Wu, J., 1969: *J. Fluid Mech.*, **35**, 531.

Chapter 6

MODES OF TURBULENCE IN STABLY STRATIFIED ENVIRONMENTS: PART II

We continue here with analysis of buoyancy-driven mixed layer turbulence beneath a stably stratified environment. In order to close the problem, it is necessary to determine the buoyancy fluxes at the bottom and top of the mixed layer. Surface layer analysis allows approximation of the former through a transfer coefficient expression, i.e.,

$$(\overline{w'b'})_0 = C_T V_0(b_0 - \bar{b}), \tag{6.1}$$

where V_0 is the low-level wind speed and C_T is a dimensionless transfer coefficient, a function of surface roughness, air-surface buoyancy difference, and (weakly) wind speed.

Determination of the buoyancy flux at the interface is the most difficult part of the problem. From laboratory experiments (Deardorff et al. 1969), it was found that the buoyancy flux at the interface is linearly related to the surface buoyancy flux, i.e.,

$$(\overline{w'b'})_- = -k(\overline{w'b'})_0, \tag{6.2}$$

where $k \sim 0.2$. Substituting (6.1) and (6.2) into (5.2), integrated over the depth of the mixed layer, we obtain the mixed layer heating rate as

$$\frac{d\bar{b}}{dt} = \frac{(1+k)C_T V_0(b_0 - \bar{b})}{z_i} \,. \tag{6.3}$$

Upon substituting (6.1), (6.2), and (5.4) into (5.6) we also obtain an expression for the rate of growth of the mixed layer, i.e.,

$$\frac{dz_i}{dt} = \frac{k C_T V_0(b_0 - \bar{b})}{b_{eo} + N^2 z_i - \bar{b}} \,. \tag{6.4}$$

Here $\bar{b}$ and z_i are the only unknowns. These equations can be solved analytically.

Further insight can be gained by considering the energetics. The turbulence kinetic energy equation may be written as

$$\frac{\partial E}{\partial t} = \overline{w^{P}rimeb'} - \epsilon - \frac{\partial}{\partial z}\left(\frac{\overline{w'p'}}{\rho_0} + \frac{\overline{w'V'^2}}{2}\right) \,. \tag{6.5}$$

Statistical steady state is assumed, so the first term is neglected. By integrating the remaining terms over the mixed layer depth, we get

$$\int_0^{z_i} (\overline{w'b'})dz = \int_0^{z_i} \epsilon\, dz \,. \tag{6.6}$$

It is assumed that the flux in parentheses in the RHS of (6.5) vanishes at the boundries. The $\overline{w'p'}$ term may, however, include a loss due to gravity waves at the top of the boundary layer. It is generally thought to be small, although Carruthers and Hunt (1986) predict from linear theory a loss of up to 10% of the total dissipation in the layer.

By substituting the empirical relation (6.2) into (6.6) and integrating the LHS across the linear flux profile, we obtain an expression for the average dissipation

$$\int_0^{z_i} \epsilon\, dz = \frac{1-k}{2}(\overline{w'b'})_0\, z_i \,. \tag{6.7}$$

If k were equal to 1, the flux at the surface and at the interface would be equal and opposite and, from (6.7), there could be no dissipation. If k were equal to 0, there would be no flux at the interface and hence no entrainment.

If we compare the areas of the positive and negative regions of the buoyancy flux profile on the left panel of Fig. 6.1, it can be seen that

$$\int (\overline{w'b'})_{<0} dz = -k^2 \int (\overline{w'b'})_{>0} dz \,. \tag{6.8}$$

For a value of $k = 0.2$, this suggests that $(0.2)^2 = 4\%$ of the energy in the mixed layer goes into entrainment of "clear" air from above.

An alternative view is that of Stage and Businger (1981a, b). They decompose the buoyancy flux profile into positive and negative flux regions, each of which decreases linearly to zero through the boundary layer, as shown in the dashed line of 6.1. They then derive the relationship

$$\int (\overline{w'b'})_{<0} dz = -k \int (\overline{w'b'})_{>0} dz, \tag{6.9}$$

which implies a 20% rather than a 4% efficiency. Moeng (1987) has used conditional sampling of the results of a large eddy simulation model of the boundary layer to test both hypotheses. She separately evaluated the contributions to the above integrals of warm updrafts, warm downdrafts, cold updrafts, and cold downdrafts and found agreement with the latter interpretation in the case of a dry atmosphere. However, when the cloud base was within the mixed layer no good agreement was found for either hypothesis.

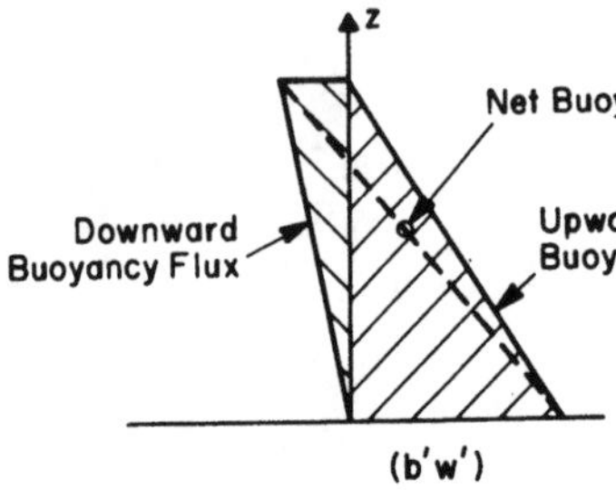

Figure 6.1: Lilly-Deardorff interpretation of buoyancy flux profile (dashed) and Stage-Businger interpretation (solid).

6.1. Breaking Gravity Waves

Wave breaking is associated with regions where the buoyancy surfaces become bent back beyond the vertical, so that the buoyancy locally decreases with height. Thus turbulent energy may be derived from buoyancy through overturning. Other ways of generating turbulence by waves are through shearing instability and in hydraulic jump-like flows produced at stability interfaces. Fritts (1984) presents a recent review of wave breaking in the upper atmosphere.

One might expect that shearing instability would always occur before overturning, since the condition for the former is that the Richardson number must be less than 1/4, while the latter requires that $Ri < 0$. Hodges (1967) showed, however, that in an environment with constant stability and wind speed, the local vertical shear and stability decrease together as wave amplitude increases, and at the point of overturning Ri decreases discontinuously from 1/2 to zero, so that overturning is evidently the preferred mode of instability. In less idealized circumstances this singularity may not strictly occur, but the location where shearing and overturning instabilities occur may be so close in time and space that only the latter are relevant.

A number of authors have proposed that the vertical energy spectrum is in a state of saturation, so that a small change in energy content produces a large change in dissipation. There is evidence to support this, (see, e.g., Fritts, 1984; Smith et al., 1987), though I don't believe the theoretical justification is yet solid. A prediction associated with this hypothesis is that

$$E_H(k_z) = CN^2 k_z^{-3}, \tag{6.10}$$

where $E_H(k_z)$ is the vertical spectrum of the horizontal energy and C is a constant, presumably of order unity.

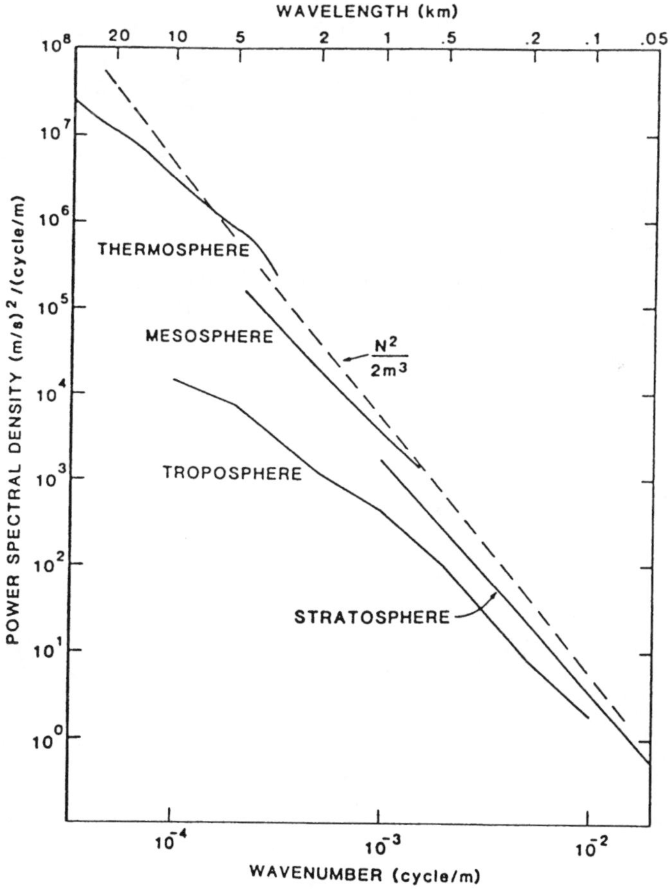

Figure 6.2: Vertical spectra of horizontal flow for various levels in the atmosphere (from Smith et al., 1987).

Observations (see Fig. 6.2) show a spectrum proportional to k_z^{-n}, where $n \sim 2.5 - 2.8$. This might be thought to be insignificantly different from the hypothesized prediction with $n = 3$. However, if $n = 3$, then the spectrum of the vertical shear must be proportional to k_z^{-1}, indicating that every decade in vertical wavenumber contributes equally to mean shear. This seems unlikely, since it suggests that instabilities would occur with equal probability on all scales. In fact they are generally observed only at length scales of order 100 m or less. This seems consistent with $n < 3$. A point in favor of the saturation prediction, however, is that the vertical spectrum does scale well with N^2. Specifically, it increases by about a factor of four going through the tropopause.

Carruthers, D. J., and J. C. R. Hunt, 1986: *J. Fluid Mech.*, **165**, 475.

Deardorff, J. W., G. E. Willis, and D. K. Lilly, 1969: *J. Fluid Mech.*, **35**, 7.

Fritts, D. C., 1984: *Rev. Geophys. Space Phys.*, **22**, 275.

Hodges, R. R., 1967: *J. Geophys. Res.*, **72**, 3455.

Moeng, C.-H., 1987: *J. Atmos. Sci.*, **44**, 1605.

Smith, S. A., D. C. Fritts, and T. E. VanZandt, 1987: *J. Atmos. Sci.*, **44**, 1404.

Stage, S. A. and J. A. Businger, 1981a: *J. Atmos. Sci.*, **38**, 2213.

Stage, S. A. and J. A. Businger, 1981b: *J. Atmos. Sci.*, **38**, 2230.

Chapter 7

SUBGRID CLOSURES IN LARGE EDDY SIMULATION: PART I

7.1 Definitions

The procedure of numerically computing the largest scales of a large Reynolds number flow, while parameterizing the subgrid scales (SGS), has come to be known as Large Eddy Simulation (LES). The purpose of this chapter is to review some aspects of SGS closures.

Early workers defined the resolved scale velocity field as the volume average

$$\overline{u}_i(x,y,z) = \int_{-\frac{\Delta}{2}}^{\frac{\Delta}{2}}\int_{-\frac{\Delta}{2}}^{\frac{\Delta}{2}}\int_{-\frac{\Delta}{2}}^{\frac{\Delta}{2}} u_i(x+\xi, y+\eta, z+\zeta)\,d\xi d\eta d\zeta, \tag{7.1}$$

where Δ is an averaging distance, originally (and normally) the spacing of variables defined on a grid. A similar expression was used for other dependent variables in the problem. The volume-averaged variables could be considered as either continuous or only defined on discrete grid volumes. In the latter case, as applied by Schumann (1975), the values vary discontinuously throughout the domain. In the continuous approach, which most workers have tended to use, volume-averaged variables do not satisfy the Reynolds hypotheses ($\overline{\overline{u}}_i = \overline{u}_i$, and $\overline{\overline{u}_i u_j'} = 0$, where $u_i' = u_i - \overline{u}_i$). We note that the volume-averaged variable defined by (7.1) has a very different physical significance than does the traditional Reynolds (ensemble)-averaged variable.

To investigate some of the detail of SGS closure, we begin by averaging an appropriate form of the momentum equation, for example,

$$\frac{\partial \overline{u}_i}{\partial t} + \frac{\partial}{\partial x_j}(\overline{u_i u_j}) + \frac{1}{\rho_0}\frac{\partial \overline{p}}{\partial x_i} - \overline{b}\delta_{i3} = 0, \tag{7.2}$$

where $\overline{u}_i$, $\overline{p}$, and $\overline{b}$ are respectively the volume-averaged velocity, pressure and buoyancy. We have neglected viscous effects. The problem is how to treat the advective term. The original approach used by Smagorinsky (1963), Lilly (1962, 1966, 1967), and Deardorff (1970) is to replace $\overline{u_i u_j}$ with $\overline{u}_i \overline{u}_j$ on the LHS and add $\partial \tau_{ij}/\partial x_j$ to the RHS, where

$$\tau_{ij} = -\overline{u_i u_j} + \overline{u}_i \overline{u}_j + \frac{\delta_{ij}}{3}\left[\overline{u_k^2} - \overline{u}_k^2\right] \tag{7.3a}$$

and

$$\overline{\pi} = \frac{\overline{p}}{\rho_0} + \frac{\overline{u_i^2} - \overline{u}_i^2}{3} \qquad (7.3b)$$

(δ_{ij} is the Kronecker symbol). The quantity τ_{ij} is referred to as the SGS stress; it has been constructed so that it is nonzero only in anisotropic flow. The momentum equation thus becomes

$$\frac{\partial \overline{u}_i}{\partial t} + \frac{\partial}{\partial x_j}(\overline{u}_i\overline{u}_j) + \frac{\partial \overline{\pi}}{\partial x_i} - \overline{b}\delta_{i3} = \frac{\partial \tau_{ij}}{\partial x_j} \; . \qquad (7.4)$$

A suitable parameterization for τ_{ij} must be used to close the problem. No parameterization is required for $\overline{\pi}$ in an incompressible flow, since it is obtained by solution of the Poisson equation derived by taking the divergence of (7.4).

7.2. Smagorinsky Closure

The earliest and best-known SGS closure, due to Smagorinsky (1963), relates stresses to fluid strains through an eddy viscosity coefficient, i.e.,

$$\tau_{ij} = 2KS_{ij}, \qquad (7.5)$$

where S_{ij} is the volume–averaged strain rate tensor,

$$S_{ij} = \frac{1}{2}\left(\frac{\partial \overline{u}_i}{\partial x_j} + \frac{\partial \overline{u}_j}{\partial x_i}\right),$$

and

$$K = (C\Delta)^2(2S_{ij}S_{ij})^{1/2}. \qquad (7.6)$$

In the above, C is an adjustable constant and Δ is the grid spacing. This type of closure is known as "first order," because the stresses are related directly, though nonlinearly, to the explicitly predicted volume-averaged variables. In second order methods, conservation equations for the stresses are derived and they are then also predicted explicitly.

From the chosen eddy viscosity form of the stress, it is easily shown that this term removes energy from the volume-averaged flow and transfers it to the implicit SGS motion field. The form and amplitude of K in (7.6) is chosen to allow a matching of this energy removal with the rate of viscous dissipation, under the assumption that the grid scale lies within a turbulent inertial subrange.

If the explicit kinetic energy equation is derived from the volume-averaged momentum equation (7.4) by multiplying it by $\overline{u}$, the RHS becomes

$$\overline{u}_i \frac{\partial \tau_{ij}}{\partial x_j} = \frac{\partial}{\partial x_j}(\overline{u}_i \tau_{ij}) - \frac{\partial \overline{u}_i}{\partial x_j}\tau_{ij} . \tag{7.7}$$

Since the first term on the RHS is the gradient of a flux, we see that any exchange of energy between explicit and SGS forms must be produced by the second term, which we designate as $-\epsilon_g$. When the Smagorinsky closure is used in this term, the result is

$$\langle \epsilon_g \rangle = \langle \frac{\partial \overline{u}_i}{\partial x_j}\tau_{ij} \rangle = \langle 2K \frac{\partial \overline{u}_i}{\partial x_j} S_{ij} \rangle = (C\Delta)^2 \langle (2S_{ij}S_{ij})^{3/2} \rangle , \tag{7.8}$$

where $\langle\ \rangle$ indicates the spatial average and we have used the fact that $S_{ij}\partial\overline{u}_i/\partial x_j = S_{ij}S_{ij}$.

If we assume that turbulence is homogeneous, we can define the constant C so that the loss of explicit energy closely approximates the energy cascaded through the inertial subrange and dissipated by viscosity. We begin by noting that for homogeneous turbulence, we can write

$$\langle 2S_{ij}S_{ij} \rangle = \langle \overline{\omega}_i^2 \rangle / 2 . \tag{7.9}$$

We can approximate the resolved enstrophy as

$$\frac{\langle \overline{\omega}_i^2 \rangle}{2} \simeq \int_0^{\pi/\Delta} k^2 E(k) dk , \tag{7.10}$$

where E(k) is the kinetic energy spectrum and π/Δ is the largest resolvable wavenumber of the finite difference approximation. Note that (7.10) is only approximate, because it assumes a sharp cutoff in Fourier space between resolved and subgrid scales, rather than a volume–average filter. We assume that the largest resolvable wave number of the simulation lies within the inertial subrange, so that the energy spectrum is given by

$$E(k) = \alpha \langle \epsilon \rangle^{2/3} k^{-5/3} , \tag{7.11}$$

where $\langle \epsilon \rangle$ is the spatially averaged viscous energy dissipation, which is assumed to be out of the implicit SGS energy. The Kolmogorov constant, α, is approximately 1.5. Using (7.11) and (7.10) in (7.9) gives

$$\langle 2S_{ij}S_{ij}\rangle = \frac{3\alpha}{4}\langle\epsilon\rangle^{2/3}\left(\frac{\pi}{\Delta}\right)^{4/3} . \qquad (7.12)$$

Now we assume that $\langle\epsilon_g\rangle$, the transfer of energy from explicit to SGS scales, is identical to $\langle\epsilon\rangle$, the dissipation of SGS energy. This is equivalent to assuming that the energy in the SGS scales remains constant. By elimination of these two quantities from (7.8) and (7.12), we obtain an expression for C, the Smagorinsky constant, as

$$C = \frac{1}{\pi}\left(\frac{2}{3\alpha}\right)^{3/4}\frac{\langle S_{ij}S_{ij}\rangle^{3/2}}{\langle(S_{ij}S_{ij})^{3/2}\rangle} . \qquad (7.13)$$

If the strain ratio is assumed unity (as has been normally done), evaluation of the expression indicates that $C \sim 0.2$. By using a more accurate version of (7.10), Lilly (1966) obtained a slightly larger value. The strain ratio should be less than unity for any statistical distribution of the six non-duplicated S_{ij} components. Its neglect could be easily justified if its components were distributed as Gaussian variables. This is not a good assumption, however, as observed flow gradient distributions always have thicker tails and larger kurtosis than do Gaussian variables. This is, therefore, an incomplete element of the theory.

As another apparent weakness, the existence of SGS stresses may seem inconsistent with the assumption of isotropic turbulence which is used to evaluate the constants. If, however, we evaluate the SGS energy as that of the inertial subrange, i.e.,

$$E \simeq \int_{\pi/\Delta}^{\infty} \alpha\langle\epsilon\rangle^{2/3}k^{-5/3}\,dk = \frac{3\alpha}{2}\left(\frac{\langle\epsilon\rangle\Delta}{\pi}\right)^{2/3} ,$$

and then substitute (7.8) for $\langle\epsilon\rangle$, we obtain

$$E \simeq \frac{3\alpha}{2}\left(\frac{C^2}{\pi}\right)^{2/3}\Delta^2\langle 2S_{ij}S_{ij}\rangle .$$

Then the ratio of stress amplitude [from (7.5) and (7.6)] to SGS energy amplitude is

$$\frac{|\tau_{ij}|}{2E} \simeq \frac{(C\pi)^{2/3}}{3\alpha} \sim 0.16 . \qquad (7.14)$$

Thus the deviation from isotropy is not large.

Deardorff (1970), using Lilly's recommended value of $C \simeq 0.2$, applied the Smagorinsky model to simulate simple shear flows and the planetary boundary layer. Although the value of $C = 0.2$ was found to work well in some circumstances, it often was set to a value as small as about 0.1 to obtain expected spectra. The smaller value was found to be necessary in flows with a strong mean shear. Evidently in such cases the mean shear overwhelmed the local stresses, indicating that the grid scale was not really within the inertial subrange.

7.3. Limitations and Adjustments to SGS Closures

A typical result of an LES is a profile of some important quantity. An example is the buoyancy flux profile between parallel plates (Fig. 7.1). This shows the contributions to buoyancy flux $\overline{wb}$ calculated from both resolvable motions and the SGS parameterization. This type of profile displays the simultaneous strength and weakness of LES. The LES is considered to be successful because in most of its domain the resolvable flux dominates the contribution from SGS parameterization. But close to the boundary, and in this case near the top of the mixed layer, LESs rely on their SGS parameterization, and fail to describe the flux when the energy-containing scales of turbulence are not much larger than the computational resolution. Increasing the resolution in just the direction normal to the boundaries is inadequate, since it only allows for unrealistically

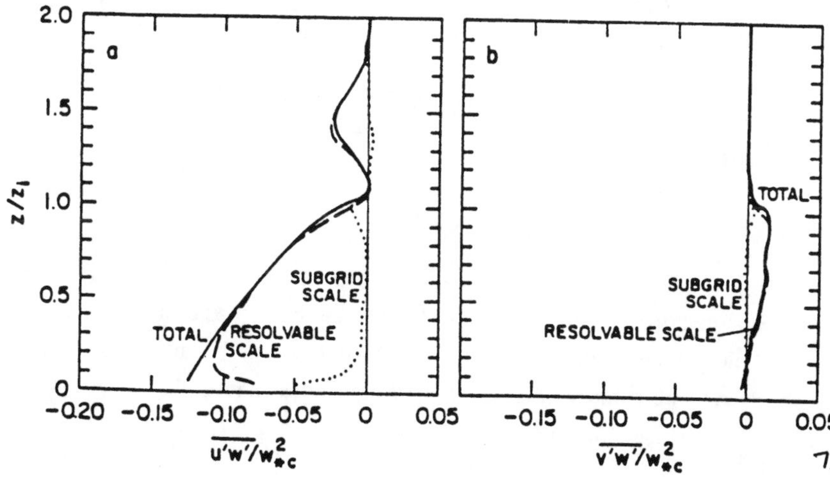

Figure 7.1: Resolvable and SGS fluxes of momentum from a boundary layer numerical simulation by Moeng (1986).

long flat eddies. With extra computational effort it may be possible to increase resolution isotropically as the boundary is approached. An optimal technique would be adaptive mesh modeling, in which the local computational resolution responds to the strength of local gradients.

In addition to the problems near the boundaries and those associated with strong shear, it has been found that stable or unstable stratification, especially the former, may reduce the scale of the inertial range to below that of the grid spacing, so that the eddy viscosity must be adjusted. A modified eddy viscosity formulation which has been widely used (Lilly, 1962) is given by

$$K_m = (C\Delta)^2 \left[2S_{ij}S_{ij} - \frac{K_h}{K_m}\frac{\partial \bar{b}}{\partial z}\right]^{1/2}, \qquad (7.15)$$

where K_m is the eddy viscosity and K_h is the eddy conductivity of buoyancy. To show how this arises, we form the equation of conservation of buoyancy,

$$\frac{\partial \bar{b}}{\partial t} + \frac{\partial}{\partial x_i}(\bar{u}_i\bar{b}) = \frac{\partial H_i}{\partial x_i}, \qquad (7.16)$$

where

$$H_i = -\overline{u_i b} + \bar{u}_i\bar{b} \quad \rightarrow \quad K_h\frac{\partial \bar{b}}{\partial x_i}. \qquad (7.17)$$

By multiplying (7.16) by the negative of the vertical coordinate, $-z$, we obtain the potential energy equation. The RHS can be rearranged as follows:

$$-z\frac{\partial H_i}{\partial x_i} = \frac{\partial(-zH_i)}{\partial x_i} - \epsilon_{gb}, \qquad (7.18)$$

where

$$\epsilon_{gb} = -K_h\frac{\partial \bar{b}}{\partial z}\ .$$

The quantity ϵ_{gb} is the energy transferred from explicit potential into SGS kinetic forms. It is negative for positive static stability, indicating (as one should expect) that turbulent kinetic energy is lost to potential energy by mixing in a stable environment. The sum of the SGS energy gains, designated here as ϵ_{gt}, is then obtained by adding ϵ_g from (7.8) to ϵ_{gb}, to obtain

$$\epsilon_{gt} = 2K_mS_{ij}S_{ij} - K_h\frac{\partial \bar{b}}{\partial z}\ . \qquad (7.19)$$

The form of the eddy viscosity in (7.15) is chosen by analogy to (7.6), leading to the final expression for ϵ_{gt},

$$\epsilon_{gt} = (C\Delta)^2 \left(2S_{ij}S_{ij} - \frac{K_h}{K_m}\frac{\partial \bar{b}}{\partial z} \right)^{3/2} . \tag{7.20}$$

Note that for large $\partial \bar{b}/\partial z$, the coefficient K_m tends toward zero. If turbulence is to be maintained,

$$\frac{\left(\partial \bar{b}/\partial z\right)}{2S_{ij}S_{ij}} \leq \frac{K_m}{K_h}, \tag{7.21}$$

which is a generalized Richardson number criterion.

One might expect that the coefficient ratio (eddy Prandtl number) could also be a function of Richardson number. In his early simulations, however, Deardorff (1970) found that a ratio of $K_m/K_h = 1/3$ produced acceptable results both for stable and unstable cases. Most workers have followed these results, but some have preferred to assume a unit eddy Prandtl number or to leave out the Prandtl number in the definition of eddy viscosity. In their renormalization group theory of turbulence, Yakhot and Orszag (1986) predict an eddy Prandtl number ~ 0.7 for a passive scalar. Moeng and Wyngaard (1988) have, however, reexamined the question on the basis of higher-resolution simulations and found general agreement with Deardorff.

Deardorff, J. W., 1970: *J. Fluid Mech.*, **41**, 453.

Lilly, D. K., 1962: *Tellus*, **14**, 148.

Lilly, D. K., 1966: *On the Application of the Eddy Viscosity Concept in the Inertial Sub-Range of Turbulence*. NCAR Ms. No. 123, National Center for Atmospheric Research, Boulder, CO.

Lilly, D. K., 1967: The representation of small-scale turbulence in numerical simulation experiments. *Proceedings of the IBM Scientific Computing Symposium on Environmental Sciences*, H. H., Goldstine, ed., pp. 195–210, IBM Form No. 320-1951.

Moeng, C.-H., 1986: *J. Atmos. Sci.*, **43**, 2886.

Moeng, C.-H. and J. Wyngaard, 1988: *J. Atmos. Sci.*, **45**, 3573.

Schumann, U., 1975: *J. Comp. Phys.*, **18**, 376.

Smagorinsky, J., 1963: *Mon. Wea. Rev.*, **91**, 99.

Yakhot, V., and S. A. Orszag, 1986: *J. Sci. Comp.*, **1**, 3.

Chapter 8

SUBGRID CLOSURES IN LARGE EDDY SIMULATION: PART II

8.1 Leonard's Analysis

We now consider a line of analysis and SGS parameterization that has had a strong influence on the field in the last decade, largely due to a paper by Leonard (1974). Instead of the simple grid-box filtering that we have already mentioned, Leonard considered a more general filter of the form

$$\overline{u_i}(x,y,z) = \iiint G(\xi,\eta,\zeta)u_i(x+\xi,y+\eta,z+\zeta)d\xi d\eta d\zeta \quad , \tag{8.1a}$$

which has a Fourier transform

$$\hat{\bar{u}}_i(k) = \hat{G}(k)\hat{u}_i(k) \quad . \tag{8.1b}$$

This general filter [G] can be specified to incorporate a number of filters, including

$$\text{(a) grid–box or boxcar filter}: G = \frac{1}{\Delta^3} \text{ for } \frac{-\Delta}{2} < \left\{\begin{matrix}\xi\\ \eta\\ \zeta\end{matrix}\right\} < \frac{\Delta}{2} \, , \tag{8.2a}$$

$$G = 0 \text{ otherwise,}$$

$$\text{with } \hat{G} = \frac{\sin(k_x\Delta/2)\sin(k_y\Delta/2)\sin(k_z\Delta/2)}{\left(\frac{k_x\Delta}{2}\right)\left(\frac{k_y\Delta}{2}\right)\left(\frac{k_z\Delta}{2}\right)} \quad ; \tag{8.2b}$$

$$\text{(b) Gaussian filter}: G = \left[\frac{6}{\pi\Delta^2}\right]^{3/2} \exp\left[-6\left(\xi^2+\eta^2+\zeta^2\right)/\Delta^2\right] \, , \tag{8.3a}$$

$$\text{with } \hat{G} = \exp\left[-k^2\Delta^2/24\right] \; ; \tag{8.3b}$$

$$\text{(c) sharp spectral filter}: G = \frac{8\sin(\pi\xi/\Delta)\sin(\pi\eta/\Delta)\sin(\pi\zeta/\Delta)}{\pi^3\xi\eta\zeta} \, , \tag{8.4a}$$

$$\text{with } \hat{G} = \begin{cases} 1 \text{ for } |k_x|,\ |k_y|,\ |k_z| < \frac{\pi}{\Delta} \, , \\ 0 \text{ otherwise} \; . \end{cases} \tag{8.4b}$$

The Stanford research group (including Leonard) usually preferred the Gaussian filter (b). For finite difference equations, the choice between filters is (up to this point) a philosophical one, since data is defined only at grid points.

Leonard wrote the filtered equations of motion in a form somewhat different from that of (7.4), i.e.,

$$\frac{\partial \overline{u}_i}{\partial t} + \frac{\partial}{\partial x_j}\left(\overline{\overline{u}_i\overline{u}_j}\right) + \frac{\partial \overline{\pi}}{\partial x_i} - \overline{b}\delta_{i3} = \frac{\partial \tau_{ij}}{\partial x_j} , \tag{8.5a}$$

$$\text{where } \tau_{ij} = -\overline{u_i u_j} + \overline{\overline{u}_i\overline{u}_j} + \frac{\delta_{ij}}{3}\left(\overline{u_k^2} - \overline{\overline{u}_k^2}\right) , \tag{8.5b}$$

$$\text{and } \overline{\pi} = \frac{\overline{p}}{\rho_0} + \frac{\overline{u_i^2} - \overline{\overline{u}_i^2}}{3} . \tag{8.5c}$$

The difference in form arises from an expansion of the nonlinear term

$$\overline{u_i u_j} = \overline{\overline{u}_i\overline{u}_j} + \overline{u_i'\overline{u}_j} + \overline{\overline{u}_i u_j'} + \overline{u_i' u_j'} . \tag{8.6}$$

It is evident that the first term on the RHS is (at least approximately) evaluable from filtered data. Leonard grouped the divergence of this product term with the explicit terms on the LHS. Thus the stress, (8.5b), may be written

$$\tau_{ij} = -\overline{u_i'\overline{u}_j} - \overline{\overline{u}_i u_j'} - \overline{u_i' u_j'} + \frac{\delta_{ij}}{3}\left(\overline{2u_k'\overline{u}_k} + \overline{u_k'^2}\right) . \tag{8.7}$$

As before, Leonard used the Smagorinsky (1963) eddy viscosity assumption as a parameterization for τ_{ij} The modified pressure/density (8.5c) is expanded into

$$\overline{\pi} = \frac{\overline{p}}{\rho_0} + \frac{1}{3}\left(\overline{2u_i'\overline{u}_i} + \overline{u_i'u_i'}\right) . \tag{8.8}$$

As with (7.3b), this is not used directly, as the pressure is obtained from a Poisson equation.

As compared to the original SGS formulation, Leonard's version seems to involve more small-scale energy and covariance being included in the parameterization and less being explicitly calculated, since $\overline{\overline{u}_i\overline{u}_j}$ is a low-pass filtered version of $\overline{u}_i\overline{u}_j$. Leonard's rationale for this reorganization of the equations is based on an analysis of the resulting energy equation, using homogeneous turbulence theory. In the volume-averaged momentum equation derived earlier, (7.4), the advective term conserves energy over a closed volume, since $\overline{u}_i\partial(\overline{u}_i\overline{u}_j)/\partial x_j = (1/2)\partial(\overline{u}_i\overline{u}_i\overline{u}_j)/\partial x_j$. This conservation does not hold for Leonard's version, however. Using homogeneous and isentropic turbulence theory, Leonard predicted that the energy loss from the resolvable scale,

$$\epsilon_{RS} = \overline{u}_i \frac{\partial}{\partial x_j}\left(\overline{\overline{u}_i\overline{u}_j}\right) , \tag{8.9}$$

is approximately 0.3 to 0.4 of the expected total loss to SGS motions. If this is true, it effectively reduces the requirement on the SGS terms, making the choice of parameterization less critical. This has strongly influenced the field of SGS.

Since Leonard used finite difference equations and a Gaussian filter, he approximated the filtered velocity product by use of a Taylor series expansion as

$$\overline{\overline{u}_i\overline{u}_j} = \overline{u}_i\overline{u}_j + \left(\frac{\Delta^2}{24}\right)\nabla^2\left(\overline{u}_i\overline{u}_j\right) , \tag{8.10}$$

where ∇^2 is the second order finite difference Laplacian.

Most recent workers at Stanford and Queen Mary's College have used Leonard's filtering scheme, but with spectral discretization and pseudo-spectral integration replacing the original grids and finite differences. The initially specified u_i variables have been "pre–filtered" to $\overline{u}_i$ by use of the spectral filter, (8.3b). Then the Leonard term in the explicit equations involves another pass with the filter on $\overline{u}_i\overline{u}_j$. The τ's are again generally evaluated from the Smagorinsky assumption.

8.2. Subsequent Developments

More recently the Leonard SGS structure has been called into question, and in some cases abandoned. Antonopoulos-Domis (1981) found from analysis of simulation results that the Leonard terms do not produce dissipative removal of energy, but instead "backscatter" energy from the smallest resolved scales to larger scales, and are therefore harmful. Also, Speziale (1985) found that the Leonard terms do not satisfy Galilean invariance.

Comparisons have been made between LES and higher-resolution direct simulation to test the accuracy of SGS terms. Results from Bardina et al. (1983)

show very little correlation between τ_{ij} and "exact" SGS stresses, except that the τ_{ij}'s produce about the right dissipation. The correlation between the "exact" and parameterized τ_i improves greatly, however, if another term is added to the Smagorinsky formulation, i.e.,

$$\tau_{ij} \;=\; \text{Smagorinsky's term} \;+\; R_{ij} \;, \tag{8.11}$$

$$\text{where } R_{ij} \;=\; -\overline{u}_i\overline{u}_j + \overline{\overline{u}}_i\overline{\overline{u}}_j \;. \tag{8.12}$$

This is equivalent to replacing $\overline{\overline{u}_i\overline{u}_j}$ on the LHS of (8.5a) by $\overline{\overline{u}_i\overline{u}_j} + \overline{u}_i\overline{u}_j - \overline{\overline{u}}_i\overline{\overline{u}}_j$, which seemingly involves adding back most of the variance removed by the Leonard assumption. By this reasoning, the improved correlation occurs simply because part of the resolved stress is added into the SGS terms. Addition of the Bardina terms also restores the Galilean invariance.

The most recent tests of SGS formulations (Piomelli, et al., 1987) show that the best results are obtained from either a Fourier cutoff filter with the original SGS τ_{ij}'s, or a Gaussian filter with the Bardina terms re-added (Leonard terms plus "correction"). These results, together with those of Moeng and Wyngaard (1988), suggest an almost circular evolution of the SGS problem over the last 15 years.

8.3. Future Work

Further testing of existing closures is in order. Many meteorological models use what is sometimes called a $1\frac{1}{2}$'th order closure, in which SGS energy is predicted with a conservation equation, but stresses and fluxes are assumed proportional to gradients, i.e.,

$$\frac{\partial E}{\partial t} + \frac{u_i\partial u_j}{\partial x_i} \;=\; \tau_{ij}\frac{\partial \overline{u}_i}{\partial x_j} - H_3 - \frac{C_\epsilon E^{3/2}}{\Delta} + \frac{\partial}{\partial x_i}\left(K_e\frac{\partial E}{\partial x_i}\right) \;, \tag{8.13a}$$

$$\tau_{ij} \;=\; 2K_m S_{ij} \;,\; H_i \;=\; K_h\frac{\partial \overline{b}}{\partial x_i} \;,\; K_m \;=\; C_\tau E^{1/2}\Delta \;, \tag{8.13b}$$

with C_ϵ, C_τ, K_e/K_m, and K_h/K_m additional constants to be prescribed. This formulation differs from a first order closure only by the inclusion of time derivative, advective, and diffusive terms in the energy equation. Experiments indicate it often gives better results, however. Deardorff (1972) used a complete second order closure for a planetary boundary layer simulation with equations for all the stresses and buoyancy fluxes. This frontal assault on the closure problem produced better performance at the capping inversion, but suffered from complexity and realization problems. Neither of these closures has been tested against direct simulation. Most recently, however (since the summer lectures), Moeng and Wyngaard (1988) compared results of a second order closure model using a Fourier cutoff filter against those using the Leonard formulations, and found the former produced much superior inertial range spectra.

Antonopoulos-Domis, M., 1981: *J. Fluid Mech.*, **104**, 55.

Bardina, J., J. Fertziger, and W. C. Reynolds, 1983: *Improved Turbulence Models*, Stanford Report #TF-19, Thermosciences Div., M.E. Dept., Stanford University.

Deardorff, J., 1972: *J. Atmos. Sci.*, **29**, 91.

Leonard, A., 1974: *Adv. Geophys.*, **18A**, 237.

Moeng, C.-H., 1986: *J. Atmos. Sci.*, **43**, 2886.

Moeng, C.-H. and J. Wyngaard, 1988: *J. Atmos. Sci.*, **45**, 3573.

Piomelli, U., P. Moin, and J. H. Fertziger, 1987: Model consistency in the LES of turbulent channel flows, *AIAA*, 87-1446.

Smagorinsky, J., 1963: *Mon. Wea. Rev.*, **91**, 99.

Speziale, C. G., 1985: *J. Fluid Mech.*, **156**, 55.

LECTURES ON TURBULENCE AND LATTICE GAS HYDRODYNAMICS

Uriel Frisch
Centre National de la Research Scientifique
Observatoire de Nice
B.P. 139, 06003 Nice Cedex, France

Presented at the Summer School on Turbulence,
National Center for Atmospheric Research
Boulder, Colorado, June 1987

This set of lectures covers the following topics: fully developed turbulence (#1-2), turbulent transport (#3), large-scale instabilities in flows lacking parity-invariance (#4), lattice gas hydrodynamics (#5-10). Introductions to the topics may be found at the beginning of Chapters 1, 3, and 5 and at the beginning of two reprints of published papers attached at the end of the lecture notes by Frisch.

Chapter 1

CURRENT BELIEFS ABOUT FULLY DEVELOPED TURBULENCE: PART I

This and the next chapter are based on material published in French [1] and in English translation [2]. To avoid duplication of existing material, the present notes and references remain sketchy. Phenomenological aspects of fully developed turbulence are not covered (see the lectures by H. Tennekes).

Incompressible three-dimensional flow is described by the Navier-Stokes (N-S) equations

$$\begin{aligned} \partial_t \mathbf{v} + \mathbf{v}\cdot\nabla\mathbf{v} &= -\nabla p + \nu\nabla^2\mathbf{v} + \boldsymbol{f}\,, \\ \nabla\cdot\mathbf{v} &= 0, \end{aligned} \tag{1.1}$$

plus initial and boundary conditions. The Reynolds number, R, is defined as the ratio of the product of the typical length scale, L, times the typical velocity, V, divided by the kinematic viscosity, ν:

$$R = \frac{LV}{\nu}. \tag{1.2}$$

Fully developed turbulence is defined as the behavior obtained in the limit $R \to \infty$. For example, in astrophysical turbulence a typical value is $R = 10^{12}$. As R tends to infinity the number of effectively excited degrees of freedom in a typical volume scales approximately as $R^{9/4}$.

There are situations, such as statistical mechanics of many interacting particles, where the large numbers of degrees of freedom can be used to greatly simplify the problem. For fully developed turbulence, the hope has been for a long time that a contracted and universal description will be found. By universal, one understands a description which is independent of the details of boundary conditions and initial conditions, except for scaling factors.

Nothing of this sort has yet been derived systematically from the N-S equations. There is, however, experimental evidence for universal laws. For example, the experiment on tidal channel flow with $R = 10^8$ by Grant et al. [3] clearly shows the Kolmogorov -5/3 scaling law for the energy spectrum over three decades (Fig. 1.1). Experimental data encompassing comparable inertial ranges while taking advantage of modern experimental and data processing techniques have recently been obtained at the Modane wind tunnel by Gagne and Hopfinger [4]. These data reveal a power-law somewhat steeper than predicted by the Kolmogorov theory (the exponent is about -1.72).

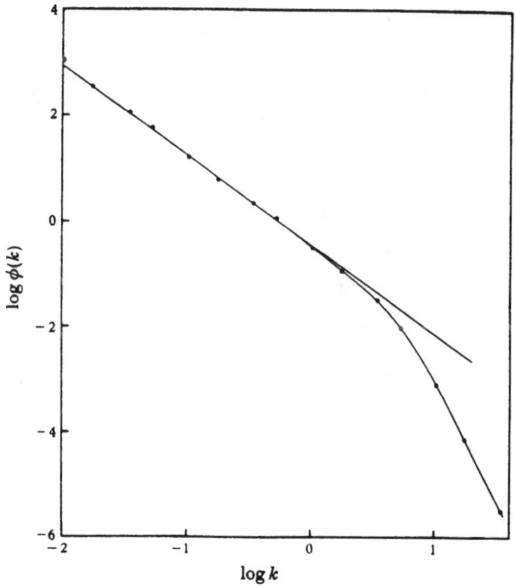

Figure 1.1: A logarithmic plot of one-dimensional energy spectrum. The straight line has a slope of $-5/3$ [3].

Understanding universality (or disproving it) without recourse to phenomenology or closure assumptions is a great challenge for theoreticians. As has been the case in high energy physics, the use of symmetries and conservation laws can shed some light on the issues. This was at the core of Kolmogorov's derivation of the -5/3 law, [5] much more than was dimensional analysis, in the present author's view. We shall now present Kolmogorov's argument using such language.

In the absence of forces or boundaries, the N-S equations are invariant under the (non-exhaustive) list of transformations: space and time translations, rotations, planar symmetries, Galilean transformations, and scaling transformations (for $\nu = 0$ or $\nu \to 0$). The scaling transformations can be parameterized as

$$r \to \lambda r, \quad \mathrm{v} \to \lambda^h \mathrm{v}, \quad t \to \lambda^{1-h} t, \tag{1.3}$$

where λ is a positive real number, and h is to be determined. The 1941 argument of Kolmogorov [K-41] can be expressed by the following equivalence principle: *In the limit $R \to \infty$, all invariance properties of the Navier-Stokes equations, possibly broken at large scales by mechanisms generating turbulence, are recovered at small scales in a statistical sense.*

In other words, the flow tries to have maximum symmetry at small scales. The exponent h is determined by the constraint of a constant rate of energy flow from large scales to small scales. Assuming that the flow of energy through a wavenumber k is independent of k, so that it equals the total rate of energy injected at high wavenumbers, fixes h to the value 1/3. This argument is independent of the details of the energy transfer, whether it is done in small steps or large steps. On dimensional grounds, the energy flux is given by $vvv\nabla$, which behaves like v^3/l; since this scales as l^{3h-1} and must be independent of l, $h = 1/3$. There is no doubt that the N-S equations admit such scale-invariant solutions, at least formally. It is well known, however, that solutions of nonlinear systems may or may not exhibit the full symmetry of the equations. The symmetry may be *broken* either externally or spontaneously.

An example is given by Burgers' equation [6]

$$\partial_t u + u\partial_x u = \nu\partial_{xx} u \ , \tag{1.4}$$

which has all the symmetries of the N-S equations, but is integrable. The transformation

$$u = -2\nu\partial_x \log\theta \tag{1.5}$$

changes Burgers' equation into the heat equation. As is well known, for $\nu \to 0$ the solutions of Burgers' equation develop shocks which break the scale-invariance. Burgers' equation is very useful for "spoiling" the game: most turbulence models which are applied to it do not work! If you have a new idea for fully developed turbulence, try it out on Burgers' equation first.

The definition of self-similarity used in the Kolmogorov theory implies that when the flow is viewed with a magnifying glass, placed *anywhere*, it always exhibits the same structure. This property is not shared by a shock-like solution where all the singularities are concentrated in isolated points. More generally, this definition of self-similarity rules out singularities concentrated on objects which are not space-filling, such as fractals. The latter are only conditionally self-similar. To take a famous example borrowed from Mandelbrot [7], the fractal nature of the coast of Britain is not revealed if we put a magnifying glass at Canterbury! Kolmogorov's assumed self-similarity has strong implications for moments. For example, the nondimensionalization of the fourth order moment through division by the square of the second order moment should be scale-independent.

Singularities provide one possible mechanism for breaking self-similarity. It is interesting to examine their presence in high Reynolds number flow. An experiment on turbulent flow in a channel exhibits intermittency in the dissipation range (Fig. 1.2).

Figure 1.2: Signal recorded by an anemometer behind grid turbulence after high-pass filtering. Data obtained by Y. Gagne (Grenoble) [4].

The turbulent velocity signal was high-pass filtered, with a cutoff frequency in the dissipation range. If the cutoff frequency is too high the signal becomes swamped by the noise. Related intermittent bursts were noticed as early as 1949 by Batchelor and Townsend [8]. Where does such dissipation range intermittency come from? It has been found that these bursts reflect the fact that the solution has singularities at complex times [9]. To understand this we make a brief digression into some basic complex analyses.

When solutions of the N-S equation are extended to complex space or time, singularities are unavoidable. An understanding of intermittency can be obtained by examining the Fourier transform of a function with isolated singularities at complex time-locations $z_j = t_j + i\tau_j$, assumed to be of power-law type:

$$v(z) \sim a_j(z - z_j)^\rho, \quad z \to z_j, \tag{1.6}$$

where ρ is anything but an integer.

The Fourier transform is given by

$$\hat{v}(w) = \int_R e^{i\omega t} v(t)dt, \tag{1.7}$$

and can be evaluated asymptotically as $\omega \to \infty$ by Laplace's method (Fig. 1.3) as

$$\hat{v}(w) \sim \frac{-2\sin(\pi\rho)\Gamma(\rho + 1)}{\omega^{\rho+1}} e^{-i\pi\rho/2} \sum_j \epsilon_j a_j e^{i\omega z_j}. \tag{1.8}$$

The sum is over singularities with positive imaginary parts. The ϵ_j's are determination factors.

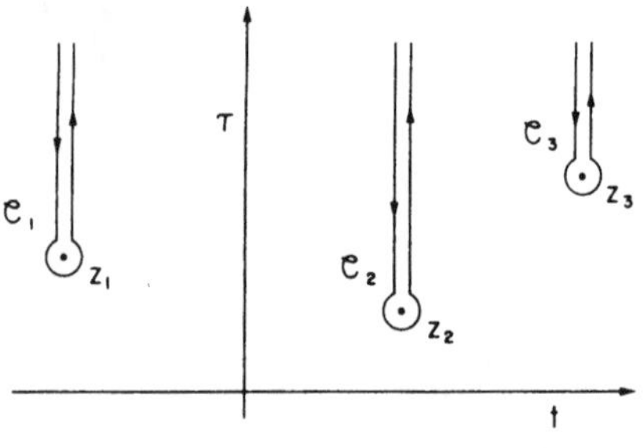

Figure 1.3: Integration contour around singularities in the complex plane [9].

The high-pass-filtered signal with cutoff $\Omega \to \infty$ is given by

$$\begin{aligned} v_\Omega(t) &= \frac{1}{2\pi} \int_{|\omega|>\Omega} e^{-i\omega t} \hat{v}(\omega) d\omega \\ &\sim \frac{\Gamma(\rho + 1)}{\Omega^{\rho+1}} \sum_j \epsilon_j a_j e^{-\Omega\tau_j} \mathrm{Re}\left[\frac{exp(-i\Omega(t - t_j) - i\pi\rho/2)}{t - z_j}\right]. \end{aligned} \tag{1.9}$$

This shows that $v(t)$ has bursts centered at t_j with amplitudes proportional to $\exp(-\Omega\tau_j)$. Since Ω is large, the asymptotic solution picks out singularities closest to the real axis. It turns out that any nonlinear dynamical system generically has singularities in the complex plane. This is not surprising since physical constraints are only strong in the real domain. For example, the energy, as a sum of squares, gives bounded terms on the real axis, but not so in the complex plane where blow-up can occur. Generically, complex singularities translate to bursts. The probe in Gagne's [4] experiment recorded a real signal displaying bursts after high-pass filtering, evidence for singularities in the complex domain.

Nonlinear systems, especially nonintegrable ones, have actually a more complicated singularity structure. Fractal clustering of complex time singularities, rather than isolated singularities (as above), is the rule. Known examples include the Hénon-Heiles [10] system, the time-independent Kuramoto-Sivashinski [11,12] equation, and the Arnold-Beltrami-Childress flow [13]. Typically, the solution near a complex singularity is given by an expansion of the form:

$$u \sim (z - z_*)^{-\alpha} F(z - z_*,\ g_1(z - z_*), g_2(z - z_*), \ldots), \tag{1.10}$$

where z_* is the location of the complex time singularity, $F(u, v, w, \ldots)$ is an analytic function of its arguments, and the g_i are nonanalytic functions of the form:

$$g_i \sim \begin{cases} (z - z_*)^{\rho_i}, & \rho_i \ complex \\ (z - z_*)\log(z - z_*). \end{cases} \tag{1.11}$$

Complex powers imply that in the neighborhood of a singularity, there are many others; the whole set of singularities generally forms a fractal set.

It is still not known if the singularities of the Navier-Stokes equations do form such a fractal set in the complex domain. Neither is it known if, as $R \to \infty$, some complex singularities migrate to the real domain. Scaling relations observed in experiments on fully developed turbulence can be interpreted by assuming this to be the case. Let us assume that this happens and that the singularities reside on a subset S of R^3. S is characterized by its Hausdorff dimension d. The singularities are characterized by a scaling exponent for the velocity:

$$|v(r + l) - v(r)| \sim l^h, \text{ for } r \in S \text{ and } l \to 0. \tag{1.12}$$

A consequence of this for the structure functions (moments of velocity increments) is:

$$\langle[\mathrm{v}(r+l)-\mathrm{v}(r)]^p\rangle \sim l^{ph} l^{3-d} \equiv l^{\zeta_p}. \tag{1.13}$$

The l^{ph} comes from the singularity scaling, and the l^{3-d} is the probability that a ball of radius l will encounter the set S. In 1941, without using his celebrated self-similarity assumption, Kolmogorov [5] derived directly from the Navier-Stokes equations that the exponent of the third-order structure function should be 1. Thus $\zeta_3 = 3h + 3 - d = 1$, so that d and h are not independent. If the velocity field is self-similar then $d = 3$, the factor l^{3-d} is unity, and we obtain the standard [5], [K-41] result $h = 1/3$.

One way to to test this fractal model is to measure the exponents ζ_p and compare with the predicted linear-plus-constant dependence on p. This is difficult because in order to measure high-order moments one needs averages over times that grow exponentially with p. In addition, real experiments sample at discrete times, while high-order structure functions are dominated by rare, sudden events which may be missed. Figure 1.4 [14] shows data from several experiments of ζ_p vs. p. It is seen that for low p the data agree with the Kolmogorov scaling, but at high p it may not even be linear.

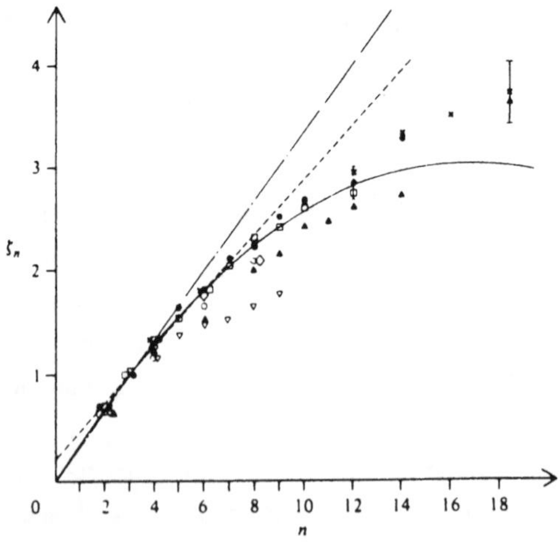

Figure 1.4: Exponent ζ_n of structure function of order n. (n is denoted p in the present text) [14].

The *multi-fractal* model of Parisi and Frisch [15] provides a possible framework for explaining such data. Exponents falling on a concave curve can be obtained from a set of singularities composed of nested, increasingly singular fractal sets, each with its own dimension and singularity exponent. The evaluation of the structure functions is done as above, but the contributions from the various singular sets must be added with an unknown (and irrelevant) weighting $d\mu(h)$:

$$\langle [\mathrm{v}(r+l) - \mathrm{v}(r)]^p \rangle \sim \int d\mu(h) l^{ph+3-d(h)}. \tag{1.14}$$

As $l \to 0$ the minimum exponent dominates, resulting in

$$\langle [\mathrm{v}(r+l) - \mathrm{v}(r)]^p \rangle \sim l^{\zeta_p}, \quad \zeta_p = \min_p (ph + 3 - d(h)). \tag{1.15}$$

Thus, the exponent ζ_p is the Legendre transform of the co-dimension $3 - d(h)$.

[1] Frisch, U., *Phys. Ser.*, **9**, 131 (1985).

[2] Frisch, U., in *Dynamical Systems; A Renewal of Mechanism*, ed. S. Diner, p. 13 (World Publishing Co.) (1986).

[3] Grant, H. L., Stewart, R. W., Moilliet, A. *J. Fluid Mech.*, **12**, 241 (1962).

[4] Gagne, Y., Thèse de Doctorat, Institut de Mécanique de Grenoble (1988).

[5] Kolmogorov, A. N., *C. R. Acad. Sci.*, USSR, **30**, 301 (1941).

[6] Burgers, J. M., *The Nonlinear Diffusion Equation* (Reidel) (1974).

[7] Mandelbrot, B. B., *Fractal Geometry of Nature* (W. H. Freeman), Rev. ed. (1982).

[8] Batchelor, G., and Townsend, A. A., Proc. Roy. Soc., **A199**, 238 (1949).

[9] Frisch, U. and Morf, R., *Phys. Rev.*, **A23**, 2673 (1981).

[10] Hénon, M., and Heiles, C., *Astron. J.*, **69**, 73 (1964).

[11] Kuramoto, Y., *Prog. Theoret. Phys. Suppl.*, **64**, 346 (1978).

[12] Sivashinski, G. I., *Acta Astronaut.* **4**, 1177 (1977).

[13] Dombre, T., Frisch, U., Greene, J. M., Hénon, M., Mehr, A., and Soward, A. M., *J. Fluid Mech.*, **167**, 353 (1986).

[14] Anselmet, F., Gagne, Y., Hopfinger, E. J., and Antonia, R. A., *J. Fluid Mech.*, **140**, 63 (1984).

[15] Parisi, G. and Frisch, U. in *Turbulence and Predictability in Geophysical Fluid Dynamics and Climate Dynamics*, School Enrico Fermi, Course 88, p. 1983 (1985).

Chapter 2

CURRENT BELIEFS ABOUT FULLY DEVELOPED TURBULENCE: PART II

In the previous chapter it was shown that singularities can break the scale-invariance, thereby providing a mechanism for intermittency corrections to the K-41 theory. In 2D, well-known theorems for ideal and viscous fluids rule out the development of real singularities in finite time from a smooth initial field. In 3D the strongest known mathematical result is that smooth fields remain smooth for at least a short time.

A simple example shows that naive phenomenological arguments can be misleading. Consider the Burgers equation with zero viscosity:

$$\partial_t u + u \partial_x u = 0 \,. \tag{2.1}$$

Under the transformation $w = -\partial_x u$ it may be rewritten in Lagrangian terms:

$$D_t w = w^2, \tag{2.2}$$

to yield:

$$w(t) = \frac{w(0)}{1 - t w(0)} \,. \tag{2.3}$$

This shows that a singularity will usually develop in finite time. It is tempting to use a similar argument for the 3D Euler equation in which w is replaced by the vorticity ω. It is a standard result that the vorticity equation can be written in the form:

$$\frac{D\omega}{Dt} = \omega \cdot \nabla \boldsymbol{u}, \tag{2.4}$$

and phenomenologically one *might* estimate the right-hand side as the square of the vorticity. With that assumption one obtains

$$\frac{D\omega}{Dt} \approx \omega^2, \tag{2.5}$$

and therefore

$$\omega(t) \approx \frac{\omega(0)}{1 - t\omega(0)}, \tag{2.6}$$

again leading to a singularity in finite time (at this level of argument the distinction between scalars and vectors is blurred). However, evidence is now presented that this need not occur.

The Taylor-Green vortex will provide such an example. The initial velocity distribution is as follows:

$$\begin{aligned} v_1 &= \sin x_1 \cos x_2 \cos x_3 \ , \\ v_2 &= \cos x_1 \sin x_2 \cos x_3 \ , \\ v_3 &= 0. \end{aligned} \tag{2.7}$$

It consists of layers of circular eddies, with those at $x_3 = \pi$ counter-rotating relative to the ones at $x_3 = 0$. For $t > 0$ the flow develops a nonvanishing v_3 component. The evolution of this field under the Euler equations ($\nu = 0$) was solved numerically by Brachet et al. [1] using a pseudo-spectral method with a resolution of 256^3 grid points.

By arguments similar to the ones presented in the previous chapter, it can be shown that in the presence of singularities in the complex space $\boldsymbol{C}^3$, the energy spectrum defined as

$$E(t, K) = \sum_{K-1 \leq |\boldsymbol{k}| < K} |\hat{\mathbf{v}}(t, k)|^2 , \tag{2.8}$$

has the asymptotic behaviour

$$E(t, K) \propto e^{-2\delta(t)K}, \quad K \to \infty, \tag{2.9}$$

to within an algebraic prefactor, determined by the nature of the singularities closest to the real domain $\boldsymbol{R}^3$. Thus, $\delta(t)$ is called the width of the analyticity band. Such exponential behaviour is indeed seen in the data of Brachet et al. (Fig. 2.1) [1].

At each time the value of δ can be estimated from the slopes in Fig. 2.1, using the asymptotic form given in (2.9). The results are plotted in Fig. 2.2.

By extrapolation the graph suggests that the singularity will take an infinitely long time to reach the real axis. This shows that the analogy with Burgers' equation fails in this case. In practice the truncation errors invalidate the solution when $\delta \leq 2\pi/K_{max}$, where K_{max} is the maximum wavenumber, but there are sufficient "good" points to make credible decreasing exponential behavior in $\delta(t)$. The simulation of Brachet et al. [1] suggests a physical reason for the

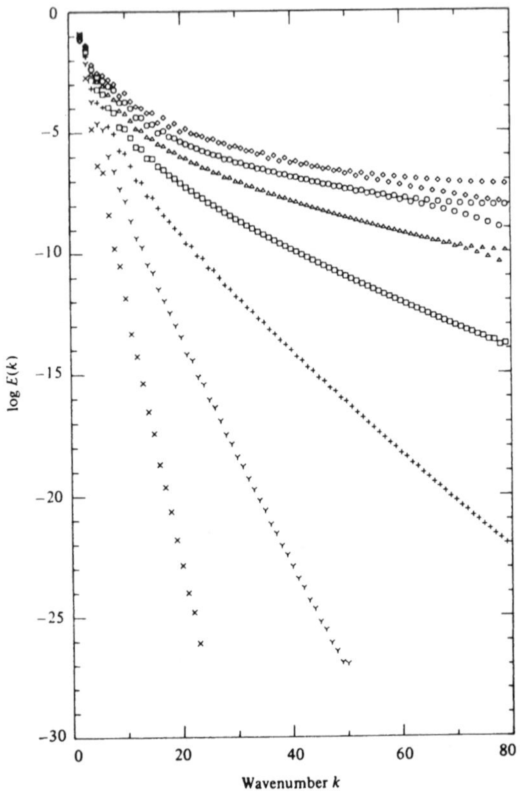

Figure 2.1: The inviscid spectrum $E(k,t)$ for the Taylor-Green vortex. Lin-log scales make exponentials appear as straight lines. The different symbols distinguish the spectra at equally spaced times from crosses at $t = 0.5$ to diamonds at $t = 3.5$ in steps of 0.5 [1].

failure of the analogy, namely, the formation of pancakes with strongly reduced nonlinearities; this may be a generic property of 3D flows. Figure 2.3 illustrates this by showing successive stages of the evolution of the inviscid Taylor-Green vortex, at $t = 0$ and $t = 2$.

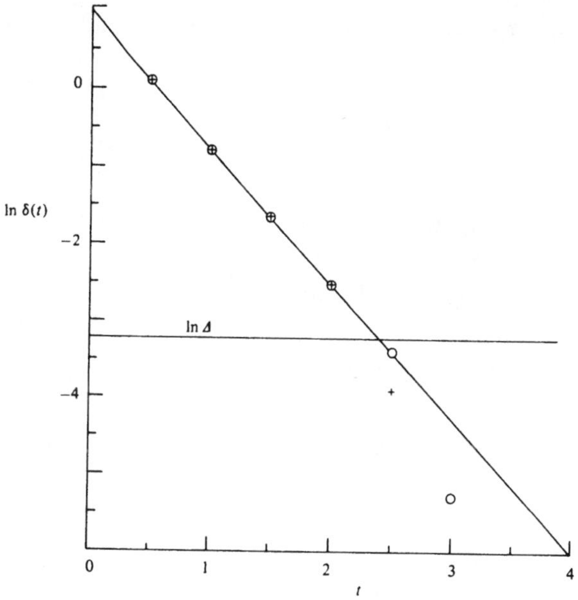

Figure 2.2: The time dependence of the width of the analyticity strip δ for the inviscid Taylor-Green vortex [1].

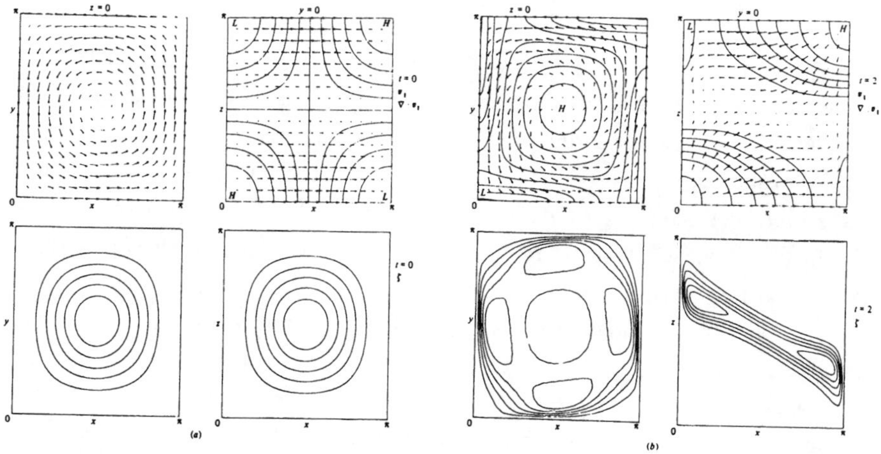

Figure 2.3: Various features for the inviscid Taylor-Green flow, at (a) $t = 0$, (a) and (b) $t = 2.0$ [1].

Another example is provided by the 2D MHD problem studied by Frisch et al. [2]. The equations are (ω and $\boldsymbol{j}$ are the curls of $\mathbf{v}$ and $\boldsymbol{b}$, respectively)

$$\begin{aligned}\partial_t\omega + \mathbf{v}\cdot\nabla\omega &= \boldsymbol{b}\cdot\nabla\boldsymbol{j}\ ,\\ \partial_t\boldsymbol{b} + \mathbf{v}\cdot\nabla\boldsymbol{b} &= \boldsymbol{b}\cdot\nabla\mathbf{v}\ ,\\ \nabla\cdot\mathbf{v} = 0, \quad \nabla\cdot\boldsymbol{b} &= 0.\end{aligned} \tag{2.10}$$

Note that despite the two-dimensionality, vorticity is not conserved, since it can be generated by the Lorentz force. Phenomenological arguments and even quite sophisticated closures predict finite-time singularities (from smooth, initial conditions). However, there is numerical evidence for the formation of quasi-one-dimensional sheets (Fig. 2.4).

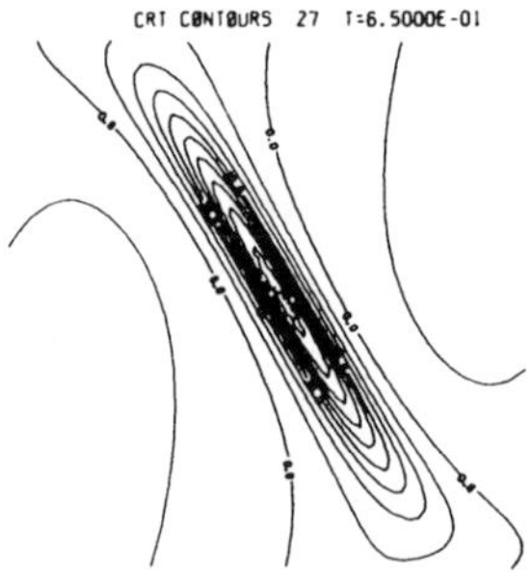

Figure 2.4: Current contours for two-dimensional ideal MHD flow [2].

To summarize: there is overwhelming evidence for complex singularities coming arbitrarily close to the real axis in the inviscid (Euler) case. It is widely believed that viscosity prevents any real singularity developing in the Navier-Stokes equations in finite time, but it is not known in general whether finite-time, real singularities develop in inviscid flow. However, in the Taylor-Green problem, straightforward extrapolation suggests that such singularities may be absent.

The presence of singularities close to the real axis would raise questions about the validity of hydrodynamics, i.e., the continuum approximation. Leray [3] suggested this as a source of randomness in turbulence. To derive hydrodynamics, one assumes the molecular mean free path λ to be much smaller than the smallest macroscopic-flow scale δ. Physically, following the ideas of K-41, we identify δ with the dissipation scale and write

$$L_0/\delta \sim R^{3/4} , \tag{2.11}$$

where L_0 is the integral ("large") scale for the turbulence. We evaluate the ratio λ/δ as follows:

$$\frac{\lambda}{\delta} = \frac{\lambda}{L_0}\frac{L_0}{\delta} \sim \frac{\lambda}{L_0}R^{3/4} \sim \frac{\lambda}{L_0}\left(\frac{L_0}{\lambda}\frac{V_0}{V_{th}}\right)^{3/4} \sim \left(\frac{\lambda}{L_0}\right)^{1/4}\left(\frac{V_0}{V_{th}}\right)^{3/4} . \tag{2.12}$$

It is here assumed that the kinematic viscosity ν can be estimated as $\nu \sim V_{th} \cdot \lambda$, where V_{th} is the thermal velocity of molecules. Note that V_0/V_{th} is approximately the Mach number, which must be small for incompressibility to hold, and therefore, unless K-41 is very strongly violated, the hydrodynamic approximation appears valid even in the limit of high Reynolds number.

It is now commonly believed, following the work of Lorentz and others, that the source of randomness in turbulence is the intrinsic stochasticity of the Navier-Stokes equations in certain regimes, i.e., "deterministic chaos." The flow-field is thought to evolve toward some attractor, with sensitive dependence on initial conditions. Large-time behavior would then appear random. One might hope that statistical properties of the system would still be predictable. However, usually in such dynamical systems there exists *more than one attractor*, and the geometry of the basins of attraction may be so complicated as to make even the statistical properties sensitively dependent on initial conditions. The problem of determining the ultimate attractor for each initial point with, say, rational coordinates, may not even be amenable to any finite algorithm.

We cannot rule out, indeed, that some of the central questions of turbulence (as they used to be formulated) are undecidable, in the sense of Gödel and Turing. If this is true, it just indicates that we are on the wrong path by insisting on a purely deterministic description of turbulence. Random perturbations, from external action or from thermal fluctuations, are not only hard to avoid in real flow; they may be essential to avoid the pitfall of "byzantine complexity" (undecidability).

[1] Brachet, A., Meiron, S., Orszag, S., Nickel, R., Morf, R., and Frisch, U., *J. Fluid Mech.*, **130**, 411 (1983).

[2] Frisch, U., Pouquet, A., Sulem, P.-L., and Meneguzzi, M., *J. Méc. Théor. Appliqu.*, special issue on "Two-dimensional turbulence," p. 191 (1983).

[3] Leray, J., *Acta Math*, **63**, 193 (1934).

Chapter 3

TURBULENT TRANSPORT OF SCALARS AND MAGNETIC FIELDS

The purpose of this chapter is to demonstrate the technique of multiple scale analysis as applied to advection of passive scalars (e.g., temperature fields) and vectors (e.g., magnetic fields); both in turn demonstrate the effect of eddy features on large-scale flow. These techniques are useful for deriving eddy diffusivities and other transport coefficients related to more exotic effects, such as the α-effect in MHD and the AKA (Anisotropic Kinetic Alpha) effect to be discussed in the next chapter. The exposition for the scalar field follows closely a paper by Papanicolaou and Pironneau [1]. The material on the transport of magnetic fields and the α-effect does not incorporate any background material or references. Such material may be found in the monographs by Moffatt [2] and by Zeldovich et al. [3].

1. Transport of Scalar Fields

We begin this study of "macrodynamics," the dynamics of large scales, by considering the diffusion of a scalar in an incompressible 2D or 3D flow. The flow domain consists of a space-time periodic array of cells whose periodicity equals ℓ in both x_1, x_2, and x_3. The (prescribed) flow velocity within the cells is denoted by $\boldsymbol{u}$ and satisfies the incompressibility condition $\nabla \cdot \boldsymbol{u} = 0$. Furthermore, the spatial-temporal average of the flow over a cell, $\langle \boldsymbol{u} \rangle$, is assumed to vanish and the characteristic value of $\boldsymbol{u}$ is denoted V.

Observing the domain at macroscopic scales, we address the question of whether or not a superimposed passively advected scalar contaminant of scale size $L \gg \ell$ will undergo diffusion. This situation is analogous to the *turbulent* transport of a passive scalar, except that a space-time periodicity, rather than random homogeneity and stationarity, have been assumed for simplicity. In the latter random case the formalism is essentially unchanged, but there are some mathematical fine points [1]. We note that when the eddy field $\boldsymbol{u}$ is time-independent and the molecular diffusivity vanishes, the scalar may not diffuse but simply "swirl around" with the flow within each periodic cell. This pathology disappears as soon as a finite molecular diffusivity is introduced.

The governing equation for the concentration Θ of the advected scalar is

$$\partial_t \Theta + \boldsymbol{u} \cdot \nabla_x \Theta = \kappa \nabla_x^2 \Theta. \tag{3.1}$$

We are interested in the behavior on scales L such that $\ell/L = \epsilon \ll 1$. Simple physical reasoning suggests that there may be an eddy diffusivity (isotropic or anisotropic) of order ℓV which is independent of ϵ. Then, the characteristic time scale associated with the large scale should be of order $L^2/\ell V = (L/\ell)^2 (\ell/V) = \epsilon^{-2} t_{\text{eddy}}$. The simplest way to deal with this situation is to use a multiscale expansion in which we keep both "fast" and "slow" variables. The fast variables are x and t while the slow variables are $X = \epsilon x$ and $T = \epsilon^2 t$. By its definition, the velocity $\boldsymbol{u}$ depends only on the fast variables and Θ depends on both fast and slow variables. Accordingly, Θ is expanded in powers of ϵ

$$\Theta(x,X,t,T) = \Theta^{(0)} + \epsilon\Theta^{(1)} + \epsilon^2\Theta^{(2)} + \cdots, \tag{3.2}$$

where all the $\Theta^{(n)}$ depend on x, X, t, and T. It is convenient to decompose $\Theta^{(0)}$ as $\langle\Theta^{(0)}\rangle + \widetilde{\Theta}^{(0)}$, where $\langle\Theta^{(0)}\rangle$ is the space-time average over the cell, depending only on X and T, and $\widetilde{\Theta}^{(0)}$ are the fluctuations. The standard multiscale technique pretends that x, X, t, and T are all independent variables to accommodate the substitution of $\Theta\,(x,X,t,T)$ for $\Theta\,(x,t)$. So, we employ the chain rule to obtain:

$$\nabla_x \leftarrow \nabla_x + \epsilon \nabla_X\,, \tag{3.3a}$$

$$\partial_t \leftarrow \partial_t + \epsilon^2 \partial_T. \tag{3.3b}$$

Using (3.3) and substituting into (3.1) yields

$$\left(\partial_t + \epsilon^2 \partial_T\right)\Theta + \boldsymbol{u}\cdot(\nabla_x + \epsilon\nabla_X)\,\Theta = \kappa\,(\nabla_x + \epsilon\nabla_X)^2\,\Theta. \tag{3.4}$$

Inserting the expansion (3.2) into the Partial Differential Equation (PDE) (3.4) and identifying the powers of ϵ gives a hierarchy of equations, the first of which is

$$O(\epsilon^0): \quad \partial_t\widetilde{\Theta}^{(0)} + \boldsymbol{u}\cdot\nabla_x\widetilde{\Theta}^{(0)} = \kappa\nabla_x^2\widetilde{\Theta}^{(0)}. \tag{3.5a}$$

The solution $\widetilde{\Theta}^{(0)}$ is easily shown to go to zero on a fast time scale (see Chapter 3 Appendix). Therefore we lose no generality by equating it to zero. The next two equations in the hierarchy are

$$O(\epsilon): \quad \underline{\partial_t\Theta^{(1)}} + \underline{\boldsymbol{u}\cdot\nabla_x\Theta^{(1)}} = \underline{-\boldsymbol{u}(x,t)\cdot\nabla_X\langle\Theta^{(0)}\rangle} + \underline{\kappa\nabla_x^2\Theta^{(1)}}\,, \tag{3.5b}$$

$$O(\epsilon^2): \quad \underline{\partial_t \Theta^{(2)}} + \underline{\boldsymbol{u} \cdot \nabla_x \Theta^{(2)}} =$$

$$-\partial_T \langle \Theta^{(0)} \rangle - \boldsymbol{u}(x,t) \cdot \nabla_X \Theta^{(1)} + \underline{\kappa \nabla_x^2 \Theta^{(2)}} + \kappa \nabla_X^2 \langle \Theta^{(0)} \rangle + \underline{2\kappa \nabla_X \cdot \nabla_x \Theta^{(1)}}. \quad (3.5c)$$

At first glance this hierarchy seems to imply a closure problem since in the above *two* equations we have *three* unknowns. Actually, (3.5b, c) will yield a closed equation for $\langle \Theta^{(0)} \rangle$ via a solvability condition. In many perturbation expansions one encounters Fredholm alternatives, that is, equations of the form $Af = g$ where the linear operator A is non-invertible, so that g has to be orthogonal to the null-space of the adjoint of A. Here, in spite of having partial differential operators, the structure is quite straightforward: we notice that all the underlined terms in (3.5b, c) have vanishing space-time averages since they are, or might be put in the form of, exact derivatives with respect to one of the fast variables. Thus the sum of *all* underlined terms must vanish. This plays the role of the usual solvability condition. All the terms of equation (3.5b) have vanishing averages and therefore do not provide a solvability condition. The solvability condition is furnished by (3.5c) and reads (summation of repeated indices henceforth understood)

$$\partial_T \langle \Theta^{(0)} \rangle + \frac{\partial}{\partial X_i} \langle u_i \Theta^{(1)} \rangle = \kappa \nabla_X^2 \langle \Theta^{(0)} \rangle. \quad (3.6)$$

The equation for the mean field is closed once $\Theta^{(1)}$ is determined from (3.5b). The solution to (3.5b) is obtained by noting that it is a PDE in fast variables x and t and that the quantity $\boldsymbol{G} = \nabla_X \langle \Theta^{(0)} \rangle$ is independent of x and t, and thus i is a constant as far as integration of the equation is concerned. Furthermore, since (3.5b) is linear in $\Theta^{(1)}$, the solution for $\Theta^{(1)}$ can be expressed as

$$\Theta^{(1)} = \boldsymbol{\chi}(x,t) \cdot \boldsymbol{G}, \quad (3.7a)$$

where the vector $\boldsymbol{\chi}$ satisfies the auxiliary equation

$$\partial_t \boldsymbol{\chi} + \boldsymbol{u} \cdot \nabla_x \boldsymbol{\chi} = -\boldsymbol{u} + \kappa \nabla_x^2 \boldsymbol{\chi}. \quad (3.7b)$$

Substituting (3.7a) into (3.6) produces a closed evolution equation for the mean field:

$$\partial_T \langle \Theta^{(0)} \rangle = \left(\kappa \nabla_X^2 + D_{ij} \frac{\partial^2}{\partial X_i \partial X_j} \right) \langle \Theta^{(0)} \rangle, \quad (3.8a)$$

with an eddy diffusivity tensor

$$D_{ij} = -\frac{1}{2}\left[\langle u_i\chi_j\rangle + \langle u_j\chi_i\rangle\right]. \qquad (3.8b)$$

We note that (3.8) represents a diffusion equation for the mean field $\langle\Theta^{(0)}\rangle$ whose diffusivity tensor may be anisotropic. This of course reflects the trivial fact that the flow is usually not invariant under arbitrary rotations. It is enough for the cellular flow above to be invariant under permutations of the coordinates to insure isotropy of second-order tensors and thereby the isotropy of the eddy diffusivity.

The diffusivity tensor is shown to be a positive definite by multiplying each component i of Equation (3.7b) by χ_j, averaging over a cell, adding the expression obtained by interchanging i and j, and integrating by parts. Thus, $D_{ij} = \kappa\langle\partial_l\chi_i\partial_l\chi_j\rangle$ is an obviously positive, definite expression. This implies that the passive scalar cannot undergo amplification (which would also violate the maximum principle).

In the special case when the molecular diffusivity κ vanishes, (3.7b) may be integrated in "Lagrangian coordinates" to yield ($\boldsymbol{a}$ denotes the starting point and the superscript L refers to Lagrangian coordinates)

$$\chi^L(\boldsymbol{a},t) = -\int_0^t \boldsymbol{u}^L(\boldsymbol{a},t')\,dt' + \text{const}, \qquad (3.9)$$

so that D_{ij} becomes

$$D_{ij} = \frac{1}{2}\int_0^t \left[\langle u_i^L(\boldsymbol{a},t)u_j^L(\boldsymbol{a},t')\rangle + i \leftrightarrow j\right] dt'. \qquad (3.10)$$

For large times and when the integral converges, this is just G.I. Taylor's expression for the eddy diffusivity as the time-integral of the Lagrangian velocity autocorrelation function.

Here a word of caution is required. In the absence of molecular diffusivity the large-scale dynamics need not be diffusive. Indeed, the study of the non-diffusive advection of a passive scalar is equivalent to the study of the Ordinary Differential Equation (ODE) for the Lagrangian trajectories of fluid particles

$$\frac{d\boldsymbol{x}}{dt} = \boldsymbol{u}(\boldsymbol{x},t). \qquad (3.11)$$

Since we assume $\nabla \cdot \boldsymbol{u} = 0$, (3.11) constitutes a *conservative dynamical system* (volume-preserving); such systems have been extensively studied and are known often to develop structures called Kolmogorov-Arnold-Moser (KAM) surfaces which prevent diffusion! We shall not dwell on such frontier questions [4–7]. With nonvanishing molecular diffusivity the problem mostly disappears. However, it will reappear if we ask about the behavior of the eddy diffusivity as $\kappa \to 0$.

The case of finite molecular diffusivity does not lead to such mathematical fine points: the existence of a finite eddy diffusivity is generally guaranteed, but its value cannot be found without solving the auxiliary equation (3.7b). This can be done numerically; since there is no ϵ left in the equation there is a single time scale, so that the problem is not "stiff." Alternatively, the equation can be solved in a low Peclet number (large κ) expansion or in a short correlation-time expansion when the field $\boldsymbol{u}$ is random. We shall not carry out such expansions here, and just mention that, to leading order, the expression for the eddy diffusivity does not involve the *helicity* of the flow, but higher order corrections do.

2. Transport of a Magnetic Field

We shall now demonstrate how the above method must be modified to study the evolution of a magnetic field embedded in the 3D periodic array described in section 1. The governing equations describing the evolution of a magnetic field in the MHD magnetic regime are (see D. Montgomery's lecture notes for explanation of the magnetic regime):

$$\partial_t \boldsymbol{b} + \boldsymbol{u} \cdot \nabla_x \boldsymbol{b} - \boldsymbol{b} \cdot \nabla_x \boldsymbol{u} = \lambda \nabla_x^2 \boldsymbol{b} \,, \tag{3.12a}$$

$$\nabla_x \cdot \boldsymbol{u} = 0, \quad \nabla_x \cdot \boldsymbol{b} = 0. \tag{3.12b}$$

We note that the term $\boldsymbol{b} \cdot \nabla \boldsymbol{u}$ has no analog in the scalar advection equation. It turns out that this new term can induce instabilities by stretching the magnetic field. If this happens, we say that we have a "dynamo."

If we anticipate an "eddy diffusivity" of order one, then the same steps presented in Section 1 can be applied to (3.12a) to produce

$$(\partial_t + \epsilon^2 \partial_T)\boldsymbol{b} + \boldsymbol{u} \cdot (\nabla_x + \epsilon \nabla_X)\boldsymbol{b} - \boldsymbol{b} \cdot (\nabla_x + \epsilon \nabla_X)\boldsymbol{u} = \lambda(\nabla_x + \epsilon \nabla_X)^2 \boldsymbol{b} \,. \tag{3.13a}$$

As before, we expand the magnetic field

$$\boldsymbol{b}(x,X,t,T) = \langle \boldsymbol{b}^{(0)} \rangle + \widetilde{\boldsymbol{b}}^{(0)} + \epsilon \boldsymbol{b}^{(1)} + \epsilon^2 \boldsymbol{b}^{(2)} \ldots , \tag{3.13b}$$

where $\langle \boldsymbol{b}^{(0)} \rangle$ only depends upon the slow variables and all the other $\boldsymbol{b}^{(n)}$'s depend on both fast and slow variables. Expanding, we obtain the following hierarchy:

$$O(\epsilon^0): \quad \underline{\partial_t \widetilde{\boldsymbol{b}}^{(0)}} + \underline{\boldsymbol{u}\cdot\nabla_x \widetilde{\boldsymbol{b}}^{(0)}} = \underline{\widetilde{\boldsymbol{b}}^{(0)}\cdot\nabla_x \boldsymbol{u}} + \underline{\langle \boldsymbol{b}^{(0)}\rangle\cdot\nabla_x \boldsymbol{u}} + \underline{\lambda\nabla_x^2 \widetilde{\boldsymbol{b}}^{(0)}}, \tag{3.14a}$$

$$\begin{aligned} O(\epsilon): \quad & \underline{\partial_t \boldsymbol{b}^{(1)}} + \underline{\boldsymbol{u}\cdot\nabla_x \boldsymbol{b}^{(1)}} + \underline{\boldsymbol{u}\cdot\nabla_X \langle \boldsymbol{b}^{(0)}\rangle} + \boldsymbol{u}\cdot\nabla_X \widetilde{\boldsymbol{b}}^{(0)} \\ & = \underline{\boldsymbol{b}^{(1)}\cdot\nabla_x \boldsymbol{u}} + \underline{\lambda\nabla_x^2 \boldsymbol{b}^{(1)}} + \underline{2\lambda\nabla_{xX}^2 \widetilde{\boldsymbol{b}}^{(0)}} , \end{aligned} \tag{3.14b}$$

$$\begin{aligned} O(\epsilon^2): \quad & \underline{\partial_t \boldsymbol{b}^{(2)}} + \partial_T \widetilde{\boldsymbol{b}}^{(0)} + \underline{\boldsymbol{u}\cdot\nabla_x \boldsymbol{b}^{(2)}} + \boldsymbol{u}\cdot\nabla_X \boldsymbol{b}^{(1)} \\ & = -\partial_T \langle \boldsymbol{b}^{(0)}\rangle + \boldsymbol{b}^{(2)}\cdot\nabla_x \boldsymbol{u} + \lambda\nabla_X^2 \langle \boldsymbol{b}^{(0)}\rangle + \underline{2\lambda\nabla_{xX}^2 \boldsymbol{b}^{(1)}} + \underline{\lambda\nabla_x^2 \boldsymbol{b}^{(2)}}. \end{aligned} \tag{3.14c}$$

In addition, we have the expanded forms of the solenoidality condition

$$\nabla_x \cdot \widetilde{\boldsymbol{b}}^{(0)} = 0, \tag{3.14a$'$}$$

$$\nabla_x \cdot \boldsymbol{b}^{(1)} + \nabla_X \cdot \widetilde{\boldsymbol{b}}^{(0)} = 0, \qquad \nabla_X \cdot \langle \boldsymbol{b}^{(0)} \rangle = 0, \tag{3.14b$'$}$$

$$\nabla_x \cdot \boldsymbol{b}^{(2)} + \nabla_X \cdot \boldsymbol{b}^{(1)} = 0. \tag{3.14c$'$}$$

The underlined terms in (3.14) have zero fast variable average. In order for the system (3.14) to be consistent, *two* solvability conditions must be satisfied:

$$\langle \boldsymbol{u}\cdot\nabla_X \widetilde{\boldsymbol{b}}^{(0)} \rangle = 0 , \tag{3.15a}$$

$$\partial_T \langle \boldsymbol{b}^{(0)} \rangle + \langle \boldsymbol{u}\cdot\nabla_x \boldsymbol{b}^{(1)} \rangle - \langle \boldsymbol{u}\nabla_x \cdot \boldsymbol{b}^{(1)} \rangle = \lambda\nabla_X^2 \langle \boldsymbol{b}^{(0)} \rangle. \tag{3.15b}$$

Equations (3.15a, b) are obtained by averaging (3.14b, c), respectively. The $-\langle \boldsymbol{u}\nabla_x \cdot \boldsymbol{b}^{(1)} \rangle$ term in (3.15b) comes from manipulating the $\boldsymbol{b}^{(2)}\cdot\nabla_x \boldsymbol{u}$ term in (3.15c) through the solenoidality condition (3.14c′).

Substituting $\boldsymbol{a} = \boldsymbol{u}$ and $\boldsymbol{b} = \widetilde{\boldsymbol{b}}^{(0)}$ into the vector identity

$$\nabla \times (\boldsymbol{a} \times \boldsymbol{b}) = (\boldsymbol{b}\cdot\nabla)\boldsymbol{a} - (\boldsymbol{a}\cdot\nabla)\boldsymbol{b} - (\nabla\cdot\boldsymbol{a})\boldsymbol{b} + (\nabla\cdot\boldsymbol{b})\boldsymbol{a}$$

gives

$$\nabla_X \times (\boldsymbol{u} \times \widetilde{\boldsymbol{b}}^{(0)}) = -(\boldsymbol{u}\cdot\nabla_X)\widetilde{\boldsymbol{b}}^{(0)},$$

since $\boldsymbol{u}$ is independent of X. The solvability conditions can thus be recast as

$$\nabla_X \times \langle \boldsymbol{u} \times \widetilde{\boldsymbol{b}}^{(0)} \rangle = 0 \,, \tag{3.16a}$$

$$\partial_T \langle \boldsymbol{b}^{(0)} \rangle = \nabla_X \times \langle \boldsymbol{u} \times \boldsymbol{b}^{(1)} \rangle + \lambda \nabla_X^2 \langle \boldsymbol{b}^{(0)} \rangle . \tag{3.16b}$$

A physical interpretation for (3.16) can be obtained by rewriting (3.12) in the form:

$$\partial_t \boldsymbol{b} = \nabla_x \times (\boldsymbol{u} \times \boldsymbol{b}) + \lambda \nabla_x^2 \boldsymbol{b}; \tag{3.17}$$

$\boldsymbol{u} \times \boldsymbol{b}$ is the so-called electromotive force. The decomposition (3.13b) implies that the first term of the electromotive force having a nonzero average is $\boldsymbol{u} \times \boldsymbol{b}^{(1)}$. However, the solvability condition (3.16a) requires that this term have a curl which has zero average. Equation (3.14a) determines $\widetilde{\boldsymbol{b}}^{(0)}$ in terms of $\langle \boldsymbol{b}^{(0)} \rangle$ and the velocity field. In general the solvability condition (3.16a) will not be satisfied, indicating that our choice of scaling is inconsistent. There is, however, one instance when solvability is guaranteed, namely, when the velocity field is *parity-invariant*. This means that the velocity field is invariant under $\boldsymbol{x} \to -\boldsymbol{x}$ and $\boldsymbol{u} \to -\boldsymbol{u}$ (fast variables only!). Note that parity-invariance is a broader concept than nonvanishing helicity: the former implies the latter but not vice versa. We now show that this implies the vanishing of the mean electromotive force. For a *prescribed* $\langle \boldsymbol{b}^{(0)} \rangle$, as we perform a parity transform on $\boldsymbol{u}$, (3.14a) remains intact (because both $\boldsymbol{u}$ and ∇_x change sign). Thus $\widetilde{\boldsymbol{b}}^{(0)}$ does not change. Therefore any product of components of $\boldsymbol{u}$ and $\widetilde{\boldsymbol{b}}^{(0)}$ changes sign. This implies the vanishing of the average of $\boldsymbol{u} \times \widetilde{\boldsymbol{b}}^{(0)}$.

Equation (3.16b) is closed once $\boldsymbol{b}^{(1)}$ is determined. The procedure for determining $\boldsymbol{b}^{(1)}$ is as follows:

(i) Since $\langle \boldsymbol{b}^{(0)} \rangle$ is independent of both x and t the integration of (3.14a) can be effected as if $\langle \boldsymbol{b}^{(0)} \rangle$ were a constant vector. Furthermore, since $\boldsymbol{b}^{(1)}$ is linear in $\langle \boldsymbol{b}^{(0)} \rangle$, the solution for $\boldsymbol{b}^{(1)}$ will be proportional to $\langle \boldsymbol{b}^{(0)} \rangle$: $b_i^{(1)} = C_{ij}(x,t) \langle b_j^{(0)} \rangle$.

(ii) The solution for $\widetilde{\boldsymbol{b}}^{(0)}$ can then inserted in (3.14b) to determine $\boldsymbol{b}^{(1)}$. Since (3.14b) is linear in $\nabla_X \boldsymbol{b}^{(0)}$, its solution $\boldsymbol{b}^{(1)}$ will once again be proportional to slow space derivatives of $\boldsymbol{b}^{(0)}$. Thus we have $b_i^{(1)} = \Gamma_{ijk}(x,t) \partial_{X_j} \langle b_k^{(0)} \rangle$.

(iii) Inserting $\boldsymbol{b}^{(1)}$ into the second solvability condition, (3.16b) yields a closed equation involving slow second-order space derivatives of $\langle \boldsymbol{b}^{(0)} \rangle$.

In the isotropic case, the resulting equation has the form

$$\partial_T \langle \boldsymbol{b}^{(0)} \rangle = (\beta + \lambda) \nabla_X^2 \langle \boldsymbol{b}^{(0)} \rangle \tag{3.18}$$

(where β is the magnetic analog of the scalar eddy diffusivity). Contrary to the scalar case the eddy diffusivity $\beta + \lambda$ need not be positive and can drive instabilities.

Next, we consider the transport of a magnetic field when the flow *lacks parity*. This will lead to a new form of the large-scale equation demonstrating the so called α-effect. We derive the corresponding equation by a variant of our multiscale formalism, distributing the ϵ's somewhat differently. We now suspect that there may be large-scale effects on shorter $O(\epsilon)$ time scales, so we try

$$\partial_t \leftarrow \partial_t + \epsilon \partial_T$$

$$\partial_x \leftarrow \partial_x + \epsilon \partial_X .$$

Equation (3.12a) becomes

$$(\partial_t + \epsilon \partial_T) \boldsymbol{b} + \boldsymbol{u} \cdot (\nabla_x + \epsilon \nabla_X) \boldsymbol{b} - \boldsymbol{b} \cdot (\nabla_x + \epsilon \nabla_X) \boldsymbol{u} = \lambda (\nabla_x + \epsilon \nabla_X)^2 \boldsymbol{b} \tag{3.19}$$

The corresponding hierarchy is

$$O(\epsilon^0): \quad \underline{\partial_t \widetilde{\boldsymbol{b}}^{(0)}} + \underline{\boldsymbol{u} \cdot \nabla_x \widetilde{\boldsymbol{b}}^{(0)}} - \underline{\boldsymbol{b}^{(0)} \cdot \nabla_x \boldsymbol{u}} = \langle \boldsymbol{b}^{(0)} \rangle \cdot \underline{\nabla_x \boldsymbol{u}} + \underline{\lambda \nabla_x^2 \widetilde{\boldsymbol{b}}^{(0)}} \tag{3.20a}$$

$$\begin{aligned} O(\epsilon): \quad & \underline{\partial_t \boldsymbol{b}^{(1)}} + \underline{\partial_T \widetilde{\boldsymbol{b}}^{(0)}} + \partial_T \langle \boldsymbol{b}^{(0)} \rangle + \underline{\boldsymbol{u} \cdot \nabla_X \langle \boldsymbol{b}^{(0)} \rangle} + \boldsymbol{u} \cdot \nabla_X \widetilde{\boldsymbol{b}}^{(0)} \\ & + \underline{\boldsymbol{u} \cdot \nabla_x \boldsymbol{b}^{(1)}} = \underline{\boldsymbol{b}^{(1)} \cdot \nabla_x \boldsymbol{u}} + \underline{\lambda \nabla_x^2 \boldsymbol{b}^{(1)}} + \underline{2 \lambda \nabla_{xX}^2 \widetilde{\boldsymbol{b}}^{(0)}}. \end{aligned} \tag{3.20b}$$

As with the previous cases, the underlined terms have zero fast variable average. From (3.20b) the solvability condition is given by

$$\partial_T \langle \boldsymbol{b}^{(0)} \rangle = \nabla_X \times \langle \boldsymbol{u} \times \widetilde{\boldsymbol{b}}^{(0)} \rangle . \tag{3.21}$$

However, from (3.20a), $\widetilde{\boldsymbol{b}}^{(0)}$ is linear in $\langle \boldsymbol{b}^{(0)} \rangle$, and thus the mean electromotive force $\langle \boldsymbol{u} \times \widetilde{\boldsymbol{b}}^{(0)} \rangle$ will be proportional to the mean magnetic field $\langle \boldsymbol{b}^{(0)} \rangle$:

$$\langle \boldsymbol{u} \times \widetilde{\boldsymbol{b}}^{(0)} \rangle_i = \alpha_{ij} \langle \boldsymbol{b}^{(0)} \rangle_j . \tag{3.22}$$

In the isotropic case, $\alpha_{ij} = \alpha \delta_{ij}$, and (3.10) becomes

$$\partial_T \langle \boldsymbol{b}^{(0)} \rangle = \alpha \nabla_X \times \langle \boldsymbol{b}^{(0)} \rangle . \tag{3.23}$$

Applying a spatial Fourier transform to (3.23) gives

$$\partial_T \langle \widehat{\boldsymbol{b}}^{(0)} \rangle = i\alpha \boldsymbol{k} \times \langle \widehat{\boldsymbol{b}}^{(0)} \rangle. \tag{3.24}$$

Seeking eigensolutions in the form $\langle \widehat{\boldsymbol{b}}^{(0)} \rangle = \boldsymbol{C} e^{\xi t}$ gives $\xi_1 = 0$ and $\xi_{2,3} = \pm k\alpha$. Thus the small-scale structures produce an instability in the mean magnetic field with growth rate proportional to wavenumber. This is what is usually referred to as the α-effect.

Various remarks can be made:

a) Inclusion in the α-effect equation (3.23) of an additional diffusion term can be achieved by modification of the assumed scaling for the molecular diffusivity: instead of an $0(1)$ diffusivity, we take an $0(\epsilon^{-1})$ diffusivity. In the isotropic case, we then obtain

$$\partial_T \langle \boldsymbol{b}^{(0)} \rangle = \alpha \nabla_X \times \langle \boldsymbol{b}^{(0)} \rangle + \lambda \nabla_X^2 \langle \boldsymbol{b}^{(0)} \rangle. \tag{3.25}$$

The corresponding eigenvalues of (3.25) are $\xi_1 = -\lambda k^2$ and $\xi_{2,3} = \pm \alpha k - \lambda k^2$. Therefore, if $\alpha k - \lambda k^2 > 0$, an amplification in $\langle \boldsymbol{b}^{(0)} \rangle$ is still possible, provided the wavenumber k is sufficiently small to render diffusion ineffective.

b) The form of (3.23) and (3.25) can be guessed by Landau-type symmetry arguments. Performing a Reynolds decomposition of the induction equation (3.12a) gives a "mean" equation

$$\partial_t \boldsymbol{B} = \nabla_X \times \langle \boldsymbol{u} \times \boldsymbol{b}' \rangle + \lambda \nabla_X^2 \boldsymbol{B}, \tag{3.26}$$

where $\boldsymbol{B} = \langle \boldsymbol{b} \rangle$ and $\boldsymbol{b}' = \boldsymbol{B} - \langle \boldsymbol{b} \rangle$ (analogous to, respectively, $\boldsymbol{b}^{(0)}$ and the fluctuating field above). Expanding $\langle \boldsymbol{u} \times \boldsymbol{b}' \rangle$ in a Taylor series in slow gradients acting on the mean field gives:

$$\langle \boldsymbol{u} \times \boldsymbol{b}' \rangle_i = \alpha_{ij} \boldsymbol{B}_j + \Gamma_{ijk} \partial_{X_j} \boldsymbol{B}_k + O(\nabla_X^2 \boldsymbol{B}). \tag{3.27}$$

Parity-invariance of $\boldsymbol{u}$ implies $\alpha_{ij} = 0$. But, for flows which do not possess parity, α_{ij} need not be zero and, to leading order, we recover (3.22).

c) A variant of the technique discussed herein can be applied to the question of viscoelasticity of turbulence [6]. When flows possess Galilean invariance, it may be shown that to leading order, the large-scale dynamics are governed by a wave equation involving second-order time and space derivatives.

3. Conclusions

We have shown here how multiscale expansions can be used to systematically analyze the large-scale dynamics of scalars and vectors passively advected by a prescribed small-scale flow. For the scalar case, we find that the large-scale behavior is always diffusive. The value of the eddy diffusivity (or the eddy diffusivity tensor in the anisotropic case) cannot usually be obtained in closed form; an auxiliary problem, involving only the small scales, must be solved. For the vector case (here a passive magnetic field), the large-scale behavior will be superficially diffusive if *parity-invariance* holds. By "superficially" we mean that it is governed by an equation involving second-order space derivatives, but without a guarantee that the eddy diffusivity is positive. For flows lacking parity-invariance, the large-scale magnetic field will usually be subject to a destabilizing α-effect involving first-order space derivatives.

We finally want to stress that essentially the same multiscale analysis which we have applied to the large-scale dynamics of a passively advected magnetic field can also be applied to the problem of the *eddy viscosity* of a small-scale flow $\boldsymbol{u}$ subject to a weak large-scale perturbation $\boldsymbol{w}$. The governing equation is the linearized Navier-Stokes equation

$$\partial_t \boldsymbol{w} + \boldsymbol{u} \cdot \nabla_x \boldsymbol{w} + \boldsymbol{w} \cdot \nabla_x \boldsymbol{u} = -\nabla p' + \nu \nabla_x^2 \boldsymbol{w} , \qquad (3.28a)$$

$$\nabla_x \cdot \boldsymbol{u} = 0, \quad \nabla_x \cdot \boldsymbol{w} = 0. \qquad (3.28b)$$

It is clear that its structure is very similar to that of (3.12a, b). The results for large-scale velocity perturbations are essentially the same as above: with parity-invariance, we get superficially diffusive behavior, but the eddy viscosity may be negative; without parity-invariance, a new type of instability may be present, the AKA-effect. This is discussed in the next chapter.

Appendix to Chapter 3

The demonstration that $\langle(\widetilde{\Theta}^{(0)})^2\rangle \to 0$ as $t \to \infty$ is given here.

Since $\widetilde{\Theta}^{(0)}$ is ℓ-periodic in x, it admits a Fourier series representation ($\boldsymbol{k} = (k_1, k_2, k_3)$)

$$\sum_{\boldsymbol{k}} \widehat{\Theta}_k^{(0)} e^{2i\pi \boldsymbol{k}\cdot\boldsymbol{x}/\ell}. \tag{A3.1}$$

The $\boldsymbol{k} = 0$ coefficient is absent (since $\widetilde{\theta}$ has zero space average); thus, we have the inequality

$$\sum_{\boldsymbol{k}} k^2 |\widehat{\Theta}_k^{(0)}|^2 \geq \sum_{\boldsymbol{k}} |\widehat{\Theta}_k^{(0)}|^2. \tag{A3.2}$$

Furthermore, using here angular brackets to denote space averaging only, and invoking Parseval's identity $\langle(\widetilde{\Theta}^{(0)})^2\rangle = \sum_{\boldsymbol{k}} |\widehat{\Theta}_{\mathbf{k}}^{(0)}|^2$ and $\langle(\nabla_x \widetilde{\Theta}^{(0)})^2\rangle = \sum_{\boldsymbol{k}} (2\pi k/\ell)^2 |\widehat{\Theta}_{\boldsymbol{k}}^{(0)}|^2$, we obtain the following "Poincaré inequality":

$$\langle(\nabla_x \widetilde{\Theta}^{(0)})^2\rangle \quad \geq \quad (2\pi/\ell)^2 \langle(\widetilde{\Theta}^{(0)})^2\rangle. \tag{A3.3}$$

Multiplying the scalar diffusion equation by $\widetilde{\Theta}^{(0)}$, taking the average over a cell and integrating by parts gives:

$$\frac{1}{2}\partial_t \langle(\widetilde{\Theta}^{(0)})^2\rangle \quad \leq \quad -\kappa(2\pi/\ell)^2 \langle(\widetilde{\Theta}^{(0)})^2\rangle, \tag{A3.4}$$

so that

$$\langle(\widetilde{\Theta}^{(0)})^2\rangle \leq C \exp(-2\kappa(2\pi/\ell)^2 t) \to 0 \quad \text{as} \quad t \to \infty. \tag{A3.5}$$

[1] Papanicolaou, G. and Pironneau, O., "On the asymptotic behavior of motions in random flows", *Stochastic Nonlinear Systems*, Arnold and Lefever, eds., Springer (1981).

[2] Moffatt, H. K., *Magnetic Field Generation in Electrically Conducting Fluid*, Cambridge Univ. Press (1978), 343 pp.

[3] Zeldovich, Ya. B., Ruzmaikin, A. A., Sokoloff, D. D., *Magnetic Fields in Astrophysics*, Gordon and Breach (1983), 363 pp.

[4] Lichtenberg, A. J. and Lieberman, M. A., *Regular and Stochastic Motion*, Springer (1983), 499 pp.

[5] Sagdeev, R. Z., Usikov, D. A., and Zaslavsky, G. M., *Nonlinear Physics*, Harwood (1988), 675 pp.

[6] Aref, H., *J. Fluid Mech.*, **143**, 1 (1984).

[7] Dombre, T., Frisch, U., Greene, J. M., Hénon, M., Mehr, A., and Soward, A. M., *J. Fluid Mech.*, **167**, 353 (1986).

[8] Frisch, U., She, Z. S., and Thual, O., *J. Fluid Mech.*, **168**, 221 (1986).

Chapter 4

LARGE-SCALE INSTABILITY IN 3D FLOWS LACKING PARITY-INVARIANCE

The present notes are meant to supplement material presented in the appended paper by Frisch, She, and Sulem [1].

The α-effect, a large-scale MHD instability, has been known for about 20 years and is usually associated with the presence of helicity. An analog to the α-effect is known in ordinary fluid dynamics for compressible flows, but was only noticed recently for incompressible three-dimensional, anisotropic flow. As we shall see, a new kind of large-scale instability may exist for incompressible three-dimensional, anisotropic flows which lack *parity-invariance* (invariance under simultaneous reversal of position and velocity vectors).

This Anisotropic Kinetic Alpha (AKA) effect can be derived by a small Reynolds number expansion using a separation of scales. The full formal expansion requires six levels of expansion and will not be presented here (see the Appendix of Ref. [1]).

A small-scale flow, $\boldsymbol{u}^0(\boldsymbol{r},t)$ is driven by a force, $\boldsymbol{f}(\boldsymbol{r},t)$ which is space- and time-periodic. The flow has characteristic length, time, and velocity scales l_0, t_0, and V_0, and a small-scale Reynolds number, $R = l_0V_0/\nu$, and obeys the Navier-Stokes equations (incompressibility conditions are henceforth omitted for brevity):

$$\frac{\partial u_i^0}{\partial t} + \frac{\partial}{\partial x_j}(u_i^0 u_j^0) = -\frac{\partial p^0}{\partial x_i} + \nu\nabla^2 u_i^0 + f_i. \tag{4.1}$$

The basic flow is perturbed to a flow $\boldsymbol{u}$ having a nontrivial large-scale component

$$\boldsymbol{u} = \boldsymbol{w} + \widetilde{\boldsymbol{u}}, \tag{4.2}$$

where $\langle \boldsymbol{u} \rangle = \boldsymbol{w}$ and the angular brackets denote averaging over the small scales. $\boldsymbol{w}$ has scales $L \gg l_0$ and $T \gg t_0$ and may be considered as constant over the small scales l_0 and t_0 . The small-scale flow, $\widetilde{\boldsymbol{u}}$, is advected by the large-scale flow

$$\frac{\partial \widetilde{u}_i}{\partial t} + w_j\frac{\partial \widetilde{u}_i}{\partial x_j} + \frac{\partial}{\partial x_j}\left(\widetilde{u}_i\widetilde{u}_j\right) = -\frac{\partial \widetilde{p}}{\partial x_i} + \nu\nabla^2\widetilde{u}_i + f_i. \tag{4.3}$$

Therefore, the average small-scale Reynolds stresses,

$$R_{ij} = \left\langle \tilde{u}_i \tilde{u}_j \right\rangle , \tag{4.4}$$

will develop a dependence on the mean field and contribute to the large-scale dynamics. The equation for the large-scale velocity components can be derived by introducing slow variables, expanding in Reynolds number, and using separation of scales to give

$$\frac{\partial w_i}{\partial t} + \frac{\partial}{\partial x_j}\left(w_i w_j + R_{ij}\right) = -\frac{\partial p}{\partial x_i} + \nu \nabla^2 w_i . \tag{4.5}$$

This equation is in terms of the slow variables.

The linear regime may be considered when the mean field is weak. The average Reynolds stresses can be expanded in powers of $\boldsymbol{w}$

$$R_{ij} = \left\langle u_i^0 u_j^0 \right\rangle + w_l \left[\frac{\partial \left\langle \tilde{u}_i \tilde{u}_j \right\rangle}{\partial w_l} \right]_{w_l = 0} . \tag{4.6}$$

The large-scale flow satisfies

$$\frac{\partial w_i}{\partial t} = \alpha_{ijk} \frac{\partial w_k}{\partial x_j} - \frac{\partial p}{\partial x_i} + \nu \nabla^2 w_i , \tag{4.7}$$

with

$$\alpha_{ijk} = -\left[\frac{\partial \left\langle \tilde{u}_i \tilde{u}_j \right\rangle}{\partial w_k} \right]_{w_k = 0} . \tag{4.8}$$

The tensor α_{ijk} will vanish for many situations, including parity-invariance, isotropy, and a random flow that is δ-correlated in time. The AKA effect vanishes with isotropy because α_{ijk} is symmetric in i and j and there are no nonvanishing, isotropic third-order symmetric tensors. The only known instance when α vanishes, in both MHD and hydrodynamics, is when parity-invariance holds. The AKA effect had not been seen before because assumptions used to derive the equation, made for convenience and not for necessity, caused the effect to disappear. Many closure methods assuming isotropy or δ-correlated in time miss the AKA effect, but the full Direct Interaction Approximation (DIA) method of Kraichnan [2] should show it. We shall here derive it by multiscale techniques without resort to closure.

A simple example of a flow that lacks parity-invariance and demonstrates the AKA effect by avoiding the known pitfalls is obtained using the following force lacking parity-invariance:

$$\begin{aligned} f_1 &= \frac{\nu V_0\sqrt{2}}{l_0^2}\cos\left(\frac{y}{l_0}+\frac{\nu t}{l_0^2}\right), \\ f_2 &= \frac{\nu V_0\sqrt{2}}{l_0^2}\cos\left(\frac{x}{l_0}-\frac{\nu t}{l_0^2}\right), \\ f_3 &= f_1+f_2. \end{aligned} \tag{4.9}$$

The basic flow is obtained from the linearized Navier-Stokes equations (to leading order in the Reynolds number)

$$\begin{aligned} u_1^0 &= V_0\cos\left(\frac{y}{l_0}+\frac{\nu t}{l_0^2}-\frac{\pi}{4}\right), \\ u_2^0 &= V_0\cos\left(\frac{x}{l_0}-\frac{\nu t}{l_0^2}+\frac{\pi}{4}\right), \\ u_3^0 &= u_1^0+u_2^0. \end{aligned} \tag{4.10}$$

For low Reynolds numbers of the *basic* flow (as we henceforth assume), the small-scale flow is essentially a Stokes flow with forcing and advection by the large-scale (quasi-uniform) flow $\mathbf{w}$; it thus satisfies the following equation (to leading order):

$$\frac{\partial \tilde{u}_i}{\partial t}+w_j\frac{\partial \tilde{u}_i}{\partial x_j} = \nu\nabla^2\tilde{u}_i+f_i. \tag{4.11}$$

This can be solved with Fourier transforms or a Galilean transformation to calculate the average small-scale Reynolds stresses.

$$\begin{aligned} R_{11} &= R_{13} = \frac{V_0^2}{2}-\frac{V_0^2 l_0 w_2}{2\nu}+O(w^2), \\ R_{22} &= R_{23} = \frac{V_0^2}{2}+\frac{V_0^2 l_0 w_1}{2\nu}+O(w^2), \\ R_{33} &= R_{11}+R_{22}, \\ R_{12} &= 0. \end{aligned} \tag{4.12}$$

The nonvanishing components of α_{ijk} for the linear AKA effect are therefore

$$\alpha_{112}=\alpha_{132}=\alpha_{332}=-\alpha_{221}=-\alpha_{231}=-\alpha_{321}=-\alpha_{331}=\alpha=\frac{V_0^2 l_0}{2\nu}. \tag{4.13}$$

The equation for the large-scale motion can be solved now that R_{ij} is known. The most unstable modes depend only on z, so large-scale perturbations depending only on z satisfy

$$\frac{\partial w_1}{\partial t} = \alpha \frac{\partial w_2}{\partial z} + \nu \frac{\partial^2 w_1}{\partial z^2} , \tag{4.14}$$

$$\frac{\partial w_2}{\partial t} = -\alpha \frac{\partial w_1}{\partial z} + \nu \frac{\partial^2 w_2}{\partial z^2} . \tag{4.15}$$

The solution is of the form

$$\psi = w_1 + i w_2 = e^{ikz} e^{(\alpha k - \nu k^2)t} , \tag{4.16}$$

so that wavenumbers satisfying $kl_0 < \frac{R^2}{2}$ are unstable.

The scale of the instability increases as the Reynolds number decreases. The large-scale flow that results from this instability is a helical standing wave, with circular polarization and exponentially growing amplitude. This is an example of a Beltrami flow.

Numerical experiments were performed to confirm the AKA effect. Pseudo-spectral simulations with 32^3 points used forcing at $k = 6$ and $R = \sqrt{1/2}$, the only linearly unstable mode being $k = 1$. At such moderately small Reynolds numbers, subleading order effects are significant (about 40 percent) but the qualitative picture is present. Note that much smaller Reynolds number calculations require very fine resolution to get a separation of scales. The resolution goes up as R^{-2} and the computational effort rises as R^{-8} . The linear AKA effect is clearly seen from the numerical simulation results in Fig. 4.1 for times up to about $t = 3.5$. The small initial energy in the $k = 1$ mode decays at first, because stable and unstable modes are competing. The exponential growth (Beltrami runaway) then takes over until the large scales have enough energy that the linear equations are no longer valid.

Beyond $t = 3.5$, we observe saturation. Physically, this is due to the feedback effect of the large-scale flow. Against a strong background flow, it becomes more difficult for the prescribed force to produce small-scale flow. Thus the small-scale Reynolds stresses should be decreasing. Indeed, when we solve (4.3) ignoring the nonlinear term $\partial(\widetilde{u}_i\widetilde{u}_j)/\partial x_j$ (which is irrelevant at low Reynolds number), we find that the Reynolds stresses have a nonlinear dependence on the mean flow $\boldsymbol{w}$, with a w^{-2} behavior for large w's. The nonlinear equations for

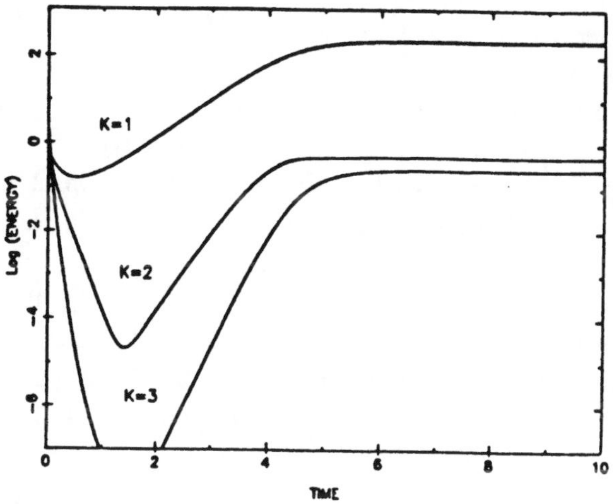

Figure 4.1: Full numerical simulation of 3D Navier Stokes equation showing AKA instability and saturation [1].

the large-scale dynamics incorporating the correct Reynolds stresses (in their full nonlinear dependence) are given below in nondimensional form:

$$\begin{aligned} \frac{\partial w_1}{\partial t} + \frac{1}{2}\frac{\partial}{\partial z}\left[\frac{1}{1 + Rw_2 + \frac{1}{2}(Rw_2)^2}\right] &= \frac{1}{R}\frac{\partial^2}{\partial z^2}w_1\,, \\ \frac{\partial w_2}{\partial t} + \frac{1}{2}\frac{\partial}{\partial z}\left[\frac{1}{1 - Rw_1 + \frac{1}{2}(Rw_1)^2}\right] &= \frac{1}{R}\frac{\partial^2}{\partial z^2}w_2, \end{aligned} \tag{4.17}$$

where $R = V_0 l_0/\nu$. Here the length scale is l_0 and the velocity scale is V_0. A naive dimensional analysis of these equations suggests that, in the nonlinear regime, the amplitudes of the w's should be $O(R^{-1})$, their spatial scale $O(R^{-2})$, and their temporal scale $O(R^{-3})$. Thus, for small R, we expect very strong fields to form eventually on very large scales.

We now report some recent results of Frisch et al. [3] concerning the possibility of an *inverse cascade*. It is known that the α-effect in MHD is the main motor of the "inverse helicity cascade" (see D. Montgomery's lecture notes). Could it be that in ordinary incompressible 3D flows there is an inverse cascade driven by the AKA effect? The possibility of an inverse cascade of helicity had already been considered by Brissaud et al. [4]. Later, André and

Lesieur [5] showed that within an isotropic closure framework the possibility is ruled out. However, the AKA effect is inherently anisotropic, so the question is worth reexamining.

We assume that there is a whole range of linearly unstable wavenumbers below the cutoff wavenumber $k_s = R^2/(2l_0)$ for linear stability. If there is an inverse cascade we expect to see, with increasing time, the peak of excitation migrates to lower and lower wavenumbers, while becoming increasingly strong. Such a cascade cannot just follow from linear growth of unstable modes, since they would forever be dominated by the most unstable one. For the AKA effect the linear growth rate is $\alpha k - \nu k^2$, and the most unstable mode is $k = k_s/2$. A *necessary* condition for the establishment of an inverse cascade is that the linearly unstable mode with the *smallest* nonvanishing k should eventually dominate, in spite of its lower *linear* growth rate.

A first numerical experiment was set up, looking for such evidence. It had modes $k = 1$ through 4 unstable, with maximum growth rate at $k = 2$. The results are shown in Fig. 4.2, where only modes 1, 2, and 9 are shown, for clarity. Up to about $t = 0.5$, we obtain an exponential growth in which mode 2 is growing much faster than mode 1, as predicted by the linear theory. After that time, nonlinear effects become significant and at about $t = 0.6$, we observe a kind of saturation reminiscent of what we saw when there was only a single linearly unstable mode. This is, however, just a plateau from which the system quickly escapes. The most salient feature is that mode $k = 1$ "leapfrogs" mode $k = 2$ around $t = 0.9$. Eventually the system goes to an approximately steady state in which mode $k = 1$ significantly dominates over mode $k = 2$.

We stress that we have not yet seen direct evidence for an inverse cascade, but just one necessary symptom. Experiments using much higher resolution and very considerable computational resources are needed to confirm the existence of an inverse cascade.

Note added in proof. Numerical integration and theoretical analysis of the set of nonlinear PDEs (4.17) have demonstrated the existence of the inverse cascade. Cases have been studied with up to 300 linearly unstable modes [6].

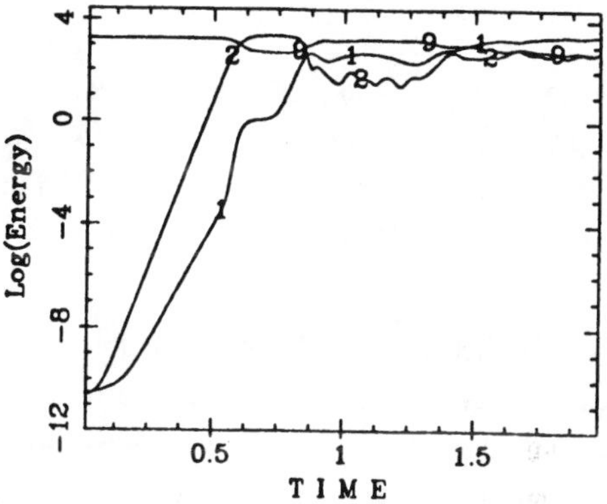

Figure 4.2: AKA instability with two linearly unstable modes ($k = 1$ and $k = 2$). Mode $k = 2$ is the linearly fastest growing, but eventually mode $k = 1$ "leapfrogs" it, an indication that we may see the beginning of an inverse cascade [3].

[1] Frisch, U., She, Z. S., and Sulem, P.-L., *Physica*, **28D**, 382 (1987).

[2] Kraichnan, R. H., *J. Math. Phys.* **2**, 124 (1961).

[3] Frisch, U., Scholl, H., She, Z. S., and Sulem, P.-L., A new large-scale instability in three-dimensional anisotropic incompressible flows lacking parity-invariance, to appear in Proc. IUTAM Symposium on Fundamental Aspects of Vortex Motion, Tokyo, Sept. 1987 *Fluid Dyn. Res.*, in press.

[4] Brissaud, A., Frisch, U., Léorat, J., Lesieur, M., and Mazure, A., *Phys. Fluids*, **16**, 1366 (1973).

[5] André, J. C. and Lesieur, M., *J. Fluid Mech.*, **81**, 187 (1977).

[6] Sulem, P.-L., She, Z. S., Scholl, H., and Frisch, U., Generation of large-scale structures in three-dimensional flows lacking parity-invariance. *J. Fluid Mech.* (1989), in press.

Chapter 5

LATTICE GAS HYDRODYNAMICS: INTRODUCTION

These lecture notes, written by participants in the 1987 Summer School on Turbulence, should be considered as supplementing the paper "Lattice gas hydrodynamics in two and three dimensions" by Frisch et al. [1], *found at the end of this series of lectures by Frisch.*

A novel technique that has recently been applied to the study of hydrodynamics is that of Cellular Automata (CA). CAs are discrete states attached to the nodes of a regular lattice with an updating rule involving only a small number of neighbors. Figures 5.1, 5.2, and 5.3 illustrate well-known fluid mechanical problems that have been simulated using this technique. Figure 5.1 shows a snapshot of a two-dimensional Karman vortex street behind a flat plate [2,4]. Figure 5.2 shows a snapshot of the recent three-dimensional lattice gas calculations behind a circular plate, as performed by J.-P. Rivet et al. [3]. Figure 5.3 shows the evolution of a Rayleigh-Taylor instability [5]. For historical perspective on CAs and their lattice gas version, the reader is referred to Section 1 [1].

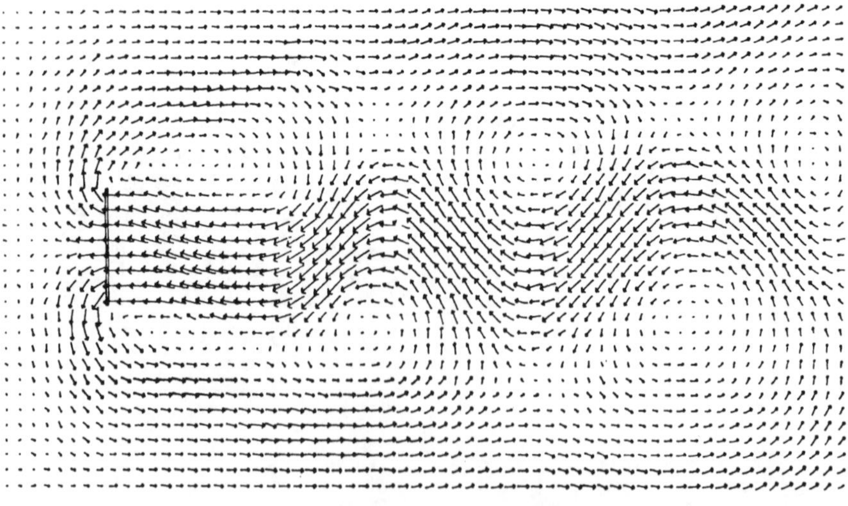

Figure 5.1: Karman vortex street behind a flat plate simulated on FPS 164 at École Normale Supérieure [2,4].

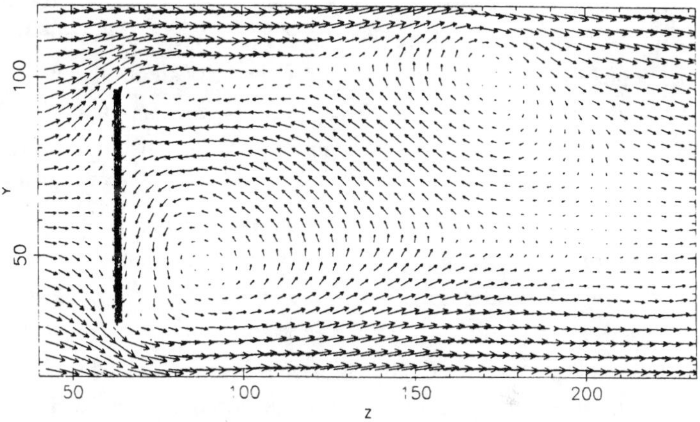

Figure 5.2: A snapshot of a recent three-dimensional lattice gas calculation of flow behind a circular plate [3]. Shown is a detail of an axial cut of the velocity field. Note that the axial symmetry is broken.

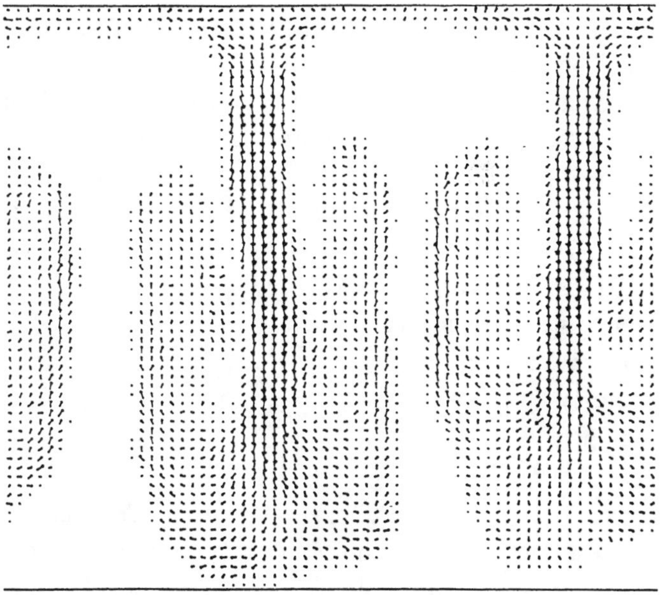

Figure 5.3: Rayleigh-Taylor instability simulated by Clavin et al. [5].

There are many precedents in physics of simple, discrete models able to correctly capture complexities of the real world. A prominent example is the Ising model, which is a very poor substitute for the intricate atomic interactions in a ferromagnet, but still manages to capture the essence of critical phase transitions. This is indeed revealed by Renormalization Group analysis (fully exploiting the physical idea of spin-blocking), which demonstrates that the real world and the Ising model differ only in "irrelevant" details.

Similarly, there are many models leading to fluid dynamics. Traditionally, turbulence theories have attempted to model phenomena through statistical averages of the velocity field. This field is, however, itself an average over microscopic motion. Is it conceivable that something can be gained by "undoing" such averages, reverting either to the full microscopic world (molecular dynamics), or to artificial, discrete CA models emulating this world in its macroscopic consequences? First, we must try to understand what the key ingredients are which lead from the microscopic laws of motion to fluid dynamics. In the real world, equilibria are parameterized by three thermodynamic quantities: the density ρ, the fluid velocity $\boldsymbol{u}$, and the temperature T, associated with the conservation of mass, momentum, and energy, respectively. The fluid dynamical equations are obtained by patching together "local" equilibria, changing slowly in space and time. It was already stressed by Maxwell that the details of the intermolecular forces are unimportant in deriving his celebrated law for velocity distribution and fluid dynamical equations. Such details only affect the values of the transport coefficients. In addition to the conservation laws, basic symmetries of Newtonian physics seem to play a crucial role in determining the form of the fluid dynamical equations. These include:

- continuous spatial-temporal translations,
- arbitrary 3D rotations,
- time reversal,
- space reversal (parity),
- Galilean transformations.

For example, the last essentially provides a unique determination of the nonlinear term $\boldsymbol{u} \cdot \nabla \boldsymbol{u}$. If we were to insist on keeping all these symmetries, there is little we could do, short of staying in the real world. The key question is to see whether *discrete* models can be constructed in which one or more of these invariance properties are by necessity relaxed, without losing fluid dynamical behavior. Certainly, discrete translational invariance in space and time does not preclude continuous invariances on large scales. Less obvious is that discrete rotations will not violate the requirement of isotropy. Even less obvious is the fact that Galilean invariance is not violated macroscopically. These issues will be addressed in following chapters.

Finally, it may be asked why we insist on making discrete CA models. The fact that the Ising model has played an essential role in the understanding of ferromagnetics was certainly one of the original motivations of Hardy, Pomeau, and de Pazzis when they introduced the first lattice gas model in the early seventies [6, 7]. For computational implementation, discrete models are appealing in view of the Boolean structure of computers. One of the most promising trends in computer architecture is toward parallelism and even *massive* parallelism. In the latter, the number of concurrent processors can, in principle, grow indefinitely, as is the case when processors interact only with nearest neighbors. This provides additional reasons to explore the possibilities of CA models of fluid dynamics.

[1] Frisch, U., d'Humières, D., Hasslacher, B., Lallemand, P., Pomeau, Y., and Rivet, J.-P., *Complex Systems*, **1**, 649 (1987). (Also attached at the end of lectures by Frisch.)

[2] d'Humières, D., Pomeau, Y., and Lallemand, P., *C. R. Acad. Sci.* Paris II, **301**, 1391 (1985).

[3] Rivet, J.-P., Hénon, M., Frisch, U., and d'Humières, D., *Europhys. Lett.* **7**, 231 (1988).

[4] d'Humières D. and Lallemand, P., *Complex Systems*, **1**, 598 (1987).

[5] Clavin, P., Lallemand, P., Pomeau, Y., and Searby, J., *J. Fluid Mech.*, **188**, 437 (1988).

[6] Hardy, J., Pomeau, Y., and de Pazzis, O., *J. Math. Phys.*, **14**, 1746 (1973).

[7] Hardy J. and Pomeau, Y., *J. Math. Phys.*, **13**, 1042 (1972).

Chapter 6

LATTICE GAS HYDRODYNAMICS: MICRODYNAMICS

The Navier-Stokes equations are themselves isotropic, and it is important that a lattice gas model preserve this property. Specifically, we hope to recover at the macroscopic level an equation of the form

$$\partial_t(\rho u)_\alpha + \partial_\beta(P_{\alpha\beta}) = \text{viscous terms}, \tag{6.1}$$

where the symmetrical momentum flux tensor is

$$P_{\alpha\beta} = p\delta_{\alpha\beta} + \rho u_\alpha u_\beta . \tag{6.2}$$

The most general parity-invariant form for $P_{\alpha\beta}$ up to second order in $\boldsymbol{u}$ (one may think of this as a Taylor-series for the limit when flow speeds are much less than particle speeds) is

$$P_{\alpha\beta} = \left(T^0_{\alpha\beta} + T_{\alpha\beta\gamma\delta}u_\gamma u_\delta\right). \tag{6.3}$$

In 2D, assuming $T_{\alpha\beta\gamma\delta}$ to be symmetric pairwise in (α,β) and in (γ,δ), a sufficient condition for isotropy is that $T_{\alpha\beta\gamma\delta}$ be invariant under $\pi/3$-rotations (the full proof was given in the lecture (Frisch et al. [1, Section 6]). This motivates the choice (in 2D) of the "FHP" lattice, based on equilateral triangles ([1], Fig. 3).

In 3D, however, no regular crystal is fully isotropic (the icosahedron is isotropic but cannot be packed). Hence, any attempt to implement lattice-gas dynamics on a regular 3D grid will lead, in the macroscopic limit, to extra terms inconsistent with the Navier-Stokes equations.

If one excludes random or quasi-lattices, then a four-dimensional representation must be sought. This has led to the introduction of the Face-Centered-Hyper-Cubic lattice (FCHC), defined as

$$\text{FCHC} = \left\{(x_1,\ x_2,\ x_3, x_4)\epsilon \boldsymbol{Z}^4 \mid x_1 + x_2 + x_3 + x_4 = 2\sigma \quad (\sigma\epsilon\boldsymbol{Z})\right\}. \tag{6.4}$$

The FCHC lattice has the following properties:

1. Any node has 24 nearest neighbors.
2. FCHC is invariant under coordinate permutations, and reversal of any or several coordinates.

3. FCHC is invariant under the isometry $x_\alpha \to x_\alpha - \sigma$.
4. It follows from 2 and 3 that the tensor, $T_{\alpha\beta\gamma\delta}$, governing the macroscopic stress-strain relation is isotropic.

In practice, one works with the *pseudo-4D* model, with a width of only one node in the fourth direction (i.e., formally assuming unit periodicity). Implementation of this lattice leads to the usual Navier-Stokes equations. In macroscopic terms, FCHC conserves density and momentum. The fourth conserved quantity associated with the four-velocity can be shown to behave effectively as a passive scalar in the incompressible limit, and hence induces no spurious dynamical effects.

Having chosen a lattice with the appropriate symmetries, consider now the equations of motion for particles on such a lattice. Each node ("gridpoint") has b cells, where b is the number of bits per node, and a cell is an "in"- or "out"-port leading to or from another node. The "microdynamics" on the lattice are described by a Boolean field, which is the cellular automaton analog of Hamilton's equations of motion in classical statistical mechanics.

Let $n_i(t_*, \boldsymbol{r}_*)$ be the Boolean field at discrete time t_* and node $\boldsymbol{r}_*$ (starred quantities being discrete), where 1 denotes "occupied" and 0 "vacant" for cell i. Let $\boldsymbol{r}_* + \boldsymbol{c}_i$ be the set of nearest neighbors of the node $\boldsymbol{r}_*$. One may think of the updating rule for the CA at each time-step in two stages: collision and propagation. For instance, in the HPP case (a square 2D lattice), ([1], Fig. 1), the rule is given by

$$\begin{aligned} n_i(t_* + 1, \boldsymbol{r}_* + \boldsymbol{c}_i) = {} & (n_i \wedge \neg(n_i \wedge n_{i+2} \wedge \neg n_{i+1} \wedge \neg n_{i+3})) \\ & \vee (n_{i+1} \wedge n_{i+3} \wedge \neg n_{i+3} \wedge \neg n_i \wedge \neg n_{i+2}), \end{aligned} \tag{6.5}$$

where $\wedge$, $\vee$, $\neg$ are the Boolean symbols for "and", "or" and "not," respectively. This can also be expressed in the arithmetical form

$$\begin{aligned} n_i(t_* + 1, \boldsymbol{r}_* + \boldsymbol{c}_i) = {} & n_i(t_*, \boldsymbol{r}_*) + n_{i+1}n_{i+3}(1 - n_i)(1 - n_{i+2}) \\ & - n_i n_{i+2}(1 - n_{i+1})(1 - n_{i+3}), \end{aligned} \tag{6.6}$$

which is reminiscent of the Boltzmann equation.

It is shown in the next chapter how the macrodynamical equations of motion can be derived from the above by taking ensemble averages.

[1] Frisch, U., d'Humières, D., Hasslacher, B., Lallemand, P., Pomeau, Y., Rivet, J.-P., *Complex Systems*, 1, 649 (1987). (Also attached at the end of the lecture notes by Frisch.)

Chapter 7

LATTICE GAS HYDRODYNAMICS: COLLISIONS AND EQUILIBRIA

In order to write the microdynamical equation for less elementary lattices we need to introduce a more compact notation. This notation will allow the use of both probabilistic and deterministic collision rules. In all cases the collision rules must ensure exact conservation laws for mass and momentum. Here is an example of a nondeterministic collision rule. On the triangular FHP lattice, a "head-on" collision with exactly two incoming particles at 0° and 180° can have two results: outgoing particles at either 60° and 240°, or 120° and 300°. If only one of these possibilities is chosen, "chirality" is introduced into the model, i.e., the model is not invariant under reflection. Similarly, on the FCHC lattice used for the pseudo-4D model, head-on collisions have 11 possible choices for outgoing channels. An advantage of probabilistic collision rules is that they may eventually make it easier to prove rigorous results, using tools from ergodic theory. Indeed, the possibility of there being several non-communicating cycles of lattice configurations, spoiling ergodicity, is then easily ruled out.

It is useful to make a distinction between random variables and their assignment. When we assign a Boolean variable n_i some value, the assignment will be denoted by s_i. The subscript i runs from 1 to b, where b is the number of neighbors, and thus the number of bits for each site. The HPP lattice has $b = 4$, the FHP lattice has $b = 6$, and the FCHC lattice has $b = 24$. Further, we shall denote the in-state, that before the collision, by $s = \{s_i\}$, and the out-state, that after the collision, by s'. For the deterministic system each s leads to a well-defined s', while in the probabilistic case transition probabilities are used. Define $A(s \to s')$ as the probability of going to state s from state s'. We shall also need the Boolean variable

$$\xi_{ss'}, \qquad \langle \xi_{ss'} \rangle = A(s \to s') \tag{7.1}$$

to represent individual realizations of the probabilistic collision process. For a given s there is a unique s' such that $\xi_{ss'} = 1$. At each r_* and t_* a new $\xi_{ss'}$ is chosen with the same statistics.

The conservation laws are expressed by the equations

$$\sum_i (s_i' - s_i) A(s \to s') a_i = 0, \quad \forall s, s', \tag{7.2}$$

where

$$a_i = \begin{cases} 1 & \text{for all } i \\ c_{i\alpha}, & \alpha = 1, \ldots, D \end{cases}, \tag{7.3}$$

and D is the spatial dimension of the lattice. The vector $\boldsymbol{c}_i$ connects each node to its nearest neighbor in the i-direction. The case $a_i = 1$ corresponds to mass conservation, while $a_i = c_{i\alpha}$ corresponds to momentum conservation. We also require that there be no other independent set of a_i's for which (7.2) is satisfied, as this would imply the existence of *spurious* conservation laws. One way to detect spurious conservation laws is to use the linearized Boltzmann approximation ([1, Section 8.2]): the null modes of the linearized collision operator correspond to the individual scalar conservation laws.

The most general microdynamical equation for a lattice gas is

$$n_i(t_* + 1, \boldsymbol{r}_* + \boldsymbol{c}_i) = \sum_{s,s'} s_i' \xi_{ss'} \prod_j n_j^{s_j} (1 - n_j)^{1 - s_j}. \tag{7.4}$$

This can be verified by inspection. The product term is the Boolean equivalent of a delta-function; it is zero unless the pattern of s's matches the pattern of n's. Thus the only term appearing in the sum over s is the one consistent with the state $n_i(t_*, \boldsymbol{r}_*)$. The $\xi_{ss'}$ gives the out-state, and the s_i' says that the final n_i is the relevant s_i'.

So far, the only nondeterministic element is in ξ. Now, as is usual in statistical mechanics, we introduce an *ensemble* of initial conditions, each with its own probability

$$P(t = 0, s(\cdot)) \geq 0, \quad \sum_{s(\boldsymbol{r}_*)} P = 1, \tag{7.5}$$

where $s(\cdot) = \{s(\boldsymbol{r}_*)\}$ is a configuration in the set $\Gamma = \{s(\cdot)\}$, the set of all configurations. Probabilities now enter twice: in the choice of initial conditions, and in the nondeterministic collision rules. The equation of evolution for P (the Liouville equation) is

$$P(t_* + 1, Ss'(\cdot)) = \sum_{s(\cdot) \in \Gamma} \prod_{\boldsymbol{r}_*} A(s \to s') P(t_*, s(\cdot)), \tag{7.6}$$

where the streaming operator S is defined by:

$$S : s_i(\boldsymbol{r}_*) \mapsto s_i(\boldsymbol{r}_* - \boldsymbol{c}_i). \tag{7.7}$$

We can now take averages with respect to this probability distribution. The most useful quantities are the mean populations

$$N_i(t_*,\boldsymbol{r}_*) = \langle n_i(t_*,\boldsymbol{r}_*)\rangle, \quad \langle(\cdot)\rangle = \sum_{s(\cdot)}(\cdot)P. \tag{7.8}$$

The relevent hydrodynamic quantities, the density per node $\rho(t_*,\boldsymbol{r}_*)$, and the mass current per node $\boldsymbol{j}(t_*,\boldsymbol{r}_*)$, can be constructed from the mean populations:

$$\begin{aligned} \rho(t_*,\boldsymbol{r}_*) &= \sum_i N_i \\ \boldsymbol{j}(t_*,\, r_*) &= \rho\boldsymbol{u} = \sum_i \boldsymbol{c}_i N_i, \end{aligned} \tag{7.9}$$

where $\boldsymbol{u}$ is the hydrodynamic velocity. Note that, depending on the lattice, the number of nodes per unit volume may be different from 1. Eventually we shall want to extract equations for the large-scale behavior of ρ and $\boldsymbol{j}$. The latter will be derived from a simple form of the local conservation equations:

$$\begin{aligned} \sum_i N_i(t_* + 1,\boldsymbol{r}_* + \boldsymbol{c}_i) &= \sum_i N_i(t_*,\boldsymbol{r}_*)\ , \\ \sum_i \boldsymbol{c}_i N_i(t_* + 1,\boldsymbol{r}_* + \boldsymbol{c}_i) &= \sum_i \boldsymbol{c}_i N_i(t_*,\boldsymbol{r}_*). \end{aligned} \tag{7.10}$$

These equations are exact equations, and they express that at each node the collisions conserve mass and momentum. Since the nodes are not summed over, they represent as many conservation relations as there are nodes.

Now that we have a probabilistic formulation, the next step is to look for equilibrium distributions. An equilibrium distribution can be guessed by analogy with ordinary statistical mechanics, and then checked to verify that it is indeed a steady-state solution of the Liouville equation. In the real world, the difficulties in finding the equilibrium distribution come from correlations between particles, which occur due to finite range interactions. Since for the lattice gas the range of interaction is zero, we expect that the equilibrium distribution will factorize into distributions at each site:

$$P(s(\cdot)) = \prod_{\boldsymbol{r}_*} p(s(\cdot)). \tag{7.11}$$

Of course the equilibrium state must be translation-invariant, so p is independent of r_*. To choose p we ask what is the most general Boolean distribution at a node. There are b variables at each node, and if they are all independent, then

$$p(s(\cdot)) = \prod_i N_i^{s_i}(1-N_i)^{1-s_i}. \tag{7.12}$$

The next step is to substitute this "guesstimated" distribution into the Liouville equation to see if it is correct. Since the distribution is translation-invariant we can ignore the streaming operator S. In addition, we only need to check the distribution at a single node. Now, at each node there are b unknowns, N_i, but there are 2^b equations. It so happens that the above distribution is indeed the equilibrium distribution, provided an additional assumption is made: semi-detailed balance. Recall that for any transition probability

$$\sum_{s'} A(s \to s') = 1, \quad \forall s. \tag{7.13}$$

Semi-detailed balance is the statement that

$$\sum_{s} A(s \to s') = 1, \quad \forall s'. \tag{7.14}$$

This holds trivially for one-to-one deterministic collisions. In the non-deterministic case, semi-detailed balance is equivalent to the statement that if there is equal probability of all states before the collision, then there will be equal probability of all states after the collision. A stronger condition, which is unnecessary here, is detailed balance, also called micro-reversibility:

$$A(s \to s') = A(s' \to s). \tag{7.15}$$

With the assumption of semi-detailed balance one can prove that the following three statements are equivalent:

a. The N_i's are solutions of the 2^b Liouville equations.

b. The N_i's are solutions of the b equations

$$\sum_{s,s'} (s_i' - s_i) A(s \to s') \prod_j N_j^{s_j}(1-N_j)^{1-s_j} = 0. \tag{7.16}$$

c. The N_i's are given by

$$N_i = \frac{1}{1 + \exp(h + \boldsymbol{q} \cdot \boldsymbol{c}_i)}, \quad h, \boldsymbol{q} \text{ arbitrary}. \tag{7.17}$$

Although it cannot be proved that this is the only equilibrium distribution, all numerical simulations on decent-size lattices ($\sim 16 \times 16$ or larger) show quick relaxation to this distribution. From the above results follows immediately a *universality theorem*: Any lattice gas satisfying semi-detailed balance has universal equilibria with mean populations depending only on density $\rho = \sum_i N_i$, and mass current $\boldsymbol{j} = \sum_i \mathbf{c}_i N_i$, independent of the detailed collision laws.

[1] Frisch, U., d'Humières, D., Hasslacher, B., Lallemand, P., Pomeau, Y., Rivet, J.-P., *Complex Systems*, 1, 649 (1987). (Also attached at the end of the lecture notes by Frisch.)

Chapter 8

LATTICE GAS HYDRODYNAMICS: FROM MICRODYNAMICS TO THE NAVIER-STOKES EQUATIONS

We have a field of Boolean variables $n_i(t_*,\boldsymbol{r}_*)$ describing the state of the b links connected to each node on a D-dimensional lattice and a set of velocity vectors $\boldsymbol{c}_i$ having a common modulus c. Averaging over an ensemble of configurations of the field yields the mean populations

$$N_i(t_*,\boldsymbol{r}_*) = \langle n_i(t_*,\boldsymbol{r}_*)\rangle, \tag{8.1}$$

and the mean density and mass current

$$\rho(t_*,\boldsymbol{r}_*) = \sum_i N_i(t_*,\boldsymbol{r}_*), \qquad \rho\boldsymbol{u} = \sum_i \boldsymbol{c}_i N_i(t_*,\boldsymbol{r}_*). \tag{8.2}$$

The conservation relations for mass and momentum are

$$\begin{aligned} \sum_i N_i(t_* + 1,\boldsymbol{r}_* + 1) &= \sum_i N_i(t_*,\boldsymbol{r}_*), \\ \sum_i \boldsymbol{c}_i N_i(t_* + 1,\boldsymbol{r}_* + 1) &= \sum_i \boldsymbol{c}_i N_i(t_*,\boldsymbol{r}_*). \end{aligned} \tag{8.3}$$

Energy conservation has not been introduced separately because it is implied by mass conservation, since all $\boldsymbol{c}_i$ have the same magnitude. For physical systems where a separate macroscopic energy conservation law must be imposed, a multi-speed lattice model can be used. The most common multi-speed models include "rest" particles of zero speed. For example, we can generalize the HPP model to include diagonally traveling particles, with speed $\sqrt{2}$, and rest particles [1]. Collision laws may then be defined such that mass, momentum and energy are conserved independently. However, this lattice has insufficient symmetry to converge to the Navier-Stokes equations, so one has to resort to multi-speed variants of the FHP triangular model to obtain a nontrivial energy variable. We shall not pursue the matter further.

The assumption of semi-detailed balance ([2, Section 2.4]) leads to universal equilibrium solutions with mean populations given by the Fermi-Dirac distribution

$$N_i = f_{FD}(h + \boldsymbol{q}\cdot\boldsymbol{c}_i)\ , \qquad \text{where} \qquad f_{FD}(x) = \frac{1}{1 + e^x}. \tag{8.4}$$

The invariants of the lattice determine the properties of the equilibria, as we shall see below. The Lagrange multipliers, $h(\rho,\boldsymbol{u})$ and $\boldsymbol{q}(\rho,\boldsymbol{u})$, are calculable in principle through the relations

$$\begin{aligned} \rho &= \sum_i f_{FD}(h + \boldsymbol{q}\cdot\mathbf{c}_i), \\ \rho\boldsymbol{u} &= \sum_i \mathbf{c}_i f_{FD}(h + \boldsymbol{q}\cdot\mathbf{c}_i). \end{aligned} \tag{8.5}$$

In general, solutions are not known in closed form. Still, for hydrodynamic velocities small compared to particle velocities, the equilibria can be calculated perturbatively. We expand h and $\boldsymbol{q}$ in powers of $\boldsymbol{u}$, expand the Fermi-Dirac distribution and apply the constraints for average mass and momentum. The derivation ([2, Section 4.2]) and the resulting equilibrium mean populations to second order are

$$N_i^{eq} = \frac{\rho}{b} + \frac{\rho D}{c^2 b} c_{i\alpha}u_\alpha + \rho G(\rho) Q_{i\alpha\beta}u_\alpha u_\beta + O(u^3),$$

where

$$G(\rho) = \frac{D^2}{2c^4 b}\frac{b-2\rho}{b-\rho} \qquad \text{and} \qquad Q_{i\alpha\beta} = \mathbf{c}_{i\alpha}\mathbf{c}_{i\beta} - \frac{c^2}{D}\delta_{\alpha\beta}. \tag{8.6}$$

This is, of course, just an expansion of N_i in powers of $\boldsymbol{u}$. The first two terms in the expansion could have been deduced simply by noting that N_i will have to have *some* expansion in powers of $\boldsymbol{u}$, and requiring that it satisfy the mass and momentum constraints together with the lattice symmetries. However, the quadratic term could not have been found this way; it depends on N_i having a Fermi-Dirac distribution. Some features of that term could still have been guessed. When $Q_{i\alpha\beta}$ is summed over i it must give zero to avoid perturbing the mass constraint. Similarly, when it is multiplied by $c_{i\alpha}$ and summed over i it must give zero to avoid perturbing the momentum equation.

That $G(\rho) = 0$ when $\rho = b/2$ could also have been guessed. Consider a "duality" transformation which interchanges particles and holes. If the equilibrium is independent of the details of the interaction laws (as is ensured by semi-detailed balance), then $N_i \to 1-N_i$ under a duality transformation, and the velocity $\boldsymbol{u}$ will reverse. When $\rho = b/2$, this implies the vanishing of quadratic terms in the velocity.

The *macrodynamical* equations can now be constructed by gluing together local equilibria of the above form. Assume that ρ and $\boldsymbol{u}$ are changing on a spatial scale ϵ^{-1}, that density is $O(1)$ and that $\boldsymbol{u}$ is small compared to the particle speed. We expect the following phenomena

a. relaxation to local equilibria on a time scale independent of ϵ (say, just a few collision times),
b. density perturbations propagating as sound waves on time scale ϵ^{-1}, because the velocity of sound will be $O(1)$ and the distance will be $O(\epsilon^{-1})$,
c. diffusive effects on time scale ϵ^{-2}, because the length scale is ϵ^{-1} and the microscopic viscosity is $O(1)$.

If $N_i^{(0)}$ is the mean equilibrium population based on the local ρ and $\boldsymbol{u}$, then for small ϵ the actual mean population $N_i(t,\boldsymbol{r})$ may be expanded in powers of ϵ

$$N_i(t,\boldsymbol{r}) = N_i^{(0)}(t,\boldsymbol{r}) + \epsilon N_i^{(1)}(t,\boldsymbol{r}) + O(\epsilon^2), \tag{8.7}$$

where the time and space variables are now treated as continuous. The Boolean conservation relations for mass and momentum can now be expanded in powers of ϵ, assuming that ρ and $\boldsymbol{u}$ can be smoothly interpolated to continuous space-time ([2, Section 5]). To leading order, $O(\epsilon)$, the result is a mass continuity equation ([2], (5.6)) and an Euler-like momentum equation ([2] (5.7)). Care must be taken in deriving the second-order approximations to incorporate diffusion effects: all the finite difference must then be expanded to second order.

The *macrodynamical* momentum equation ([2], (5.16)) has a strong resemblance to the Navier-Stokes equation, but with the discrete rotational symmetries still entering via a fourth-order tensor, $T_{\alpha\beta\gamma\delta}$. For those lattices having suitable symmetries (see the chapter on the crystallographic aspects), the macrodynamical equations can be written in a form which brings out their similarities with the equations of fluid dynamics, namely,

$$\begin{aligned} \partial_t \rho + \partial_\beta(\rho u_\beta) &= 0 \ , \\ \partial_t(\rho u_\alpha) + \partial_\beta P_{\alpha\beta} &= \partial_\beta S_{\alpha\beta} + O(\epsilon u^3) + O(\epsilon^2 u^2) + O(\epsilon^3 u), \end{aligned} \tag{8.8}$$

where the momentum flux tensor $P_{\alpha\beta}$ is

$$P_{\alpha\beta} = c_s^2\rho\left(1 - g(\rho)\frac{u^2}{c^2}\right)\delta_{\alpha\beta} + \rho g(\rho) u_\alpha u_\beta, \tag{8.9}$$

and the viscous stress tensor $S_{\alpha\beta}$ is

$$S_{\alpha\beta} = \left(\nu_c(\rho) + \nu_p\right)\left(\partial_\alpha(\rho u_\beta) + \partial_\beta(\rho u_\alpha) - \frac{2}{D}\delta_{\alpha\beta}\partial_\gamma(\rho u_\gamma)\right). \tag{8.10}$$

c_s^2, ν_p, and $g(\rho)$ are defined as follows ([1], (6.17)):

$$c_s^2 = \frac{c^2}{D}, \qquad \nu_p = -\frac{c^2}{2(D+2)}, \qquad g(\rho) = \frac{D}{D+2}\frac{b-2\rho}{b-\rho}. \tag{8.11}$$

From (8.9) we can identify the pressure

$$p = c_s^2\rho\left(1 - g(\rho)\frac{u^2}{c^2}\right). \tag{8.12}$$

The expression for $S_{\alpha\beta}$ in (8.10) is the stress-strain relation for a Newtonian fluid having kinematic viscosity $\nu = \nu_c + \nu_p$ and *zero bulk viscosity*. The kinematic viscosity has two contributions. The "collision viscosity" ν_c depends on the details of the collision rules and is positive [3]. The "propagation viscosity" ν_p is negative; it describes the enhancement of velocity gradients due to propagation on the discrete mesh.

In the hydrodynamic limit ($\epsilon \to 0$), the resulting equations ([2] (7.13)) differ from the Navier-Stokes equations only by the presence of a constant and uniform factor $g(\rho_0)$ in the advection term. This reflects the lack of Galilean invariance at the lattice level. The vanishing of the advection term when $\rho_0 = b/2$ reflects a duality invariance that does not appear in the real world. Still, for $\rho_0 < b/2$ the ordinary incompressible Navier-Stokes equations are recovered by a simple rescaling of time and viscosity:

$$t \leftarrow \frac{t}{g(\rho_0)} \qquad \nu \leftarrow g(\rho_0)\nu. \tag{8.13}$$

A tricky point, in the real world as well in lattice gases, relates to the incompressible limit. At low Mach numbers M, the density differs from a constant and uniform background, equal to ρ_0, by small fluctuations $O(M^2)$. To leading order these may be consistently ignored everywhere, except in the pressure term.

Two strategies are available for calculating the viscosity, the fluctuation-dissipation approach, or noisy hydrodynamics ([2, Section 8.1]). In the next chapter we will use the lattice analog of the Boltzmann approximation to calculate the velocity explicitly ([2, Section 8.2]).

[1] d'Humières, D., Lallemand, P., and Frisch, U., *Europhys. Lett.*, **2**, 291 (1986).

[2] Frisch, U., d'Humières, D., Hasslacher, B., Lallemand, P., Pomeau, Y., Rivet, J.-P., *Complex Systems*, 1, 649 (1987). (Also attached at the end of the lecture notes by Frisch.)

[3] Hénon, M., *Complex Systems*, 1, 762 (1987).

Chapter 9

LATTICE GAS HYDRODYNAMICS: REYNOLDS NUMBER, NOISY HYDRODYNAMICS

This chapter covers three topics:

1. evaluation of the viscosity for the lattice gas,
2. maximum achievable Reynolds numbers,
3. the influence of thermal noise.

With regard to the determination of viscosity of the lattice gas, we will describe it in general terms, and supply some detail only on the nontrivial parts of the calculation. Viscosity can be estimated by means of various approximations; in most instances, it has been found that the Boltzmann approximation is both appropriate and accurate. The resulting predictions are in very good accord with simulation measurements. The potency of this approach may be related to the possibility that the Boltzmann approximation is not only a low density approximation, but may represent the leading order form of a $1/N$ type expansion. For the lattice gas, the number N would be the number of cells per node, b.

The Boltzmann approximation reads as follows in terms of the mean populations:

$$\begin{aligned} N_i(t_* + 1,\ \boldsymbol{r}_* + \boldsymbol{c}_i) &= \sum_{ss'} s_i' A(s \to s') \prod_j N_j^{s_j}(1 - N_j)^{1-s_j} \\ &\equiv N_i^{eq} + \Delta_i\ , \end{aligned} \tag{9.1}$$

where

$$\Delta_i = \sum_{ss'} (s_i' - s_i) A(s \to s') \prod_j N_j^{s_j}(1 - N_j)^{1-s_j}. \tag{9.2}$$

When the equilibrium values $N_i = N_i^{eq}$ are substituted into this expression, it is easy to check that the collision terms vanish, as we must expect. To evaluate the viscosity, we use as before multiscale expansion in space and in time (to calculate $N_i^{(1)}$, we need only Taylor-expand to linear order). A trick used in Chapman-Enskog procedures is useful here, whereby the lowest-order equation is used to reexpress the time derivatives in terms of space derivatives. One of the resulting equations takes the form:

$$\frac{D}{bc^2} Q_{i\alpha\beta}\partial_{i\alpha}(\rho u_\beta) = \sum_j \mathcal{A}_{ij} N_j^{(1)}, \tag{9.3}$$

where

$$\mathcal{A}_{ij} \equiv \left[\frac{\partial \Delta_i}{\partial N_j}\right]_{N_i=\rho/b} . \tag{9.4}$$

This looks like 24 equations in as many unknowns. However, symmetries allow a vast simplification. We may write, following Hénon [1],

$$N_i^{(1)} = \psi Q_{i\alpha\beta}\partial_{i\alpha}(\rho u_\beta). \tag{9.5}$$

The system then reduces to a single equation for the scalar function ψ. This permits a closed-form expression for the viscosity (sum of propagation and collision viscosities):

$$\begin{aligned} \nu &= \frac{c^2}{2(D+2)} \frac{\mu}{1-\mu}, \\ \mu &= \frac{D}{4(D-1)} \frac{1}{bc^4} \sum_{ss'} A(s \to s') d^{p-1}(1-d)^{b-p-1} \sum_{\alpha\beta} (Y_{\alpha\beta} + Y'_{\alpha\beta})^2, \end{aligned} \tag{9.6}$$

where

$$\begin{aligned} d &= \rho_0/b = \text{reduced density}, \\ p &= \sum_i s_i , \\ Y_{\alpha\beta} &= \sum_i \left(s_i c_{i\alpha} c_{i\beta} - \frac{pc^2}{D} \delta_{\alpha\beta} \right) . \end{aligned} \tag{9.7}$$

The last quantity has the form of the deviator of a kind of tensor of inertia.

The most interesting part of the present expression for the viscosity relates to the dependence on $Y_{\alpha p}$. Note that if the tensor of $Y_{\alpha\beta} + Y'_{\alpha\beta}$ is isotropic for a given collision rule, there is zero contribution to the viscosity. An example is the "head-on collision with spectator," introduced by D. Levermore [2]

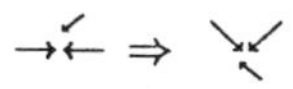

To get small viscosities, we wish to match in- and out-states to minimize the deviators, subject to the necessary conservation properties. This can be accomplished with optimization algorithms for the "perfect matching" problem. (Actually, we are considering a variant, the "perfect mismatch", since we wish to minimize the sum $Y_{\alpha\beta} + Y'_{\alpha\beta}$ and not the difference.)

In this manner, a collision table can be worked out which best minimizes the viscosity [3]. In 2D the best-known model is FHP-III ([4], Sec. 2.2), which includes all possible collisions in a seven-bit model (six with speed-one and one particle at rest).

What are the implications for the resulting Reynolds numbers? The natural unit of length is the lattice constant, and the time unit is the update interval. Recall from the last chapter that the variables which produce the Navier-Stokes equations are scaled as follows:

$$\boldsymbol{r} = \epsilon^{-1}\boldsymbol{r}_1 \ , \ \boldsymbol{u} = \epsilon \boldsymbol{U} \ , \ \nu = g(\rho_o)\nu' . \tag{9.8}$$

These are the variables in which the Reynolds number R must be formed. Hence,

$$R = \ell_0 u_0 \frac{g(\rho_0)}{\nu(\rho_0)} \ . \tag{9.9}$$

u_0 is characteristic velocity in natural units of the lattice. It is useful to bring out the Mach number. Thus we write $u_0 = c_s M$ and obtain:

$$R = M\ell_0 R_*(\rho_0), \quad R_* \equiv \frac{c_s g(\rho_0)}{\nu(\rho_0)} . \tag{9.10}$$

For a real fluid, there is evidence from several sources which indicates that compressibility corrections to incompressible dynamics, which scale like $O(M^2)$, are not prominent for $M \lesssim 0.3$. In lattice gases, the scaling is the same, but the constants seem to be more favorable, probably because of the presence of the $g(\rho)$ factor. Typical values chosen for the Mach number M in lattice gas simulations are thus in the range 0.3 to 0.5.

Note that the function $g(\rho_0)$ is universal for a given lattice, provided that semi-detailed balance is retained. Increasing ℓ_0 is expensive computationally. So, it is worthwhile to expend effort to minimize the viscosity by the strategy outlined previously.

The "Reynolds coefficient" R_*^{max}, the maximum of R_* over the density, is shown here for various models:

	FHP-I	optimal FCHC (pseudo-4D)
R_*^{max}	0.387	7.57
d_{max}	0.187	0.33.

It is necessary to check consistency conditions for our formulation, in terms of the required scale separation (i.e., we must check that our ϵ is acceptable). A necessary condition is that the dissipation scale be large compared with the lattice constant. We can check this in the following way. In a turbulent flow, the ratio of dissipation scale η and integral scale ℓ_0 is:

$$\frac{\eta}{\ell_0} \sim CR^{-m} \begin{cases} m = 3/4 & \text{(3D), according to Kolmogorov 1941} \\ m = 1/2 & \text{(2D), according to Batchelor-Kraichnan.} \end{cases} \quad (9.11)$$

So, we find for the dissipation scale the following estimates:

$$\eta = \begin{cases} C(MR_*^{max})^{-1/2}\ell_0^{1/2} = C(MR_*^{max})^{-1}R^{1/2} & \text{(2D)} \\ C(MR_*^{max})^{-1/4}\ell_0^{1/4} = C(MR_*^{max})^{-1}R^{1/4} & \text{(3D).} \end{cases} \quad (9.12)$$

We stress again that the dissipation scale η is here measured in lattice constants (=meshes). As we aim for high Reynolds numbers, η will be quite large, thereby ensuring the separation of scale required for the validity of the hydrodynamic approximation. Actually, for computational efficacy, it is of interest to *reduce* η by increasing R_*^{max}: a gain of a factor 2 on R_*^{max} amounts to a computational gain of a factor 8 in two dimensions and of a factor 16 in three dimensions.

We have stated that a separation of scale between micro- and macro-worlds is necessary for hydrodynamics, but it may not be sufficient. The following analysis has been worked out in collaboration with V. Yakhot and S. Orszag. We shall show that at high Reynolds numbers there is actually a kind of breakdown of the hydrodynamics in the dissipation range: microscopic noise terms must be added to the incompressible Navier-Stokes equations. The subsequent analysis applies equally well in lattice gases and in the real world, provided the density is O(1) (with the mean free path as unit length).

We already know from the Kolmogorov 1941 theory in 3D and from the Batchelor-Kraichnan theory in 2D that incompressible fluid velocity fluctuations are very small at small scales. Hydrodynamic pressure fluctuations are even smaller since they scale similarly to the square of the former. Maybe they become so small that microscopic fluctuations dominate. Using the mean free path as unit length and the thermal velocity as unit speed, we obtain kinematic viscosities ν which are $O(1)$ (we here ignore subtle divergence effects which could be present in 2D). The dissipation scale η, in any dimension is characterized by its turnover time $t_\eta \sim \eta/\mathrm{v}_\eta$ equal to the viscous diffusive time $t_\eta^{diss} \sim \eta^2/\nu \sim \eta^2$.

Thus $v_\eta \sim 1/\eta$ and $t_\eta \sim 1/\eta^2$. Typical hydrodynamic (macroscopic) turbulent fluctuation of the pressure (actually of pressure/density) is

$$p_\eta^{macro} \sim v_\eta^2 \sim 1/\eta^2. \tag{9.13}$$

Microscopic fluctuations must be evaluated, averaged over a space-time domain of spatial extent η and temporal extent t_η. Spatially there is essentially no microscopic coherence, so that relative fluctuations can be estimated by the usual $1/\sqrt{N}$ argument ($N = \eta^D$ being the number of particles in a box of size η). The temporal coherence of pressure fluctuations is the time necessary for a sound wave to propagate over a distance η, that is $t_\eta^{sound} \sim \eta$. Thus there are $O(\eta)$ such coherence times in t_η. This results in an additional lowering factor of $1/\sqrt{\eta}$ in the relative spatio-temporal fluctuations. Hence the expected relative density (or pressure) fluctuations of microscopic origin should be

$$p_\eta^{micro} \sim \frac{1}{\eta^{D/2}} \cdot \frac{1}{\eta/2} \sim \eta^{(-\frac{1+D}{2})}. \tag{9.14}$$

Comparisons of (9.13) and (9.14) show that in two dimensions the microscopic noise in the pressure swamps the hydrodynamic signal (recall that η measured in units of mean free path is large!). This argument of course applies only in "Flatland," a country which is 2D, even microscopically. In three dimensions, we find that microscopic fluctuations of the pressure are just about equal to the hydrodynamic signal at the dissipation scale.

A random noise term of microscopic origin should therefore be added to the Navier-Stokes equations to accommodate the microscopic fluctuations, ([4, Section 8.1]). This additional noise term has the form of an inhomogeneous forcing term which is a gradient. Therefore, it does not directly affect the dynamics of the velocity field in the incompressible limit. The above analysis can be modified to estimate fluctuations of the incompressible (solenoidal) velocity which are of microscopic origin [5]. Such fluctuations are very small and become relevant only in the far dissipation range. Still, at moderately small Mach numbers M, pressure fluctuations of microscopic origin will contaminate solenoidal velocity fluctuations (by $O(M^2)$ terms). We can therefore expect a breakdown of the usual noiseless hydrodynamics not very far out in the dissipation range. Such questions require further study. Here, we just mention that noise contamination should have implications for the predictability issue in strongly turbulent flows.

[1] Hénon, M., *Complex Systems*, **1**, 762 (1987).

[2] Levermore, D., private communication (1986).

[3] Rivet, J.-P., Hénon, M., Frisch, U., d'Humières, D., *Europhys. Lett.*, **7**, 231 (1988).

[4] Frisch, U., d'Humières, D., Hasslacher, B., Lallemand, P., Pomeau, Y., Rivet, J.-P., *Complex Systems*, **1**, 649 (1987). (Also reprinted at the end of the lectures by U. Frisch.)

[5] Ruelle, D., *Phys. Lett.*, **72A**, 81 (1979).

Chapter 10

LATTICE GAS HYDRODYNAMICS: SOFTWARE AND HARDWARE IMPLEMENTATIONS. CONCLUDING REMARKS

The following is an outline of implementation questions [1,2].

Boundary conditions are easy to implement at the microscopic level. As already stressed by Maxwell, walls are not microscopically "flat": molecules are diffusely reflected in such a way that the average velocity vector is (usually) zero at the wall. In practice, in lattice gas calculations, one implements the no-slip condition by bouncing back the particles from the boundary nodes. The free-slip boundary condition is obtained by a specular reflection. In- and out-flow conditions, which can be quite tricky to implement in traditional methods, are easily handled in lattice gas calculations through particle injection and removal; in this way one can design a lattice gas wind tunnel. Since boundary nodes usually represent a very small fraction of all the nodes, the computational overhead from incorporating boundaries is typically only a few percent.

The density (i.e., mean number of particles per node) is an important parameter that affects the viscosity (and hence the maximum achievable Reynolds number) of a lattice gas CA. The maximum hydrodynamic velocity must be adjusted with care. If it chosen to be too small, the Monte Carlo noise will swamp the macroscopic signal and the Reynolds number will be comparatively small. If the velocity is too large (i.e., the Mach number is too high) compressibility effects and spurious higher-than-quadratic nonlinearities cannot be neglected. Adjusting the hydrovelocity is an art (that is, not easily codified by rigid rules). In traditional floating-point calculations using, for example, spectral methods, comparable craftsmanship is required in adjusting the viscosity: too large, resolution is wasted; too small, truncation errors become severe.

In order to be confident of the Reynolds number at which one is operating the CA, it is desirable to know the value of the viscosity. It can be calculated using the Boltzmann approximation, or measured by simulating the decay of a single Fourier-mode, either a compressible mode (sound-wave) or a shearing mode. Comparison with the expected exponential damping (time-variation-like $\exp(-\nu k^2 t)$) gives an estimate for the viscosity ν.

To extract the hydrovelocity, space-time averaging over "mesoscopic" domains is required. By "mesoscopic" we understand large compared to microscopic distances (the lattice constant), and small compared to macroscopic distances (the integral scale for the velocity and the dissipation scale for the vorticity and other space-derivatives). Time averaging is often found to be unnecessary for the velocity. Space averaging can be done by just adding the microscopic velocities of all the nodes within mesoscopic cells, or by fancier filtering techniques, e.g., Fourier-space filtering which is particularly useful for getting derivatives.

In software simulations of lattice gases, the updating is done in two steps: collision (including boundary-node update) and propagation. There are two basic strategies for software implementation of a lattice gas CA. The first (storage by node) consists in storing the nb bits corresponding to the b different permitted microscopic velocities of n nodes into the same memory word. The second strategy (storage by velocity) consists in storing in one memory, word bits pertaining to identical microscopic velocities of different nodes, usually consecutively located on the lattice. Storage by node allows an easy solution for the collision phase: the b-bit out-state (after collision) of a given node is fetched from a look-up table of $b \times 2^b$ bits with the in-state as address; several tables are used if the collision rules involve a random element. With storage by node, propagation requires the moving of the various bits of a word into different directions; this may be operation-intensive. Storage by velocity renders propagation very easy: it just amounts to a shift; however, collisions become more difficult to implement by a look-up table, since the b bits of a given node must first be assembled from different memory words. General purpose computers may not have the hardware to do this in a small number of clocks. An alternative to look-up tables is "collision by Boolean logic." This is best illustrated by considering the HPP lattice. The collision rule is then that the states $(1,0,1,0)$ and $(0,1,0,1)$ are to be interchanged and that all other states are unaffected. A brute-force Boolean logic implementation is provided by the r.h.s. of (6.5). This requires $4 \times 13 = 52$ operations. However, an equivalent formulation of the HPP collision rule is "if the first bit is different from the second bit and the second bit is different from the third bit and the third bit is different from the fourth bit, then negate all the bits, otherwise leave them as they stand." Using the *XOR* and the *AND* logical operations, this can be implemented in only nine operations. In general, one can start from a brute-force logical expression of the collision operator and apply "Boolean reduction," that is, find an equivalent Boolean expression involving a minimum number of operations. Unfortunately, Boolean reduction is a very complex problem (somewhat like the traveling salesman problem), and no efficient systematic Boolean reduction schemes are known. Choosing among the various strategies above depends very much on specific architectural details of the machine on which the software implementation is done. For example, if the machine has a very large common memory, as on CRAY-2s and some CRAY-XMPs, very efficient implementations of the pseudo-4D model can be made, using an optimized collision look-up table with 2^{24} entries [3].

Existing computers (particularly the so-called supercomputers) are usually optimized for floating-point calculations. For CA calculations it may be better and/or cheaper to use specially designed hardware. This is the case with the MIT Cellular Automaton Machine (CAM) [4], and with École Normale Supérieure's Réseau d'Automates Programmable (RAP), [2]. RAP-1, which has been optimized for 2D lattice gas calculations on 256×512 nodes with up to 16 bits per node, runs at a speed which is a sizeable fraction of the best known CRAY-1 software implementation (depending on the specific model). It displays hydrodynamic phenomena in real time (50 complete lattice updates per second). The cost of its off-the-shelf hardware components is about one thousand dollars. Larger 2D machines, also operating in real time, but able to handle lattices several thousands by several thousands, are now being built. Their cost may be compared to that of a mini-computer. Based on the acquired experience with 3D simulations, designs for 3D hardware are now being considered.

We mention that lattice gas models can be constructed to handle free-boundary flows, multi-phase flows, and combustion [5, 6]. Keeping a sharp interface between two non-mixing species can be tricky in traditional methods. Introducing a microscopic level allows easy solutions. One way to keep molecules A and B segregated is to have a majority rule, such as $2A + B \to 3A$ and $2B + A \to 3B$. Segregation with interfacial tension has recently been achieved by Rothman and Keller [7]. Combustion phenomena can be simulated by having chemical reactions between species releasing heat (this requires a multi-speed model). Another potentially interesting domain of application of lattice gas methods is cavitation, because of its two-phase flow aspects and also because there are connected problems of sound generation.

Lattice gas methods are now becoming competitive with floating-point methods for certain applications; for example, the simulation of transition to full three-dimensionality in flows past obstacles with symmetries [3]. Still, unrestrained enthusiasm would be inappropriate. Floating-point methods are usually a lot more flexible than lattice gas methods, considerable inventiveness being required for the latter, whenever new phenomena are to be introduced (e.g. surface tension). Furthermore, floating-point methods have benefited from about half a century of development. It is thus desirable to develop mixed approaches combining the advantages of both methods.

What can we hope to learn about turbulence from lattice gas simulations? This may depend on how crucial it is to have realistic boundaries. Complex boundaries are easy to implement in lattice gas calculations. High Reynolds number simulations with simple or trivial (periodic) boundary conditions are presently most efficiently achieved by *spectral methods*. One view common among statistical turbulence theorists is that turbulence at a very high Reynolds number is sufficiently universal and that it forgets about the details of the boundaries. According to this view, we may simulate the essential features of very high Reynolds number turbulence by running on a supercomputer a spectral code with periodic boundary conditions at the highest possible Reynolds numbers.

Another view is that without boundaries we may not even be capturing the fine-scale structure correctly, i.e., we are throwing out the baby. At transitional and moderately high Reynolds numbers, boundaries are clearly essential. We cannot rule out that this remains true at very high Reynolds numbers. The traditional picture of fully developed turbulence has the fine-scale structure generated by a cascade process, starting at the most energetic scales. We know, however, that vorticity, a key to understanding the fine-scale structure, usually originates from detached boundary layers generated at the walls. Putting vorticity "by hand" at large scales, as in Taylor-Green simulations, may or may not make a difference. Experimental results obtained in Chicago on turbulent convection at Rayleigh numbers up to 6×10^{12} tend to support the view that keeping realistic boundaries is essential [8]. This may, however, be so because the Chicago experiment has not yet attained a fully developed turbulence regime.

Turbulence is at this moment an undefinable concept and is likely to remain so as long as it is considered interesting. We must be prepared to accept that turbulence presents a mixture of universal and non-universal features. It may even have a complexity comparable to that of biological structures. We do not know what will provide the "crucial" data needed for the theory. Simulations (floating-point and lattice gas) allow a very detailed analysis of moderately turbulent flows. Experiments have no difficulty achieving very high Reynolds numbers, but flow control and visualization can become major challenges. All avenues should be tried.

[1] d'Humières, D., and Lallemand, P., *Complex Systems*, **1**, 598 (1987).

[2] Clouqueur, A., and d'Humières, D., *Complex Systems*, **1**, 584 (1987).

[3] Rivet, J.-P., Hénon, M., Frisch, U., and d'Humières, D., *Europhys. Lett.*, **7**, 231 (1988).

[4] Margolus, N., Toffoli, T., and Vichniac, G., *Phys. Rev. Lett.*, **56**, 1694 (1986).

[5] Clavin, P., d'Humières, D., Lallemand, P., and Pomeau, Y., *C. R. Acad. Sci.* Paris II, **303**, 1169 (1986).

[6] Clavin, P., Lallemand, P., Pomeau, Y., and Searby, J., *J. Fluid Mech.*, **188**, 437 (1988).

[7] Rothman, D., and Keller, J., *J. Stat. Phys.*, **52**, 1119 (1988).

[8] Castaing, B., Gunaratne, G., Heslot, F., Kadanoff, L., Libchaber, A., Thomae, S., Wu, X.-Z., Zalesky, S., and Zanetti, G., *Preprint*, Res. Institutes, Univ. Chicago, 1988.

LARGE-SCALE FLOW DRIVEN BY THE ANISOTROPIC KINETIC ALPHA EFFECT

Retyped from Physica, **28D**, *382 (1987), Elsevier Science Publishers B.V., North-Holland Physics Publishing Division*

U. Frisch[1], Z.S. She[1,2], and P. L. Sulem[1,3]

[1] *CNRS, observatoire de Nice, B.P. 139, 06003, Nice Cedex, France*

[2] *Nanjing University, Nanjing, People's Republic of China*

[3] *School of Mathematical Sciences, Tel Aviv University, Israel*

Three-dimensional incompressible flows lacking parity-invariance (reversal of coordinates) can display a large-scale instability analogous to the α-effect of magnetohydrodynamics, but essentially anisotropic. At low Reynolds number R, the unstable modes have wavenumbers $\mathcal{O}(R^2)$. A specific example is given which leads to the growth of a very strong large-scale Beltrami flow. Nonlinear saturation of the instability is obtained by feed-back on the small-scale flow. All the results are illustrated by full simulations of the three-dimensional Navier-Stokes equations.

1. Introduction

The α-effect [1,2] is a large-scale Magnetohydrodynamic (MHD) instability, usually associated with *helical* flow (that is a flow with non-vanishing velocity–vorticity correlation). It is believed to play an important role in the generation of large-scale magnetic fields in cosmical objects [3]. When the α-effect is present the average $\boldsymbol{B}$ of a weak large-scale magnetic field satisfies

$$\partial_t \boldsymbol{B} = \alpha \nabla \times \boldsymbol{B} + \eta \nabla^2 \boldsymbol{B}, \qquad \nabla \cdot \boldsymbol{B} = 0. \tag{11.1)}$$

At sufficiently large scales the $\eta\nabla^2\boldsymbol{B}$ is irrelevant compared to the $\alpha\nabla\times\boldsymbol{B}$ term. The latter leads to the growth of a Lorentz-force-free field with growth-rate αk (k is the wavenumber). The α-effect has a known analog for *compressible* ordinary (non-MHD) fluid dynamics [4]. For incompressible statistically *isotropic* and helical flow, it has been shown that the average $\boldsymbol{w}$ of a weak large-scale perturbation satisfies [5]

$$\partial_t \boldsymbol{w} = \beta \nabla \times \nabla^2 \boldsymbol{w} + \gamma \nabla^2 \boldsymbol{w},$$
$$\nabla \cdot w = 0. \tag{11.2}$$

At large scales the $\beta\nabla \times \nabla^2 \boldsymbol{w}$ term is dominated by the $\gamma\nabla^2 \boldsymbol{w}$ term; thus no instability is obtained. Sufficiently *anisotropic* two-dimensional and three-dimensional incompressible flows are know to have large-scale instabilities of the negative-viscosity type; the latter contrary to the α-effect, have a growth-rate proportional to the square of the wavenumber [6,7]. The possibility of α-like effects in incompressible three-dimensional flows in the presence of convection or large-scale shear was noticed recently [8,9].

As we shall see, there exists actually a *kinetic* (i.e., non-MHD) large-scale instability governed by a first order operator (like the α-effect) in incompressible, three-dimensional anisotropic flows lacking *parity-invariance.* Parity-invariance is here defined as invariance under simultaneous reversal of the position and velocity vectors (with respect to a suitable center), in a deterministic or statistical sense, depending on the context. Lack of parity-invariance is a broader concept than helicity (obviously parity-invariant flows have zero helicity) and could in some instances (e.g., primordial turbulence) have its origin in parity nonconservation of electroweak interactions.

The paper is organized as follows. In Section 2, we derive the Anisotropic Kinetic Alpha (AKA) effect in the linear regime. We explain why the assumptions usually made (for convenience rather than for basic need) in deriving the MHD α-effect will rule out the AKA effect. In Section 3, we give an explicit example of a time-dependent basic flow displaying large-scale instability by the AKA effect. This is illustrated by a numerical experiment. The nonlinear saturation of the AKA-driven instability is discussed in Section 4 and also illustrated by numerical experiments. Concluding remarks are present in Section 5. All the derivations in this paper are essentially of an asymptotic nature, the expansion parameters being the scale-separation and the Reynolds number. Still, we have chosen to make the presentation as intuitive as possible, leaving formal expansions for the Appendix.

2. The AKA Effect: The Linear Regime

We assume that the basic (small-scale) flow $\boldsymbol{u}^{(0)}(\boldsymbol{r},t)$ is driven by a time-dependent space-and time-periodic (or random homogeneous and stationary) force $\boldsymbol{f}(\boldsymbol{r},t)$. It satisfies the incompressible Navier-Stokes equation (henceforth the incompressibility condition is always understood)

$$\partial_t u_i^{(0)} + \partial_j \left(u_i^{(0)} u_j^{(0)}\right) = -\partial_i p^{(0)} + \nu \nabla^2 u_i^{(0)} + f_i. \tag{11.3}$$

We denote l_0 and t_0 the characteristic spatial and temporal scales of the basic flow, and by V_0 the typical velocity amplitude. $R = l_0 V_0/\nu$ will be called the (small-scale) Reynolds number. In the sequel angular brackets denote space-time (or ensemble) averaging over the small scales; we assume $\langle f \rangle = 0$. We now perturb the basic flow $\boldsymbol{u}^{(0)} \to \boldsymbol{u}$, where $\boldsymbol{u}$ has nontrivial large-scale component

$$\boldsymbol{w} = \langle \boldsymbol{u} \rangle. \tag{11.4}$$

$\boldsymbol{w}$ is assumed to vary on a scale $L \gg l_0$ and time $T \gg t_0$. The small-scale flow $\widetilde{\boldsymbol{u}} = \boldsymbol{u} - \boldsymbol{w}$ is advected by the mean flow, and therefore satisfies

$$\partial_t \widetilde{u}_i + w_j \partial_j \widetilde{u}_i + \partial_j (\widetilde{u}_i \widetilde{u}_j) = -\partial_i \widetilde{p} + \nu \nabla^2 \widetilde{u}_i + f_i. \tag{11.5}$$

Here $\boldsymbol{w}$ may be considered uniform and constant. Thus the average small-scale Reynolds stresses,

$$R_{ij} = \langle \widetilde{u}_i \widetilde{u}_j \rangle, \tag{11.6}$$

become $\boldsymbol{w}$-dependent and, thereby, contribute to the large-scale dynamics of the mean field. The latter satisfies

$$\partial_t w_i + \partial_j (w_i w_j + R_{ij}) = -\partial_i p + \nu \nabla^2 w_i. \tag{11.7}$$

In order to actually write down this equation, we must go through the following steps: (i) solve (11.5) for the perturbed small-scale field, using for example, an expansion in powers of the small-scale Reynolds number (when the latter is small) or numerical tools; (ii) calculate the mean small-scale Reynolds stress R_{ij}; (iii) substitute into (11.7) which will henceforth be referred to as the AKA equation. For a more systematic multi-scale derivation of the AKA equation, see the Appendix.

When the mean field is weak, as we shall assume in the remainder of this section, the average Reynolds stresses may be Taylor-expanded

$$R_{ij} = \langle u_i^{(0)} u_j^{(0)} \rangle + w_l \left[\frac{\partial \langle \widetilde{u}_i \widetilde{u}_j \rangle}{\partial w_l} \right]_{w=0} + \mathcal{O}(w^2). \tag{11.8}$$

Note that, now, the perturbed small-scale field $\widetilde{u}$ is very close to $\boldsymbol{u}^{(0)}$ and may be obtained from a linearized version of (11.5).

Thus, the large-scale flow satisfies an equation, which well be referred to as the linear AKA equation

$$\partial_t w_i = \alpha_{ijl}\partial_j w_l - \partial_i p + \nu\nabla^2 w_i, \tag{11.9}$$

with

$$\alpha_{ijl} = -\left[\frac{\partial\langle\widetilde{\boldsymbol{u}}_i\widetilde{\boldsymbol{u}}_j\rangle}{\partial w_l}\right]_{w=0} \tag{11.10}$$

There are many important instances where the tensor α_{ijl} vanishes. When the basic flow is parity-invariant, this vanishing occurs because (11.10) has an odd number of velocities. When the basic flow is random isotropic (that is invariant under genuine rotations, not including parity), vanishing occurs because the tensor α_{ijl} is by construction symmetrical in i and j and there exists no non-vanishing third order isotropic tensor with such symmetry. When the basic flow is time-independent α_{ijl}, calculated perturbatively in powers of the Reynolds number, vanishes to leading order. Vanishing also occurs when the basic flow is random and δ-correlated in time. Finally, vanishing occurs for the ABC flows [10]. It must be stressed that in MHD, none of the above assumptions, excepting parity-invariance, imply the vanishing of the α-effect.

Once the α_{ijl} tensor has been calculated (perturbatively or otherwise), it is straightforward to analyze the stability of the mean-field equation (11.9). In two dimensions, stability always holds. Indeed, using a stream function representation, it is easily shown that the first order operator involving the α and pressure terms is anti-Hermitian and can only produce dispersive effects. In three dimensions, however, large-scale instabilities are possible, with a growth rate proportional to the wavenumber, just as in MHD. A specific example will be given in Section 3.

3. A Specific Example of AKA Instability

A simple deterministic flow that lacks partiy-invariance (but has no helicity) and displays the AKA effect is obtained with the force

$$\begin{aligned} f_1 &= \frac{\nu V_0\sqrt{2}}{l_0^2}\cos\left(\frac{y}{l_0}+\frac{\nu t}{l_0^2}\right), \\ f_2 &= \frac{\nu V_0\sqrt{2}}{l_0^2}\cos\left(\frac{x}{l_0}-\frac{\nu t}{l_0^2}\right), \\ f_3 &= f_1 + f_2. \end{aligned} \tag{11.11}$$

For small Reynolds number, the corresponding basic flow is easily obtained from the Navier-Stokes equation (11.3); it is given to leading order in R by

$$\begin{aligned} u_1^{(0)} &= V_0 \cos\left(\frac{y}{l_0} + \frac{\nu t}{l_0^2} - \frac{\pi}{4}\right), \\ u_2^{(0)} &= V_0 \cos\left(\frac{x}{l_0} - \frac{\nu t}{l_0^2} + \frac{\pi}{4}\right), \\ u_3^{(0)} &= u_1^{(0)} + u_2^{(0)}. \end{aligned} \tag{11.12}$$

Henceforth, we limit ourselves to the case where the Reynolds number $R = l_0 V_0/\nu$ is small. We may thus, to leading order linearize (11.5) with respect to the small-scale (perturbed) flow $\widetilde{\boldsymbol{u}}$:

$$\partial_t \widetilde{u}_i + w_j \partial_j \widetilde{u}_i = \nu \nabla^2 \widetilde{u}_i + f_i. \tag{11.13}$$

Furthermore, as we shall see, the smallness of the Reynolds number will provide us with the required scale-separation between the basic flow and the AKA-unstable modes. It is now a simple matter to solve (11.13) and calculate the average small-scale Reynolds stresses. We give both the full expressions and their expansions for small $\boldsymbol{w}$, appropriate for the linear AKA effect.

$$\begin{aligned} R_{11} &= \frac{V_0^2}{2 + 2\frac{l_0}{\nu} w_2 + \left(\frac{l_0}{\nu} w_2\right)^2} \\ &= \frac{V_0^2}{2} - \frac{V_0^2 l_0}{2\nu} w_2 + \mathcal{O}(w^2), \\ R_{22} &= \frac{V_0^2}{2 - 2\frac{l_0}{\nu} w_1 + \left(\frac{l_0}{\nu} w_1\right)^2} \\ &= \frac{V_0^2}{2} + \frac{V_0^2 l_0}{2\nu} w_1 + \mathcal{O}(w^2), \\ R_{13} &= R_{11}, \\ R_{23} &= R_{22}, \\ R_{12} &= 0, \\ R_{33} &= R_{11} + R_{22}. \end{aligned} \tag{11.14}$$

In the remainder of this section we concentrate on the linear AKA effect. The nonvanishing components of the tensor α_{ijl}, given by (11.10), are now

$$
\begin{aligned}
\alpha_{112} &= \alpha_{132} = \alpha_{312} = -\alpha_{221} = -\alpha_{231} \\
&= -\alpha_{321} = -\alpha_{331} = \alpha_{332} = \alpha \\
&= \frac{V_0^2 l_0}{2\nu} = \frac{1}{2} R V_0 .
\end{aligned}
\tag{11.15}
$$

Thus weak large-scale perturbations depending only on z (here, the most unstable modes) satisfy

$$
\begin{aligned}
\partial_t w_1 &= \alpha \frac{\partial}{\partial z} w_2 + \nu \frac{\partial^2}{\partial z^2} w_1, \\
\partial_t w_2 &= -\alpha \frac{\partial}{\partial z} w_1 + \nu \frac{\partial^2}{\partial z^2} w_2 .
\end{aligned}
\tag{11.16}
$$

It follows that there are solutions of the form

$$
\psi = w_1 + \mathrm{i} w_2 = e^{ikz} e^{(\alpha k - \nu k^2)t} . \tag{11.17}
$$

When the wavenumber k satisfies

$$
k l_0 < \frac{1}{2} R^2 , \tag{11.18}
$$

such modes are unstable. The resulting large-scale flow is a circularly polarized plane wave with vanishing nonlinear self-interaction (Beltrami flow). We shall refer to this type of exponential growth as *Beltrami runaway*.

Several remarks are in order. Beltrami runaway has also been observed by Herring [11] in simulations of three-dimensional convection at very low Prandtl numbers; it is there caused by a buoyancy-driven large-scale instability. By (11.16), in a system which is unbounded (in the z-direction), the AKA instability is present for arbitrary low Reynolds numbers of the basic flow. This instability is thus much stronger than negative viscosity instabilities of, e.g., the Kolmogorov flow [6,12]; the latter require a sufficiently large number to overcome molecular diffusion. The AKA instability (as well as the α-effect in MHD) is reminiscent of the Darrieus-Landau instability of flame fronts whose growth rate has a similar functional dependence [13]. Finally, we notice that although the basic flow $\boldsymbol{u}^{(0)}$ depends only on x and y, the perturbation is z-dependent, so that the instability is genuinely three-dimensional. In the example above the AKA instability is a soft-mode instability made possible by the translational invariance in z. It is not a phase instability [14], since the latter would require a preexistent non-trivial structure in z.

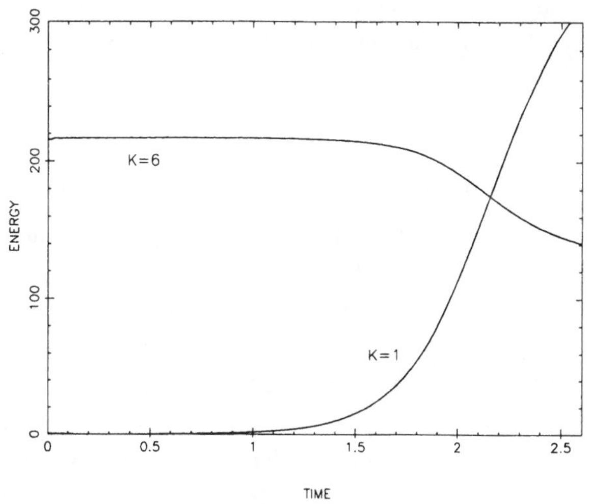

Figure 11.1: Full numerical simulation of 3-D Navier-Stokes equation showing linear AKA instability and the beginning of saturation. The force is applied at $k = 6$, giving a basic flow of Reynolds number $R = \sqrt{2/3}$. Only the modes with wavevector $(0, 0, \pm 1)$ are linearly unstable.

To illustrate the AKA instability, we have performed a full simulation of the three-dimensional Navier-Stokes equation, including both the small and the large scales. We used a spectral method [15] with $(32)^3$ collocation points. The flow was forced at wavenumber $k_0 = 6$ with $V_0 = \sqrt{216}$ and $\nu = 3$, giving a Reynolds number $R = \sqrt{2/3}$, such that only the modes with wavevector $(0, 0, \pm 1)$ are linearly unstable. Figure 11.1, showing the evolution of the energies in the modes of wavenumber one and six, clearly displays the AKA instability. The growing mode is circularly polarized, just as predicted. The measured growth rate of mode one is about 40 percent higher than predicted by (11.17). The latter gives α only to leading order in R; in fact there are corrections of relative strength $\mathcal{O}(R^2)$. The agreement can be improved by decreasing the Reynolds number; however the required resolution goes up as R^{-2} and the required computational resources as R^{-8}. Presently each simulation takes about one hour of CRAY1 so that substantial improvement in the leading-order agreement is at the moment ruled out. Still, we believe that the simulation gives a convincing check on the existence of an AKA instability.

Finally, we note the beginning of the feedback of the growing mode on the basic flow, as evidenced by the depletion of the energy at $k = 6$ beyond time $t = 2$. This will lead to the saturation of the instability discussed in the next section.

4. Nonlinear Saturation of the AKA Instability

When there is a single unstable large-scale mode, such as in the example of Section 3, saturation of the AKA instability cannot come about by self-interaction of this mode. Indeed, the latter vanishes for incompressible flows. Saturation can however take place by feedback on the small-scale flow, i.e., depletion of the average small-scale Reynolds stresses. Indeed, an unchanged small-scale flow if it must do so against a strong quasi-uniform large-scale flow. A similar reason explains the reduction of the α-effect in MHD [16,17]. For the example presented in Section 3, we can work out a quantitative theory of the feedback loop. We make the same assumptions as in Section 3, except that the large-scale mean field $\boldsymbol{w}$ is now strong. The mean field, being only z-dependent has vanishing nonlinear self-interaction. We now use the finite-$\boldsymbol{w}$ expression (11.14) of the Reynolds stresses in the mean field equation (11.7) to obtain the nonlinear AKA equation (for a more systematic derivation, see Appendix)

$$\begin{aligned} \partial_t w_1 + \frac{\partial}{\partial z}\left(\frac{V_0^2}{2 + 2\frac{l_0}{\nu}w_2 + \left(\frac{l_0}{\nu}w_2\right)^2}\right) &= \nu\frac{\partial^2}{\partial z^2}w_1, \\ \partial_t w_2 + \frac{\partial}{\partial z}\left(\frac{V_0^2}{2 - 2\frac{l_0}{\nu}w_1 + \left(\frac{l_0}{\nu}w_1\right)^2}\right) &= \nu\frac{\partial^2}{\partial z^2}w_2. \end{aligned} \tag{11.19}$$

As in Section 3, the Reynolds number $R = l_0V_0/\nu$ is our expansion parameter. In order for all the terms in (11.19) to be relevant, the mean field $\boldsymbol{w}$ must have amplitude $V = \mathcal{O}(V_0R^{-1})$, spatial scale $L = \mathcal{O}(l_0R^{-2})$, and temporal scale $T = \mathcal{O}(R^{-3}l_0/V_0)$. Consequently, we expect the large-scale Reynolds number LV/ν to be $\mathcal{O}(R^{-2})$. In spite of its large value, such a Reynolds number is irrelevant since the large-scale self-interaction vanishes. A detailed study of (11.19) is beyond the scope of this paper. The situation is somewhat simpler if we assume a finite periodicity (say $2\pi L$), large compared to l_0 in the z-direction. If $R < R_c = \sqrt{2l_0/L}$, there is no instability. At $R = R_c$ a *subcritical* bifurcation occurs, and the mean flow jumps to a finite-amplitude steady branch [18].

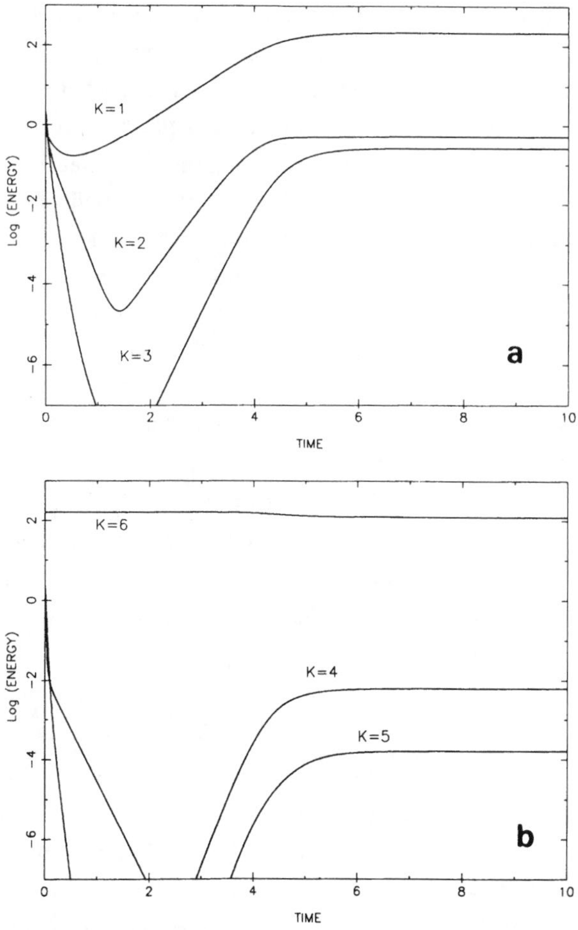

Figure 11.2: Saturation of AKA instability by the feedback mechanism. (a) Energies in modes $k = 1$ to 3, (b) energies in modes $k = 4$ to 6. Forcing is at wavenumber 6. Reynolds number of basic flow is $R = \sqrt{1/2}$. Note the Beltrami runaway phase from $t = 1$ to $t = 4$, followed by feedback saturation leading to a slight depletion of the $k = 6$ modes.

When there is more than one unstable wavenumber, (11.19) must be modified to include all the unstable modes which are not necessarily z-dependent. We shall not elaborate on these matters. Here, we shall just illustrate the saturation mechanism, using, as in Section 3, a full simulation. Forcing wavenumber is again six, $V_0 = \sqrt{162}$, $\nu = 3$, and $R = \sqrt{1/2}$. Figure 11.2 shows the evolution of the energies in wavenumbers 1 through 3 (a) and 4 through 6 (b). A log-scale is used for the energies so as to bring out the exponential growth phases. The most conspicuous feature is that after a phase of exponential Beltrami runaway, the large-scale flow saturates. The small-scale energy also tends to a steady state, but the amplitude of the basic flow ($k = 6$) is somewhat depleted. A few additional observations are in order. At very early times, the $k = 1$ modes are decaying (from a small nonvanishing initial value); this is due to viscous dissipation during a phase where the AKA effect has not yet built up (this takes of the order of one small-scale eddy-turnover-time). During the Beltrami runaway phase, the $k = 2$ and $k = 3$ modes are growing exponentially at rates twice and thrice the rate for the $k = 1$ modes. This happens because these modes are *slaved* to the $k = 1$ modes: they are linearly stable but receive an input from nonlinear interactions of the $k = 1$ modes. The nonlinear interaction is now in the sense of (11.19) (i.e., mediated by feedback), and not in the sense of the original Navier-Stokes equation.

Finally, we explore the possibility of a different kind of nonlinear saturation, namely by self-interaction of the large-scale modes when there are several unstable modes. It is then conceivable that energy transfer to stable modes can check the growth of the unstable modes. The relevant large-scale equation is (see also the Appendix)

$$\partial_t w_i + \partial_j (w_i w_j) = \alpha_{ijl} \partial_j w_l - \partial_i p + \nu \nabla^2 w_i. \tag{11.20}$$

Here, the tensor α_{ijl} is given by its linear theory expression (11.10). Assuming once more that the Reynolds number is small (without necessarily restricting ourselves to the flow considered in Section 3), we find that the components of the tensor α_{ijl} are $\mathcal{O}(V_0 R)$. Thus relevance of all the terms in (11.20) requires a mean field amplitude $V = \mathcal{O}(V_0 R)$, and scales $L = \mathcal{O}(l_0 R^{-2})$, and $T = \mathcal{O}(R^{-3} l_0 / V_0)$. Therefore, the large-scale Reynolds number is now $\mathcal{O}(1)$. A noteworthy feature of (11.20) is that any explicit reference to the driving force has disappeared, so that the equation is invariant under Galilean transformations

[19]. Therefore, it is conceivable the solutions of (11.20) display viscoelasticity with respect to perturbations of scale $\gg L$ [20]. In the absence of a detailed theory, this is of course speculative.

We made various numerical experiments, trying to find nonlinear saturation by self-interaction. For this we modified the force so as to destabilize several large-scale modes. In some experiments we find a plateau in the temporal evolution of the large-scale modes, which may be due to self-interaction. However, eventually the flows undergo Beltrami runaway.

5. Conclusion

We have found that flows lacking parity-invariance can have a three-dimensional large-scale instability leading to strongly coherent Beltrami structures. Lack of parity-invariance, not helicity, is the basic ingredient of the AKA instability. This observations applies also to the α-effect in MHD [21]. The growth rate of the AKA instability being proportional to the wavenumber, it will overcome diffusive damping at sufficiently large scales; so the critical Reynolds number for the AKA instability is zero in an infinitely extended medium. Anisotropy is required for the AKA instability, but there is no threshold as shown by the following argument [19]. Let the total force driving the flow be of the form $\boldsymbol{f}_1 + \epsilon \boldsymbol{f}_2$, where $\boldsymbol{f}_1$ is a random isotropic force, $\boldsymbol{f}_2$ is the deterministic anistropic force used in the specific example of Section 3, and ϵ is a small parameter which controls the amount of anisotropy. Only the anisotropic term contributes to the α_{ijl} tensor in the linear AKA equation (11.9), so that the AKA effect is still present (albeit at scales $\mathcal{O}(\epsilon^{-2})$ larger). We also stress that an AKA effect requires that a uniform large-scale field produces a change in the average small-scale Reynolds stresses; thus, some mechanism is necessary for breaking the Galilean invariance (external forces, rigid boundaries, etc.).

We have corroborated our theoretical analysis with numerical simulations clearly demonstrating the three-dimensional instability. Further supporting evidence can probably be obtained experimentally. In two dimensions, a magnetically driven (time-dependent) cellular flow in an electrolyte can be used to display the large-scale dispersive waves predicted by the theory. To obtain the three-dimensional AKA instability, three-dimensional forcing is needed. In three dimensions there may be a similar effect in porous media lacking isotropy and parity-invariance (say, a school of fish); we expect corrections to Darcy's law proportional to the wavenumber of the large-scale flow, which may be drag-reducing. The astrophysical and geophysical environment with its huge range of scales also provides a natural setting for AKA effects.

It is clear that much work remains to be done on the nonlinear aspects of the AKA instability. What happens when there is a whole range of unstable modes? Does the large-scale nonlinear feedback loop process described by (11.19) lead to an *inverse cascade*, as is the case in MHD [17]? An inverse cascade has been conjectured to exist in ordinary helical hydrodynamics [22]. Preliminary results indicate that inverse cascade effects may be present in anisotropic forced turbulence lacking parity-invariance: when there are several unstable wavenumbers, say $k = 1, 2, 3, 4$, the energy is the smallest wavenumber ($k = 1$) eventually dominates the energy in the fastest growing mode (say $k = 2$) [23].

Finally, we note that in the modelling of strong turbulence, the effect of small-scale motion is usually represented by second order eddy-viscosity terms. Are there situations where first order AKA-type terms should be included? Is there a connection between the "Beltrami-runaway" observed in the context of our present investigation and the Beltramization observed in channel flow simulation data [24]?

Acknowledgements

We are grateful to R H. Kraichnan, H. K. Moffatt, and V. Yakhot for illuminating discussions. The simulations were done on the CRAY1 of the Centre de Calcul Vectoriel pour la Recherche. This work was supported by CNRS (ATP Dynamique des Fluids Geophysique et Astrophysique) and by EEC grant ST-2J-0029-1-F.

APPENDIX

The aim of this Appendix is to derive systematically the various asymptotic regimes discused in the paper. Our expansion parameter is the Reynolds number R of the basic (small-scale) flow, which is assumed to be *small*. We write the Navier-Stokes equation in the following form

$$\begin{aligned} \partial_\tau u_i + R\partial_j(u_i u_j) &= -R\partial_i p + \nabla^2 u_i + f_i, \\ \partial_j u_j &= 0. \end{aligned} \qquad (A.1)$$

We assume that the force $\boldsymbol{f}(\boldsymbol{x}, \tau)$ has zero divergence and is periodic in space and time and that $\langle \boldsymbol{f} \rangle = 0$, where the angular brackets denote averaging over the periodicity.

The characteristic spatial scale and amplitude of the basic flow $\boldsymbol{u}^{(0)}$ are chosen to be $O(1)$ (in R). The natural time-scale for the small-scale flow is here also chosen to be $O(1)$. Consequently, the time-variable τ is related to the time-variable t of (11.3) by $t = R\tau$. The large-scale instabilities that we are looking for are all on the spatial scale R^{-2} and on the temporal scale R^{-4}. We therefore introduce the slow variables

$$\boldsymbol{X} = R^{-2}\boldsymbol{x}, \qquad T = R^{-4}\tau. \qquad (A.2)$$

The corresponding partial derivatives are denoted ∇_i and ∂_T.

All the fields are then expressed in terms of fast and slow variables τ, $\boldsymbol{x}$, and T, $\boldsymbol{X}$. Thus, in the Navier-Stokes equation (A.1), we substitute $\partial_\tau + R^4\partial_T$ for ∂_τ and $\partial_i + R^2\nabla_i$ for ∂_i. We obtain

$$\begin{aligned} (\partial_\tau + R^4\partial_T)u_i + R(\partial_j + R^2\nabla_j)(u_i u_j) &= -R(\partial_i + R^2\nabla_i)p\ , \\ + (\partial + R^2\nabla)^2 u_i + f_i, \quad (\partial_j + R^2\nabla_j)u_j &= 0. \end{aligned} \qquad (A.3)$$

Here, ∂ denotes the gradient with respect to the $\boldsymbol{x}$ variable.

Two different amplitudes of the large-scale flow were predicted in Section 4. One regime had amplitude $O(R^{-1})$ with vanishing large-scale nonlinear interactions and saturation by feedback on the basic flow. The other regime, with large-scale amplitude $O(R)$, had "self-interactions," i.e., non-vanishing large-scale interactions. Different expansions are thus needed. The case with self-interaction is treated first in Section A.1 (the linear AKA effect is a special casé with low amplitude); the feedback case is treated in Section A.2. The contribution of order R^n to the velocity is denoted $\mathbf{v}^{(n)}$, in order to avoid confusion between $\mathbf{v}^{(0)}$ and the basic flow $\boldsymbol{u}^{(0)}$ in the absence of large-scale perturbation.

A.1. The Self-Interaction Regime

We substitute into (A.3) the expansions

$$\begin{aligned} u_i &= \mathrm{v}_i^{(0)} + R\mathrm{v}_i^{(1)} + R^2\mathrm{v}_i^{(2)} + \cdots, \\ p &= p^{(0)} + Rp^{(1)} + R^2p^{(2)} + \cdots. \end{aligned} \tag{A.4}$$

In the present regime, there should be no large-scale flow to order R^0; thus $\langle \mathrm{v}_i^{(0)} \rangle = 0$. We then identify the various powers of R. This leads to a hierarchy of equations, of which we write the first six

(R^0):

$$\partial_\tau \mathrm{v}_i^{(0)} = \partial^2 \mathrm{v}_i^{(0)} + f_i, \tag{A.5.1}$$

(R^1):

$$\partial_\tau \mathrm{v}_i^{(1)} + \partial_j \left(\mathrm{v}_i^{(0)}\mathrm{v}_j^{(0)}\right) = -\partial_i p^{(0)} + \partial^2 \mathrm{v}_i^{(1)}, \tag{A.5.2}$$

(R^2):

$$\begin{aligned} &\partial_\tau \mathrm{v}_i^{(2)} + \partial_j \left(\mathrm{v}_i^{(0)}\mathrm{v}_j^{(1)} + \mathrm{v}_i^{(1)}\mathrm{v}_j^{(0)}\right) \\ &= -\partial_i p^{(1)} + \partial^2 \mathrm{v}_i^{(2)} + 2\nabla_j \partial_j \mathrm{v}_i^{(0)}, \end{aligned} \tag{A.5.3}$$

(R^3):

$$\begin{aligned} &\partial_\tau \mathrm{v}_i^{(3)} + \partial_j \Big(\mathrm{v}_i^{(0)}\mathrm{v}_j^{(2)} + \mathrm{v}_i^{(2)}\mathrm{v}_j^{(0)} \\ &\qquad + \mathrm{v}_i^{(1)}\mathrm{v}_j^{(1)}\Big) + \nabla_j \left(\mathrm{v}_i^{(0)}\mathrm{v}_j^{(0)}\right) \\ &= -\partial_i p^{(2)} - \nabla_i p^{(0)} + \partial^2 \mathrm{v}_i^{(3)} + 2\nabla_j \partial_j \mathrm{v}_i^{(1)}, \end{aligned} \tag{A.5.4}$$

(R^4):

$$\begin{aligned} &\partial_T \mathrm{v}_i^{(0)} + \partial_\tau \mathrm{v}_i^{(4)} + \partial_j \Big(\mathrm{v}_i^{(0)}\mathrm{v}_j^{(3)} + \mathrm{v}_i^{(3)}\mathrm{v}_j^{(0)} \\ &\qquad + \mathrm{v}_i^{(1)}\mathrm{v}_j^{(2)} + \mathrm{v}_i^{(2)}\mathrm{v}_j^{(1)}\Big) + \nabla_j \left(\mathrm{v}_i^{(0)}\mathrm{v}_j^{(1)} + \mathrm{v}_i^{(1)}\mathrm{v}_j^{(0)}\right) \\ &= -\partial_i p^{(3)} - \nabla_i p^{(1)} + \partial^2 \mathrm{v}_i^{(4)} + 2\nabla_j \partial_j \mathrm{v}_i^{(2)} + \nabla^2 \mathrm{v}_i^{(0)}, \end{aligned} \tag{A.5.5}$$

(R^5):

$$\partial_T v_i^{(1)} + \partial_\tau v_i^{(5)} + \partial_j \Big(v_i^{(0)} v_j^{(4)} + v_i^{(4)} v_j^{(0)} + v_i^{(1)} v_j^{(3)} + v_i^{(3)} v_j^{(1)} + v_i^{(2)} v_j^{(2)} \Big) + \nabla_j \Big(v_i^{(0)} v_j^{(2)} + v_i^{(2)} v_j^{(0)} + v_i^{(1)} v_j^{(1)} \Big) = -\partial_i p^{(4)} - \nabla_i p^{(2)} + \partial^2 v_i^{(5)} + 2\nabla_j \partial_j v_i^{(3)} + \nabla^2 v_i^{(1)}. \quad (A.5.6)$$

The above equations must be supplemented by the expanded incompressibility conditions, namely

$$\partial_j v_j^{(0)} = 0, \quad (A.6.1)$$
$$\partial_j v_j^{(1)} = 0, \quad (A.6.2)$$
$$\partial_j v_j^{(2)} + \nabla_j v_j^{(0)} = 0, \quad (A.6.3)$$
$$\partial_j v_j^{(3)} + \nabla_j v_j^{(1)} = 0, \quad (A.6.4)$$
$$\partial_j v_j^{(4)} + \nabla_j v_j^{(2)} = 0, \quad (A.6.5)$$

Equations (A.5.1) to (A.5.6) constitute a set of *linear* PDE's in the fast variables for the unknown fields $\mathbf{v}^{(0)}, \ldots, \mathbf{v}^{(5)}$. Solvability of these equations requires that the inhomogeneous terms be orthogonal to the kernel of the adjoint operators; here, this means that they should have zero mean. Our leading-order large-scale equation will emerge as the solvability condition for (A.5.6). All lower-order solvability conditions happen to be satisfied. Before we can actually write the solvability condition for (A.5.6), we must solve (A.5.1) to (A.5.3).

The condition $\langle \mathbf{v}^{(0)} \rangle = 0$ ensures that (A.5.1) has a unique solution, depending only on the fast variables

$$\mathbf{v}^{(0)} = \mathbf{v}^{(0)}(\boldsymbol{x}, \tau), \quad (A.7)$$

which can easily be obtained by Fourier transformation.

Equation (A.5.2) for $\mathbf{v}^{(1)}$ does not explicitly involve the slow variables. Its solution is determined up to an additive field depending only on the slow variables. We thus write

$$\mathbf{v}^{(1)} = \boldsymbol{w} + \widetilde{\mathbf{v}}^{(1)}, \qquad \boldsymbol{w} = \langle \mathbf{v}^{(1)} \rangle. \quad (A.8)$$

$\tilde{\mathbf{v}}^{(1)}$ is the solution of (A.5.2) with vanishing mean independent of the slow variable. $\boldsymbol{w}$ is the "mean field."

We now consider (A.5.3) for $\mathbf{v}^{(2)}$. By (A.7), the last term in the right-hand side vanishes. When we substitute the decomposition (A.8) for $\mathbf{v}^{(1)}$, we obtain three source terms in the left-hand side, namely $\partial_j\left(\mathrm{v}_i^{(0)}\tilde{\mathrm{v}}_j^{(1)}\right)$, $\partial_j\left(\tilde{\mathrm{v}}_j^{(1)}\mathrm{v}_j^{(0)}\right)$, and $w_j\partial_j\mathrm{v}_i^{(0)}$. The first two do not depend on the slow variables and will not contribute to the leading-order large-scale equation (because they will give $\boldsymbol{X}$-independent contributions to the mean Reynolds stresses). The contribution of the last term will be denoted $\hat{\mathbf{v}}^{(2)}$; it satisfies

$$
\begin{aligned}
\partial_\tau\hat{\mathrm{v}}_i^{(2)} + w_j\partial_j\mathrm{v}_i^{(0)} &= \partial^2\hat{\mathrm{v}}_i^{(2)},\\
\partial_j\hat{\mathrm{v}}_j^{(2)} &= 0,
\end{aligned}
\tag{A.9}
$$

equations which are also easily solved by Fourier transformation.

We now write the solvability condition for (A.5.6)

$$
\begin{aligned}
\partial_T w_i + \nabla_j(w_i w_j + R_{ij}) &= -\nabla_i\langle p^{(2)}\rangle + \nabla^2 w_i,\\
\nabla_j w_j &= 0,\\
R_{ij} &= \langle \mathrm{v}_i^{(0)}\hat{\mathrm{v}}_j^{(2)} + \mathrm{v}_j^{(0)}\hat{\mathrm{v}}_i^{(2)}\rangle.
\end{aligned}
\tag{A.10}
$$

By (A.9), $\hat{\mathbf{v}}^{(2)}$ depends linearly on $\boldsymbol{w}$, thus the mean Reynolds stresses R_{ij} are themselves linear in $\boldsymbol{w}$. Writing

$$
R_{ij} = -\alpha_{ijl}w_l,
\tag{A.11}
$$

we obtain an equation equivalent to (11.20) of section 4.

The linear AKA equation (11.9) is just a special case for small $\boldsymbol{w}$ when the $w_i w_j$ contribution may be dropped.

A.2. The Feedback Regime

An essential assumption for the present regime is that the large-scale field should have vanishing nonlinear interactions to *all orders*, that is, we require

$$
\begin{aligned}
\nabla_j\left(\langle u_i\rangle\langle u_j\rangle\right) &= -\nabla_i\langle p\rangle,\\
\nabla_j\langle u_j\rangle &= 0.
\end{aligned}
\tag{A.12}
$$

We now substitue into (A.3) the expansions

$$
\begin{aligned}
u_i &= R^{-1}\mathrm{v}_i^{(-1)} + \mathrm{v}_i^{(0)} + R\mathrm{v}_i^{(1)} + R^2\mathrm{v}_i^{(2)} + \cdots, \\
p &= R^{-2}p^{(-2)} + R^{-1}p^{(-1)} + p^{(0)} + Rp^{(1)} + \cdots.
\end{aligned} \tag{A.13}
$$

This time the final equation emerges as the solvability condition for the $\mathcal{O}(R^3)$ equation; the ingredients to write it involve only the $\mathcal{O}(R^{-1}$ and $\mathcal{O}(R^0)$ equations. Below, we shall only write the relevant equations. To leading order, we obtain

(R^{-1}):

$$
\begin{aligned}
\partial_\tau \mathrm{v}_i^{(-1)} + \partial_j\left(\mathrm{v}_j^{(-1)}\mathrm{v}_i^{(-1)}\right) &= -\partial_i p^{(-2)} + \partial^2 \mathrm{v}_i^{(-1)}, \\
\partial_j \mathrm{v}_j^{(-1)} &= 0
\end{aligned} \tag{A.14}
$$

This is an "unforced" Navier-Stokes equation in the fast variables. After relaxation of fast transients, we obtain a solution depending only on slow variables:

$$
\mathbf{v}^{(-1)} = \boldsymbol{w}(\boldsymbol{X}, T). \tag{A.15}
$$

To next order, we have

(R^0):

$$
\partial_\tau \mathrm{v}_i^{(0)} + w_j \partial_j \mathrm{v}_i^{(0)} = \partial^2 \mathrm{v}_i^{(0)} + f_i. \tag{A.16}
$$

Here, there is no pressure term, since $w_j \partial_j \mathrm{v}_i^{(0)}$ and f_i have vanishing (fast) divergence. Finally, from the solvability of the $\mathcal{O}(R^3)$ equation, we obtain

$$
\begin{aligned}
\partial_T w_i + \nabla_j R_{ij} &= -\nabla_i \langle p^{(0)} \rangle + \nabla^2 w_i, \\
\nabla_j w_j &= 0, \\
R_{ij} &= \langle \mathrm{v}_i^{(0)} \mathrm{v}_j^{(0)} \rangle,
\end{aligned} \tag{A.17}
$$

where $\mathbf{v}^{(0)}$ is obtained from (A.16), which can be solved by Fourier transformation. For the special case of the flow introduced in Section 3, (A.16) and (A.17) are equivalent to (11.13) and (11.19).

We emphasize that in deriving all this, we assumed the vanishing of large-scale nonlinear interactions (A.12). If we only assume this condition to hold initially, does the solution of (A.16) and (A.17) satisfy it at all times? We have no general answer to this question. However, for the special case introduced in Section 3, the condition is clearly satisfied when the large-scale flow is only z-dependent as we assumed in Section 4.

[1] Steenbeck, M., Krause, F., and Rädler, K. H., *Z. Naturforsch.*, **21**, 369 (1966).

[2] Moffatt, H. K., *Magnetic Field Generation in Electrically Conducting Fluids*, (Cambridge Univ. Press, Cambridge, 1978).

[3] Zeldovich, Ya. B., Ruzmaikin, A. A., and Sokoloff, D. D., "Magnetic Fields in Astrophysics," vol. 3 of: *The Fluid Mechanics of Astrophysics and Geophysics*, P. H. Roberts, ed. (Gordon and Breach, New York, 1983).

[4] Moiseev,S. S., Sagdeev, R. Z., Tur, A. V., Khomenko, G. A., and Yanovskii, V. V., *Z. Eskp. Teor. Fiz.*, **85**, 1979 (1983); *Sov. Phys.* **JETP 58**, 1149 (1983).

[5] Fournier, J.-D., Sulem, P.-L., and Pouquet, A., *J. Phys.*, **A15**, 1393 (1982).

[6] Sivashinski, G. I., *Physica*, **17D**, 243 (1985).

[7] Bayly B. and Yakhot, V. *Phys. Rev.*, **A34**, 381 (1986).

[8] Moiseev, S. S., Rutkevitch, P. B., Tur, A. V., and Yanovskii, V. V., *Proceedings* Intern. Conf. Plasma Physics, Kiev, April 1987 (Naukova Dumka, Kiev, 1987), vol. 2, p. 75.

[9] Tur, A. V., Khomenko, G. A., Gvavamadze, V. V., and Chkhetiani, O. G., *Proceedings* Intern. Conf. Plasma Physics, Kiev, April 1987, (Naukova Dumka, Kiev, 1987), vol. 2, p. 203.

[10] Dombre, T., Frisch, U., Greene, J. M., Hénon, M., Mehr, A., and Soward, A., *J. Fluid Mech.*, **167**, 353 (1986).

[11] J. H. Herring, private communication.

[12] Nepomnyachtchi, A. A. , *Prikl. Math. Mekh.*, **40**, no. 5, 886 (1976).

[13] Landau, L., *Acta Physicochim.*, **19**, 77 (1944).

[14] Kuramoto, Y., *Chemical Oscillations, Waves, and Turbulence* (Springer-Verlag, Berlin, 1984).

[15] Gottlieb, D. and Orszag, S. A., *Numerical Analysis of Spectral Methods* (SIAM, Philadelphia, PA, 1977).

[16] Kraichnan, R. H., *Phys. Rev. Lett.*, **42**, 1677 (1979).

[17] Meneguzzi, M., Frisch, U., and Pouquet, A., *Phys. Rev. Lett.*, **47**, 1060 (1981).

[18] Frisch, U., She, Z. S., and Sulem, P.-L., to appear in Proc. Fifth Beer-Sheva Intern. Seminar on MHD flows and Turbulence, Israel, March 2–6, 1987.

[19] V. Yakhot, private communication.

[20] Frisch, U., She, Z. S., and Thual, O., *J. Fluid Mech.*, **168**, 221 (1986).

[21] Gilbert, A., Frisch, U., and Pouquet, A., *Geophys. Astrophys. Fluid Dynamics*, **42**, 151 (1988).

[22] Brissaud, A., Frisch, U., Léorat, J., Lesieur, M., and Mazure, A., *Phys. Fluids*, **16**, 1366 (1973).

[23] Frisch, U., Scholl, H., She, Z. S., and Sulem, P.-L., A new large-scale instability in three-dimensional anisotropic incompressible flows lacking parity-invariance, to appear in Proc. IUTAM Symp. on Fundamental Aspects of Vortex Motion, Tokyo, September 1987, *Fluid Dynamics Res.*, (1988), in press.

[24] Yakhot, V., Orszag, S. A., Yakhot, A., Panda, R., Frisch, U., and Kraichnan, R. H., Weak interactions and local order in strong turbulence, Preprint Appl. Comput. Math., Princeton Univ. (1987).

LATTICE GAS HYDRODYNAMICS IN TWO AND THREE DIMENSIONS

Uriel Frisch

CNRS, Observatoire de Nice, BP 139, 06003 Nice Cedex, France

Dominique d'Humières

CNRS, Laboratoire de Physique de l'École Normale Supérieure

24 rue Lhomond, 75231 Paris Cedex 05, France

Brosl Hasslacher

Theoretical Division and Center for Nonlinear Studies

Los Alamos National Laboratories, Los Alamos, New Mexico 87544, USA

Pierre Lallemand

CNRS, Laboratoire de Physique de l'École Normale Supérieure

24 rue Lhomond, 75231 Paris Cedex 05, France

Yves Pomeau

CNRS, Laboratoire de Physique de l'École Normale Supérieure

24 rue Lhomond, 75231 Paris Cedex 05, France

and Physique Théorique, Centre d'Études Nucléaires de Saclay, 91191 Gif-sur-Yvette, France

and

Jean-Pierre Rivet

Observatoire de Nice, BP 139, 06003 Nice Cedex, France

and École Normale Supérieure, 45 rue d'Ulm, 75230 Paris Cedex 05

Proceedings of the Workshop "Modern Approaches to Large Non-linear Systems",

Santa Fe, October 26-29 1986

Published in *Complex Systems*, **1** (1987) p. 649-707.

ABSTRACT

Hydrodynamical phenomena can be simulated by discrete lattice gas models obeying cellular automata rules (U. Frisch, B. Hasslacher, and Y. Pomeau, Phys. Rev. Lett. **56**, 1505, (1986); D. d'Humières, P. Lallemand, and U. Frisch, Europhys. Lett. **2**, 291,. (1986)). It is here shown for a class of D-dimensional lattice gas models how the macrodynamical (large-scale) equations for the densities of microscopically conserved quantities can be systematically derived from the underlying exact "microdynamical" Boolean equations. With suitable restrictions on the crystallographic symmetries of the lattice and after proper limits are taken, various standard fluid dynamical equations are obtained, including the incompressible Navier-Stokes equations in two and three dimensions. The transport coefficients appearing in the macrodynamical equations are obtained using variants of the fluctuation-dissipation theorem and Boltzmann formalisms adapted to fully discrete situations.

1. Introduction

It is known that wind or water tunnels can be indifferently used for testing low Mach number flows, provided the Reynolds numbers are identical. Indeed, two fluids with quite different microscopic structures can have the same macroscopic behaviour because the *form* of the macroscopic equations is entirely governed by the microscopic conservation laws and symmetries. Although the values of the *transport coefficients*, such as the viscosity may depend on the details of the microphysics, still, two flows with similar geometries and identical values for the relevant dimensionless transport coefficients are related by similarity.

Recently, such observations have led to a new simulation strategy for fluid dynamics: fictitious microworld models obeying discrete cellular automata rules have been found, such that two and three-dimensional fluid dynamics are recovered in the macroscopic limit.[1,2] Cellular automata, introduced by von Neumann and Ulam,[3] are constituted of a lattice, each site of which can have a finite number of states (usually coded by Boolean variables); the automaton evolves in discrete steps, the sites being simultaneously updated by a deterministic or nondeterministic rule. Typically, only a finite number of neighbours are involved in the updating of any site. A very popular example is Conway's Game of Life.[4] In recent years there has been a renewal of interest in this subject (see e.g. Ref. 5-7), especially because cellular automata can be implemented in massively parallel hardware.[8-10]

The class of cellular automata used for the simulation of fluid dynamics are here called "lattice gas models". Historically, they emerged from attempts to construct discrete models of fluids with varying motivations. The aim of *Molecular Dynamics* is to simulate the *real* microworld in order, for example, to calculate transport coefficients; one concentrates mass and momentum in discrete particles with continuous time, positions and velocities and arbitrary interactions.[11-14] Discrete velocity models, introduced by Broadwell[15] (see also Refs. 16-20), have been used mostly to understand rarefied gas dynamics; the velocity set is now finite; space and time are still continuous and the evolution is probabilistic, being governed by Boltzmann scattering rules. To our knowledge, the first lattice gas model with fluid dynamical features (sound waves) was introduced by Kadanoff and Swift;[21] it uses a master-equation model with continuous time. The first fully deterministic lattice gas model (now known as HPP) with discrete time, positions and velocities was introduced by Hardy, de Pazzis and Pomeau;[22,23] see also related work in Ref. 24. The HPP model, a presentation of which will be postponed to section 2, was introduced to analyze, in as simple a framework as possible, fundamental questions in Statistical Mechanics, such as ergodicity and the divergence of transport coefficients in two dimensions.[23] The HPP model leads to sound waves, which have been observed in simulations on the MIT cellular automaton machine.[8] The difficulties of the HPP model in coping with full fluid dynamics were overcome by Frisch, Hasslacher and Pomeau[1] for the two-dimensional Navier-Stokes equations; models adapted to the three-dimensional case were introduced by d'Humières, Lallemand and Frisch.[2] This has led to rapid development of the subject.[25–47] These papers are mostly concerned with lattice gas models leading to the Navier-Stokes equations. A number of other problems are known to be amenable to lattice gas models: Buoyancy effects,[48] Seismic P-waves,[49] Magnetohydrodynamics,[50–52] Reaction-Diffusion models,[53–55] Interfaces and Combustion phenomena,[56,57] Burgers' model.[58]

The aim of this paper is to present in detail and without unnecessary restrictions the theory leading from a simple class of D-dimensional "one-speed" lattice gas models to the continuum macroscopic equations of fluid dynamics in two and three dimensions. The extension of our approach to multi-speed models, including for example zero-velocity "rest-particles", is quite staightforward; there will be occasional brief comments on such models. We now outline the paper in some detail while emphasizing some of the key steps. Some knowledge of Nonequilibrium Statistical Mechanics is helpful for reading this paper, but we have tried to make the paper self-contained.

Section 2 is devoted to various lattice gas models and their symmetries. We begin with the simple fully deterministic HPP model (square lattice), we then go to the FHP model (triangular lattice) which may be formulated with deterministic or nondeterministic collision rules; finally, we consider a general class of (usually) nondeterministic one-speed models containing the pseudo-4-D face-centered-hypercubic (FCHC) model for use in three dimensions.[2] In this section, we also introduce various abstract symmetry assumptions, which hold for all three models (HPP, FHP, and FCHC), and which will be very useful in reducing the complexity of the subsequent algebra.

In section 3, we introduce the "microdynamical equations", the Boolean equivalent of Hamilton's equations in ordinary Statistical Mechanics. We then proceed with the probabilistic description of an ensemble of realizations of the lattice gas. At this level, the evolution is governed by a (discrete) Liouville equation for the probability distribution function.

In section 4, we show that there are equilibrium statistical solutions with no equal-time correlations between sites. Under some mildly restrictive assumptions a Fermi-Dirac distribution is obtained for the mean populations, which is *universal*, i.e. independent of collision rules. This distribution is parametrized by the mean values of the collision inbehavariants (usually, mass and momentum).

Locally, mass and momentum are discrete, but the mean values of the density and mass current can be tuned continuously, just as in the "real world". Furthermore, space and time can be regarded as continuous by considering local equilibria, slowly varying in space and time (section 5). The matching of these equilibria leads to macroscopic PDE's for the conserved quantities.

The resulting "macrodynamical equations", for the density and mass current, are not invariant under arbitrary rotations. However, in section 6 we show that the essential terms in the macroscopic equations become isotropic as soon as the lattice gas has a sufficiently large crystallographic symmetry group (as is the case for the FHP and pseudo-4-D models, but not for the HPP model).

When the necessary symmetries hold, fluid dynamical equations are derived in section 7. We consider various limits involving large scales and times and small velocities (compared to particle speed). In one limit we obtain the equations of scalar sound waves; in another limit we obtain the incompressible Navier-Stokes equations in two and three dimensions. It is noteworthy that Galilean invariance, which does not hold at the microscopic level, is restored in these limits.

In section 8 we show how to determine the viscosities of lattice gases. They can be expressed in terms of equilibrium space-time correlation functions via an adaptation to lattice gases of fluctuation-dissipation relations. This is here done with a view-point of "noisy"hydrodynamics, which also brings out the crossover pecularities of two dimensions, namely a residual weak scale-dependence of transport coefficients at large scales. Alternatively, fluctuation-disipation relations can be obtained from the Liouville equation with a Green-Kubo formalism.[43] Fully explicit expressions for the viscosities can be derived via the "Lattice Boltzmann Approximation", not needed for any earlier steps. This is a finite-difference variant of the discrete-velocity Boltzmann approximation. The latter, which assumes continuous space and time variables, is valid only at low densities, while its lattice variant seems to capture most of the finite-density effects (with the exception of two-dimensional crossover effects). Further studies of the Lattice Boltzmann Approximation may be found in Ref. 42. Implications for the question of the Reynolds number are discussed at the end of the section.

Section 9 is the conclusion. Various questions are left for the appendices: detailed technical proofs, inclusion of body forces, catalog of results for various FHP models, proof of an H-theorem for the Lattice Boltzmann Approximation (due to M. Hénon).

2. Deterministic and nondeterministic lattice gas models

2.1 The HPP model

Let us begin with a heuristic construction of the HPP model.[(22–24)] Consider a two-dimensional square lattice with unit lattice constant as shown in fig. 1. Particles of unit mass and unit speed are moving along the lattice links and are located at the nodes at integer times. Not more than one particle is to be found at a given time and node, moving in a given direction (exclusion principle). When two and exactly two particles arrive at a node from opposite directions (head-on collisions), they immediately leave the node in the two other, previously unoccupied, directions (fig. 2). These deterministic collision laws obviously conserve mass (particle number) and momentum and are the only nontrivial ones with these properties. Furthermore, they have the same discrete invariance group as the lattice.

The above definition can be formalized as follows. We take an L by L square lattice, periodically wrapped around (a nonessential assumption, made for convenience). Eventually, we shall let $L \to \infty$. At each node, labelled by the discrete vector $\mathbf{r}_\star$, there are four cells labelled by an index i, defined modulo four. The cells are associated to the unit vectors $\mathbf{c}_i$ connecting the node to its four nearest neighbours (i increases counterclockwise). Each cell $(\mathbf{r}_\star, i)$ has two states coded with a Boolean variable: $n_i(\mathbf{r}_\star) = 1$ for "occupied" and $n_i(\mathbf{r}_\star) = 0$ for "unoccupied". A *cellular automaton* updating rule is defined on the *Boolean field* $n_. = \{n_i(\mathbf{r}_\star), i = 1,...,4, \mathbf{r}_\star \in \text{Lattice}\}$. It has two steps. Step one is *collision*: at each node the four-bit states $(1,0,1,0)$ and $(0,1,0,1)$ are exchanged; all other states are left unchanged. Step two is *propagation*: $n_i(\mathbf{r}_\star) \to n_i(\mathbf{r}_\star - \mathbf{c}_i)$. This two-step rule is applied at each integer time, $t_\star$. An example of implementation of the rule, in which arrows stand for cell-occupation, is shown in figs. 1a and 1b.

Collisions in the HPP model conserve mass and momentum locally, whereas propagation conserves them globally. (Actually momentum is conserved along each horizontal and vertical line, resulting in far too many conserved quantities for physical modelling.)

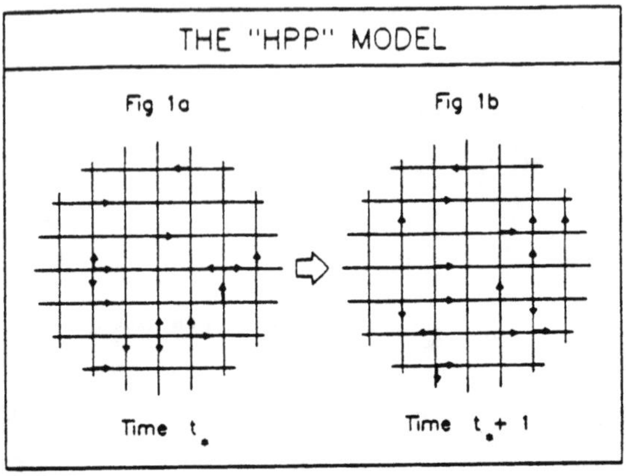

Fig. 1 The HPP model. The black arrows are for cell-occupation. In (a) and (b) the lattice is shown at two successive times.

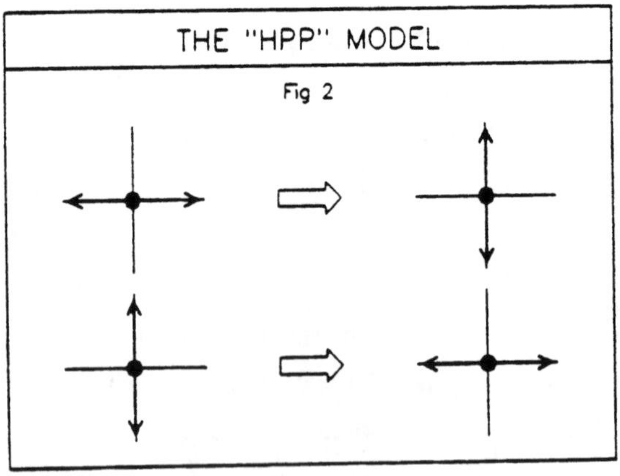

Fig. 2 Collision rules for the HPP model.

If we attribute to each particle a kinetic energy 1/2, the total kinetic energy is also conserved. Energy conservation is however indistinguishable from mass conservation and will not play any dynamical role. Models having an energy conservation law independent of mass conservation will not be considered in this paper (see Refs. 2, 29).

The dynamics of the HPP model is invariant under all discrete transformations that conserve the square lattice: discrete translations, rotations by $\pi/2$, and mirror symmetries. Furthermore, the dynamics is invariant under *duality*, that is exchange of 1's and 0's (particles and holes).

2.2 *The FHP models*

The FHP models I, II, and III (see below), introduced by Frisch, Hasslacher and Pomeau[1] (see also Refs. 25-31, 35, 38-44, 46) are variants of the HPP model with a larger invariance group. The residing lattice is triangular with unit lattice constant (fig. 3). Each node is now connected to its six neighbours by unit vectors $\mathbf{c}_i$ (with i defined modulo six) and is thus endowed with a six-bit state (or seven, cf. below). Updating involves again propagation (defined as for HPP) and collisions.

In constructing collision rules on the triangular lattice, we must consider both *deterministic* and *nondeterministic* rules. For a head-on collision with occupied "input channels" $(i,\ i+3)$, there are two possible pairs of occupied "output channels" such that mass and momentum are conserved, namely $(i+1,\ i+4)$ and $(i-1,\ i-4)$ (see fig. 4a). We can decide always to choose one of these channels; we then have a deterministic model, which is *chiral*, i.e. not invariant under mirror-symmetry. Alternatively, we can make a nondeterministic (random) choice, with equal probabilities to restore mirror-symmetry. Finally, we can make a pseudo-random choice, dependent, for example, on the parity of a time or space index.

We must also consider *spurious conservation laws*. Head-on collisions conserve, in addition to total particle number, the difference of particle numbers in any pair of opposite directions $(i,\ i+3)$. Thus, head-on collisions on a triangular lattice conserve a total of four scalar quantities. This means that in addition to mass and momentum conservation

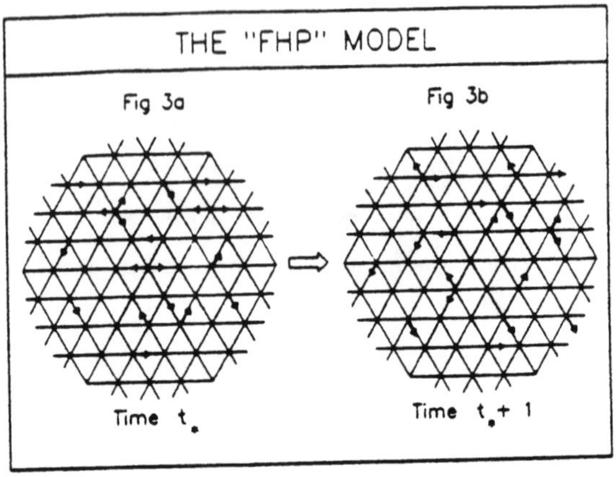

Fig. 3 The FHP model with binary head-on and triple collisions at two successive times.

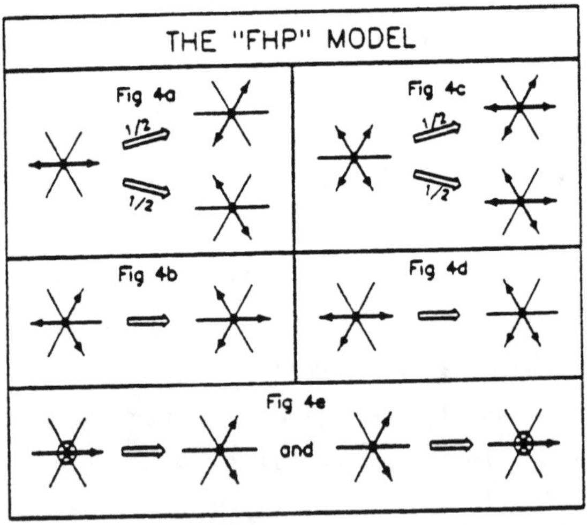

Fig. 4 Collision rules for the FHP models. (a) Head-on collision with two output channels given equal weights; (b) triple collision; (c) dual of head-on collision under particle-hole exchange; (d) head-on collision with spectator; (e) binary collisions involving one rest-particle (represented by a circle).

there is a spurious conservation law. The large-scale dynamics of such a model will differ drastically from ordinary hydrodynamics, unless the spurious conservation law is removed. One way to achieve this is to introduce triple collisions $(i,\ i+2,\ i+4) \to (i+1,\ i+3,\ i+5)$ (see fig. 4b).

Several models can be constructed on the triangular lattice. The simplest set of collision rules with no spurious conservation law, which will be called FHP- I, involves only (pseudo-random) binary head-on collisions and triple collisions. FHP-I is not invariant under duality (particle-hole exchange), but can be made so by inclusion of the duals of the head-on collisions (see fig. 4c). Finally, the set of collision rules can be saturated (exhausted) by inclusion of head-on collisions with a "spectator",(59) that is, a particle which remains unaffected in a collision; fig. 4d is an example of a head-on collision with a spectator present.

The model, FHP-II, is a seven bit variant of FHP-I including a zero-velocity "rest-particle", the additional collision rules of fig. 4e, and variants of the head-on and triple collisions of figs 4a and 4b with a spectator rest-particle. Binary collisions involving rest-particles remove spurious conservations, and do so more efficiently at low densities than triple collisions. Finally, model FHP-III is a collision-saturated version of FHP-II.(31) For simplicity we have chosen not to cover the theory of models with rest-particles in detail.

The dynamics of the FHP models are invariant under all discrete transformations that conserve the triangular lattice: discrete translations, rotations by $\pi/3$, and mirror-symmetries with respect to a lattice line (we exclude here the chiral variants of the models).

2.3 The face-centered-hypercubic 4-D and the pseudo-4-D models

Three dimensional regular lattices do not have enough symmetry to ensure macroscopic isotropy.(1,2,39) A suitable four-dimensional model has been introduced by d'Humières, Lallemand and Frisch.(2) The residing lattice is face-centered-hypercubic (FCHC), defined as the set of signed integers (x_1, x_2, x_3, x_4) such that $x_1 + x_2 + x_3 + x_4$ is even. Each node is connected via links of length $c = \sqrt{2}$ to 24 nearest neighbours, having two coordinates differing by ± 1. Thus the FCHC model has 24-bit states. The 24 possible velocity vectors

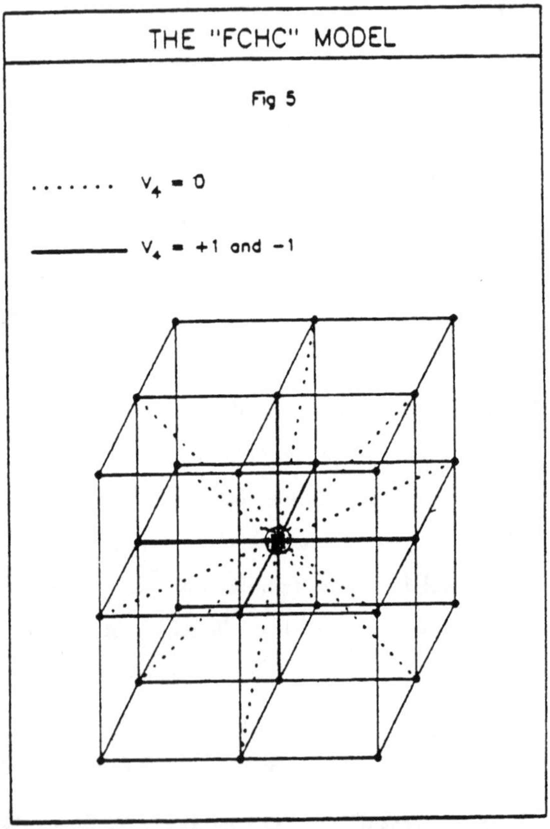

Fig. 5 The pseudo-4-D FCHC model. Only the neighbourhood of one node is shown. Along the dotted links, connecting to next-nearest neighbours, at most one particle can propagate, with component $v_4 = 0$; along the thick black links, connecting to nearest neighbours, up to two particles can propagate, with components $v_4 = \pm 1$.

are again denoted $\mathbf{c}_i$; for the index i there is no preferred ordering and we shall leave the ordering unspecified. Propagation for the FCHC lattice gas is as usual. Collision rules should conserve mass and four-momentum while avoiding spurious conservations. This can be achieved with just binary collisions, but better strategies are known.[32,33] Nondeterministic rules involving transition probabilities are needed to ensure that the collisions and the lattice have the same invariance group (precise definitions are postponed to section 2.4).

The allowed transformations of the FCHC model are discrete translations and those isometries generated by permutations of coordinates, reversal of one or several coordinates and symmetry with respect to the hyperplane $x_1 + x_2 + x_3 + x_4 = 0$.

To model 3-D fluids and maintain the required isotropies, we define the pseudo-4-D model[2] as the three-dimensional projection of an FCHC model with unit periodicity in the x_4-direction (see fig. 5). It resides on an ordinary cubic lattice with unit lattice constant. The full four-dimensional discrete velocity structure is preserved as follows. There is one communication channel to the twelve next-nearest neighbours (corresponding to the twelve velocity vectors such that v_4, the fourth component of the velocity, vanishes) and there are two communication channels to the six nearest-neighbours (corresponding respectively to velocities with $v_4 = \pm 1$). During the propagation phase, particles with $v_4 = \pm 1$ move to nearest neighbour nodes, while particles with $v_4 = 0$ move to next-nearest neighbours. The collision strategy is the same as for the FCHC model, so that four-momentum is conserved. The fourth component is not a spuriously conserved quantity because, in the incompressible limit, it does not effectively couple back to the other conserved quantities.[2]

2.4 A general class of nondeterministic models

In most of this paper we shall work with a class of models (generally nondeterministic) encompassing all the above one-speed models. The relevant common aspects of all those models are now listed: There is a regular lattice, the nodes of which are connected to nearest neighbours through links of equal length; all velocity directions are in some sense equivalent and the velocity set is invariant under reversal; at each node there is a cell associated with each possible velocity. This cell can be occupied by one particle at most; particles are indistinguishable; particles are marched forward in time by successively applying collision and propagation rules; collisions are purely local, having the same invariances as the velocity set; and collisions conserve only mass and momentum.

We now give a more formal definition of these one-speed models as cellular automata. Let us begin with the geometrical aspects. We take a D-dimensional Bravais lattice $\mathcal{L}$ in $\mathbf{R}^D$ of finite extension $O(L)$ in all directions (eventually, $L \to \infty$); the position vector $\mathbf{r}_*$ of any node of such a lattice is a linear combination with integer coefficients of D independent generating vectors.[60] We furthermore assume that there exists a set of b "velocity vectors" $\mathbf{c}_i$ having equal modulus c, the *particle speed*. $\mathbf{c}_i$ has spatial components $c_{i\alpha}$ $(\alpha = 1, ..., D)$.[1] We require the following for $\mathbf{c}_i$:

(i) for any $\mathbf{r}_* \in \mathcal{L}$, the set of the $\mathbf{r}_* + \mathbf{c}_i$'s is the set of nearest neighbours of $\mathbf{r}_*$;

(ii) any two nodes can be connected via a finite chain of nearest neighbours;

(iii) for any pair, $(\mathbf{c}_i, \mathbf{c}_j)$, there exists an element in the "crystallographic" group $\mathcal{G}$ of isometries globally preserving the set of velocity vectors, which maps $\mathbf{c}_i$ into $\mathbf{c}_j$;

(iv) for any velocity vector $\mathbf{c}_i$, we denote by $\mathcal{G}_i$ the subgroup of $\mathcal{G}$ which leaves $\mathbf{c}_i$ invariant and thus leaves its orthogonal hyperplane, Π_i, globally invariant; we assume that (a) there is no non-vanishing vector in Π_i invariant under all the elements of $\mathcal{G}_i$, and (b) the only linear transformations within the space Π_i commuting with all the elements of $\mathcal{G}_i$ are proportional to the identity.

Now, we construct the automaton. To each node $\mathbf{r}_*$, we attach a b-bit state $n(\mathbf{r}_*) = \{n_i(\mathbf{r}_*), i = 1, ..., b\}$, where the n_i's are Boolean variables. The updating of the "Boolean field", $n(.)$, involves two successive steps: collision, followed by propagation. We choose this particular order for technical convenience; after a large number of iterations it will

---

[1] In this paper Greek and Roman indices refer respectively to components and velocity labels. Summation over repeated Greek indices, but not Roman ones, is implicit.

become irrelevant which step was first.[2] Propagation is defined as

$$n_i(\mathbf{r}_*) \rightarrow n_i(\mathbf{r}_* - \mathbf{c}_i). \tag{2.1}$$

The spatial shifting by $\mathbf{c}_i$ is performed on a periodically[3] wrapped around lattice with $O(L)$ sites in any direction; eventually $L \rightarrow \infty$. Collision is the simultaneous application at each node of nondeterministic transition rules from an in-state $s = \{s_i,\, i = 1, ..., b\}$ to an out-state $s' = \{s'_i,\, i = 1, ..., b\}$. Each transition is assigned a probability $A(s \rightarrow s') \geq 0$, normalized to one ($\sum_{s'} A(s \rightarrow s') = 1\ \forall s$), and depending only on s and s' and not on the node. The following additional assumptions are made.

(v) Conservation laws: the *only* collections of b real numbers a_i such that

$$\sum_i (s'_i - s_i) A(s \rightarrow s') a_i = 0, \quad \forall s, s', \tag{2.2}$$

are linear combinations of 1 (for all i) and of $c_{i1}, ..., c_{iD}$, i.e. a_i is related to mass and momentum conservation.

(vi) Invariance under all isometries preserving the velocity set:

$$A\Big(\mathbf{g}(s) \rightarrow \mathbf{g}(s')\Big) = A(s \rightarrow s'), \quad \forall \mathbf{g} \in \mathcal{G}, \quad \forall s,\, s'. \tag{2.3}$$

(vii) Semi-detailed balance:

$$\sum_s A(s \rightarrow s') = 1, \quad \forall s'. \tag{2.4}$$

Various comments are now in order. Semi-detailed balance, also used in discrete velocity Boltzmann models,[(17)] means that if before collision all states have equal probabilities,

[2] For deterministic lattice gases, such as HPP, it is possible to bring out the reversibility of the updating rule by defining the state of the automaton at half-integer times, with particles located at the middle of links connecting nearest-neighbour nodes; updating then comprises half a propagation, followed by collision, followed by another half propagation.[(22)]

[3] Other boundary conditions at the lattice edge can also be used, for example "wind-tunnel" conditions.[(25,26,28)]

they stay so after collision. It is trivially satisfied when the collision rule is deterministic and one-to-one. There exists also a stronger assumption, detailed balance (that is $A(s \to s') = A(s' \to s)$), which will not be needed here. The HPP, FHP, and FCHP lattice gases satisfy the above assumptions (i) through (iv). The proofs are given in Appendix A. The other assumptions (v) through (vii) hold by construction with the exception of the chiral versions of FHP. The latter do not satisfy (vi) because the collision rules are not invariant under the mirror-symmetries with respect to velocity vectors. Full $\mathcal{G}$-invariance holds for the *velocity set* of the pseudo-4-D model, which is the same as for the FCHC model; however, the spatial structure is only invariant under the smaller group of the 3-D cubic lattice.

The invariance assumptions introduced above have important consequences for the transformation properties of vectors and tensors. The following definitions will be used. A tensor is said to be $\mathcal{G}$-invariant if it is invariant under any isometry in $\mathcal{G}$. A set of i-dependent tensors of order p $\{T_i = t_{i\alpha_1\alpha_2\ldots\alpha_p}, i = 1, \ldots, b\}$ is said to be $\mathcal{G}$-invariant if any isometry in $\mathcal{G}$ changing $\mathbf{c}_i$ into $\mathbf{c}_j$, changes T_i into T_j. Note that this is stronger than global invariance under the group $\mathcal{G}$. The velocity moment of order p is defined as $\sum_i c_{i\alpha_1} c_{i\alpha_2} \ldots c_{i\alpha_p}$.

We now list the transformation properties following from $\mathcal{G}$-invariance. The proofs are given in Appendix B.

P1 *Parity-invariance.* The set of velocity vectors is invariant under space-reversal.

P2 Any set of i-dependent vectors $v_{i\alpha}$, which is $\mathcal{G}$-invariant, is of the form $\lambda \mathbf{c}_{i\alpha}$.

P3 Any set of i-dependent tensors $t_{i\alpha\beta}$, which is $\mathcal{G}$-invariant, is of the form $\lambda \mathbf{c}_{i\alpha} \mathbf{c}_{i\beta} + \mu \delta_{\alpha\beta}$.

P4 *Isotropy of second order tensors.* Any $\mathcal{G}$-invariant tensor $t_{\alpha\beta}$ is of the form $\mu \delta_{\alpha\beta}$.

P5 Any $\mathcal{G}$-invariant third order tensor vanishes.

P6 *Velocity moments.* Odd order velocity moments vanish. The second order velocity moment is given by

$$\sum_i \mathbf{c}_{i\alpha} \mathbf{c}_{i\beta} = \frac{bc^2}{D} \delta_{\alpha\beta}. \tag{2.5}$$

There is, in general, no closed form expression for even order velocity moments beyond second order, with the assumptions made up to this point (cf. also section 6).

3. Microdynamics and probabilistic description

3.1 Microdynamical equations

It is possible to give a compact representation of the "microdynamics", describing the application of the updating rules to the Boolean field. This is the cellular automaton analog of Hamilton's equations of motion in Classical Statististical Mechanics. We begin with the HPP lattice gas (section 2.1). Let $n_i(t_*, \mathbf{r}_*)$, as defined in section 2.1, denote the HPP Boolean field at the discrete time t_*. With i labelling the four cells of an HPP node, the collision rule can be formulated as follows: If the in-state has i and $i+2$ empty and $i+1$ and $i+3$ occupied, then the opposite holds in the out-state; similarly, if the in-state has $i+1$ and $i+3$ empty and i and $i+2$ occupied; otherwise, the content of cell i is left unchanged. Thus, the updating of the Boolean field may be written

$$n_i(t_* + 1, \mathbf{r}_* + \mathbf{c}_i) = (n_i \wedge \neg(n_i \wedge n_{i+2} \wedge \neg n_{i+1} \wedge \neg n_{i+3})) \vee (n_{i+1} \wedge n_{i+3} \wedge \neg n_i \wedge \neg n_{i+2}), \tag{3.1}$$

where the whole r.h.s. is evaluated at t_* and $\mathbf{r}_*$. The symbols $\wedge$, $\vee$, and $\neg$ stand for **AND**, **OR**, and **NOT**, respectively. It is known that any Boolean relation can be recoded in arithmetic form ($\wedge$ becomes multiplication, $\neg$ becomes one minus the variable, etc.). In this way we obtain

$$n_i(t_* + 1, \mathbf{r}_* + \mathbf{c}_i) = n_i(t_*, \mathbf{r}_*) + \Delta_i(n). \tag{3.2}$$

The "collision function" $\Delta_i(n)$, which can take the values ± 1 and 0, describes the change in $n_i(t_*, \mathbf{r}_*)$ due to collisions. For the HPP model, it depends only on i and on the set of n_j's at t_* and $\mathbf{r}_*$, denoted n; it is given by

$$\Delta_i(n) = n_{i+1}n_{i+3}(1 - n_i)(1 - n_{i+2}) - n_i n_{i+2}(1 - n_{i+1})(1 - n_{i+3}). \tag{3.3}$$

Equation (3.2) (with $\Delta_i(n)$ given by (3.3)) will be called the *microdynamical* HPP equation.

It holds for arbitrary i (modulo four), for arbitrary integer t_*, and for arbitrary $\mathbf{r}_* \in \mathcal{L}$ ($\mathcal{L}$ designates the lattice).

It is easy to extend the microdynamical formalism to other models. For FHP-I (section 2.2), we find that the collision function may be written (i is now defined modulo six)

$$\begin{aligned}\Delta_i(n) = \xi_{t_*\mathbf{r}_*} & n_{i+1}n_{i+4}(1-n_i)(1-n_{i+2})(1-n_{i+3})(1-n_{i+5}) \\ +(1-\xi_{t_*\mathbf{r}_*}) & n_{i+2}n_{i+5}(1-n_i)(1-n_{i+1})(1-n_{i+3})(1-n_{i+4}) \\ & -n_in_{i+3}(1-n_{i+1})(1-n_{i+2})(1-n_{i+4})(1-n_{i+5}) \\ & +n_{i+1}n_{i+3}n_{i+5}(1-n_i)(1-n_{i+2})(1-n_{i+4}) \\ & -n_in_{i+2}n_{i+4}(1-n_{i+1})(1-n_{i+3})(1-n_{i+5}).\end{aligned} \tag{3.4}$$

Here, $\xi_{t_*\mathbf{r}_*}$ denotes a time- and site-dependent Boolean variable which takes the value one when head-on colliding particles are to be rotated couterclockwise and zero otherwise (remember, that there are two possible outcomes of such collisions). For the theory, the simplest choice is to assign equal probabilities to the two possibilities and to assume independence of all the ξ's. In practical implementations other choices are often more convenient.

We now give the microdynamical equation for the general class of nondeterministic models defined in section 2.4. Propagation is as before. For the collision phase at a given node, it is convenient to sum over *all* 2^b in-states $s = \{s_i = 0 \text{ or } 1,\ i = 1, ..., b\}$ and 2^b out-states s'. The nondeterministic transitions are taken care of by the introduction at each time and node and for any pair of states (s, s') of a Boolean variable $\xi_{ss'}$ (time and space labels omitted for conciseness). We assume that

$$\langle \xi_{ss'} \rangle = A(s \to s'), \quad \forall s, s', \tag{3.5}$$

where $A(s \to s')$ is the transition probability introduced in section 2.4; the angular brackets denote averaging. We also assume that

$$\sum_{s'} \xi_{ss'} = 1, \quad \forall s. \tag{3.6}$$

Since the ξ's are Boolean, eq. (3.6) means that, for a given in-state s and a given realization of $\xi_{ss'}$, one and only one out-state s' is obtained. It is now clear that the microdynamical equation can be written as

$$n_i(t_\star + 1, \mathbf{r}_\star + \mathbf{c}_i) = \sum_{s,s'} s'_i \xi_{ss'} \prod_j n_j^{s_j} (1 - n_j)^{(1-s_j)}. \tag{3.7}$$

The factor s'_i ensures the presence of a particle in the cell i after the collision; the various factors in the product over the index j ensure that before the collision the pattern of n_j's matches that of s_j's. Using (3.7) and the identity

$$\sum_s s_i \prod_j n_j^{s_j} (1 - n_j)^{(1-s_j)} = n_i, \tag{3.8}$$

we can rewrite the microdynamical equation in a form that brings out the collision function

$$\begin{aligned} n_i(t_\star + 1, \mathbf{r}_\star + \mathbf{c}_i) &= n_i + \Delta_i(n) \\ \Delta_i(n) &= \sum_{s,s'} (s'_i - s_i) \xi_{ss'} \prod_j n_j^{s_j} (1 - n_j)^{(1-s_j)}. \end{aligned} \tag{3.9}$$

In the sequel it will often be useful to have a compact notation. We define the collision operator,

$$\mathcal{C} : n_i(\mathbf{r}_\star) \longmapsto n_i(\mathbf{r}_\star) + \Delta_i(n(\mathbf{r}_\star)), \tag{3.10}$$

the streaming operator,

$$\mathcal{S} : n_i(\mathbf{r}_\star) \longmapsto n_i(\mathbf{r}_\star - \mathbf{c}_i), \tag{3.11}$$

and the evolution operator, the composition of the latter

$$\mathcal{E} = \mathcal{S} \circ \mathcal{C}. \tag{3.12}$$

The entire updating can now be written as

$$n(t_\star + 1, .) = \mathcal{E} n(t_\star, .), \tag{3.13}$$

where the point in the second argument of the n's stands for all the space variables.

An interesting property of the microdynamical equation, not shared by the Hamilton equations of ordinary Statistical Mechanics, is that it remains meaningful for an *infinite* lattice, since the updating of any given node involves only a finite number of neighbours.

3.2 Conservation relations

Conservation of mass and momentum *at each node* in the collision process can be expressed by the following relations for the collision function:

$$\sum_i \Delta_i(n) = 0, \quad \forall n \in \{0,\, 1\}^b, \tag{3.14}$$

$$\sum_i \mathbf{c}_i \Delta_i(n) = 0, \quad \forall n \in \{0,\, 1\}^b, \tag{3.15}$$

where $\{0,\, 1\}^b$ denotes the set of all possible b-bit words. This implies important conservation relations for the Boolean field

$$\sum_i n_i(t_* + 1, \mathbf{r}_* + \mathbf{c}_i) = \sum_i n_i(t_*, \mathbf{r}_*), \tag{3.16}$$

$$\sum_i \mathbf{c}_i n_i(t_* + 1, \mathbf{r}_* + \mathbf{c}_i) = \sum_i \mathbf{c}_i n_i(t_*, \mathbf{r}_*). \tag{3.17}$$

3.3 The Liouville equation

We now make the transition, traditional in Statistical Mechanics, from a deterministic to a probabilistic point of view. This can be obscured by the fact that some of our models are already probabilistic. So, let us assume for a while that the evolution operator is deterministic and invertible (as is the case for HPP).

Assuming that we have a finite lattice, we define the phase space, Γ, as the set of all possible *assignments* $s(.) = \{s_i(\mathbf{r}_*),\, i = 1, ..., b,\, \mathbf{r}_* \in \mathcal{L}\}$ of the Boolean field $n_i(\mathbf{r}_*)$. A particular assignment of the Boolean field will be called a *configuration*. We now consider at time $t_* = 0$ an *ensemble* of initial conditions, each endowed with a probability $P\Big(0, s(.)\Big) \geq 0$, such that

$$\sum_{s(.)\in\Gamma} P\Big(0, s(.)\Big) = 1. \tag{3.18}$$

We let each configuration in the ensemble evolve according to the automaton updating rule, i.e. with the evolution operator $\mathcal{E}$ of eq. (3.13). The latter being, here, invertible, conservation of probability is expressed as

$$P\Big(t_\star + 1, s(.)\Big) = P\Big(t_\star, \mathcal{E}^{-1} s(.)\Big). \tag{3.19}$$

This equation is clearly the analog of the Liouville equation of Statistical Mechanics, and will be given the same name. Alternatively, the Liouville equation can be written

$$P\Big(t_\star + 1, \mathcal{S}s(.)\Big) = P\Big(t_\star, \mathcal{C}^{-1} s(.)\Big). \tag{3.20)}$$

To derive this we have used (3.12) and put the streaming operator in the l.h.s., a form that will be more convenient subsequently.

In the *nondeterministic* case, we must enlarge the probability space to include, not only the phase space of initial conditions, but the space of all possible choices of the Boolean variables $\xi(ss')$, which at each time and each node select the unique transition from a given in-state s (cf. section 3.1). Since the ξ's are independently chosen at each time, the entire Boolean field $n(t_\star, .)$ is a Markov process (with deterministic rules, this process is degenerate). What we shall continue to call the Liouville equation, is actually the Chapman-Kolmogorov equation for this Markov process, namely

$$P\Big(t_\star + 1, \mathcal{S}s'(.)\Big) = \sum_{s(.)\in\Gamma} \prod_{\mathbf{r}_\star \in \mathcal{L}} A\Big(s(\mathbf{r}_\star) \to s'(\mathbf{r}_\star)\Big) \, P\Big(t_\star, s(.)\Big). \tag{3.21}$$

This equation just expresses that the probability at $t_\star + 1$ of a given (propagated) configuration $s'(.)$ is the sum of the probabilities at $t_\star$ of all possible original configurations $s(.)$ times the transition probability. The latter is a product, because we assumed that the ξ's are chosen independently at each node. In the deterministic case $A\left(s(\mathbf{r}_\star) \to s'(\mathbf{r}_\star)\right)$ selects the unique configuration $\mathcal{C}^{-1}s'(.)$, so that eq. (3.20) is recovered.

3.4 Mean quantities

Having introduced a probablistic description, we now turn to mean quantities. For an "observable" $q(n(t_\star, .))$, which depends on the Boolean field at a single time, the mean is given by ensemble averaging over $P(t_*, s(.))$

$$\left\langle q\left(n(t_*,.)\right)\right\rangle = \sum_{s(.)\in\Gamma} q\left(s(.)\right) P\left(t_*, s(.)\right). \tag{3.22}$$

An important role will be played in the sequel by the following mean quantities: the *mean population*

$$N_i(t_*, \mathbf{r}_*) = \left\langle n_i(t_*, \mathbf{r}_*)\right\rangle, \tag{3.23}$$

the *density*, and the *mass current* (mean momentum)

$$\rho(t_*, \mathbf{r}_*) = \sum_i N_i(t_*, \mathbf{r}_*), \quad \mathbf{j}(t_*, \mathbf{r}_*) = \sum_i \mathbf{c}_i N_i(t_*, \mathbf{r}_*). \tag{3.24}$$

Note that these are mean quantities per node, not per unit area or volume. The *density per cell* is defined as $d = \rho/b$. Finally, the *mean velocity* $\mathbf{u}$ is defined by

$$\mathbf{j}(t_*, \mathbf{r}_*) = \rho(t_*, \mathbf{r}_*)\mathbf{u}(t_*, \mathbf{r}_*). \tag{3.25}$$

Note that under duality (exchange of particles and holes) ρ changes into $b - \rho$, d into $1 - d$, $\mathbf{j}$ into $-\mathbf{j}$, and $\mathbf{u}$ into the "mean hole-velocity" $\mathbf{u}_H = -\mathbf{u}d/(1-d)$.

Averaging of the microdynamical conservation relations (3.16) and (3.17) leads to conservation relations for the mean populations

$$\sum_i N_i(t_* + 1, \mathbf{r}_* + \mathbf{c}_i) = \sum_i N_i(t_*, \mathbf{r}_*), \tag{3.26}$$

$$\sum_i \mathbf{c}_i N_i(t_* + 1, \mathbf{r}_* + \mathbf{c}_i) = \sum_i \mathbf{c}_i N_i(t_*, \mathbf{r}_*). \tag{3.27}$$

4. Equilibrium solutions

It has been shown by Hardy, Pomeau, and de Pazzis[22] that the HPP model has very simple statistical equilibrium solutions (which they call invariant states) in which the Boolean variables at all the cells are independent. Such equilibrium solutions are the lattice gas equivalent of Maxwell states in Classical Statistical Mechanics and therefore are crucial for deriving hydrodynamics. There are similar results for the general class of nondeterministic models introduced in section 2.4, which are now discussed.

4.1 Steady solutions of the Liouville equation

We are interested in equilibrium solutions, that is steady-state solutions of the Liouville equation (3.21) for a finite, periodically wrapped around lattice. Collisions on the lattice are purely local (their impact parameter is zero). This suggests the existence of equilibrium solutions with no single-time spatial correlations. The lattice properties being translation-invariant, the distribution should be the same at each node. Thus we are looking for equilibrium solutions of the form

$$P\Big(s(.)\Big) = \prod_{\mathbf{r}_*} p\Big(s(\mathbf{r}_*)\Big), \tag{4.1}$$

where $p(s)$, the probability of a given state, is node-independent. Maximization of the entropy (cf. Appendix F) suggests that $p(s)$ should be completely factorized over all cells, that is, of the form

$$p(s) = \prod_i N_i^{s_i}(1 - N_i)^{(1-s_i)}. \tag{4.2}$$

Note that $N_i^{s_i}(1 - N_i)^{(1-s_i)}$ is the probability of a Boolean variable with mean N_i.

Now, we must check that there are indeed solutions of the form that we have been guessing. Substitution of $P(s(.))$ given by (4.1) with $p(s)$ given by (4.2) into the Liouville equation (3.21) leads to

$$\prod_j N_j^{s'_j}(1 - N_j)^{(1-s'_j)} = \sum_s A(s \to s') \prod_j N_j^{s_j}(1 - N_j)^{(1-s_j)}, \quad \forall s', \tag{4.3}$$

where N_i is the mean population of cell i, independent of the node and of the time.

Eq. (4.3) is a set 2^b (the number of different states) equations for b unknowns. The fact that it actually possesses solutions is nontrivial. Furthermore, these solutions can be completely described. Indeed, we have the following lemma.

Lemma. The following statements are equivalent:

(a) The N_i's are a solution of (4.3).

(b) The N_i's are a solution of the set of b equations

$$\sum_{ss'}(s'_i - s_i)A(s \to s')\prod_j N_j^{s_j}(1 - N_j)^{(1-s_j)} = 0, \quad \forall i. \tag{4.4}$$

(c) The N_i's are given by the Fermi-Dirac distribution

$$N_i = \frac{1}{1 + \exp(h + \mathbf{q} \cdot \mathbf{c}_i)}, \tag{4.5}$$

where h is an arbitrary real number and $\mathbf{q}$ is an arbitrary D-dimensional vector.

The proof of the equivalence is given in Appendix C; it makes use of semi-detailed balance and the absence of spurious invariants. The most important consequence of the lemma is the

Universality theorem. Nondeterministic lattice gas models satisfying semi-detailed balance and having no spurious invariants admit universal equilibrium solutions, completely factorized over all nodes and all cells, with mean populations given by the Fermi-Dirac distribution (4.5), dependent only on the density ρ and the mass current $\mathbf{j} = \rho\mathbf{u}$, and independent of the transition probabilities $A(s \to s')$.

The proof follows from the observation that the Lagrange multipliers h and $\mathbf{q}$ of the Fermi-Dirac distribution can be calculated in terms of the density and the mass current through the relations

$$\rho = \sum_i N_i = \sum_i \frac{1}{1 + \exp(h + \mathbf{q} \cdot \mathbf{c}_i)}, \tag{4.6}$$

$$\rho\mathbf{u} = \sum_i N_i\mathbf{c}_i = \sum_i \mathbf{c}_i \frac{1}{1 + \exp(h + \mathbf{q} \cdot \mathbf{c}_i)}. \tag{4.7}$$

For the HPP model, this set of equations is reducible to a cubic polynomial equation, so that explicit solutions are known.[22] For the FHP model, explicit solutions are known only for special cases.[61]

It is not particularly surprising, for models that have a built-in exclusion principle (not more than one particle per cell), to obtain a Fermi-Dirac distribution at equilibrium.

Note that the factorized equilibrium solutions remain meaningful on an infinite lattice. There is no proof at the moment that the only equilibrium solutions which are relevant in the limit of infinite lattices are of the above form, namely completely factorized (which then *implies* the Fermi-Dirac distribution). There is strong numerical evidence, for those models that have been simulated, that the Fermi-Dirac is the only relevant one.[8,25,27]

4.2 Low-speed equilibria

In the "real world" equilibrium distributions with different mean velocities are simply related by a Galilean transformation. Galilean invariance does not hold at the microscopic level for a lattice gas; therefore there is no simple relation between the equilibria with vanishing and nonvanishing mean velocity. For subsequent derivations of fluid dynamical equations, we shall only need equilibria with low speeds, that is with $u = |\mathbf{u}| \ll c$, the particle speed. Such equilibria can be calculated perturbatively in powers of u.

We write the equilibrium distribution as

$$N_i = f_{\text{FD}}\Big(h(\rho,\mathbf{u}) + \mathbf{q}(\rho,\mathbf{u}) \cdot \mathbf{c}_i\Big), \tag{4.8}$$

where we have used the Fermi-Dirac function

$$f_{\text{FD}}(x) = \frac{1}{1+e^x}. \tag{4.9}$$

We observe that

$$\mathbf{u} = 0 \Rightarrow N_i = \frac{\rho}{b} = d. \tag{4.10}$$

Indeed, by assumption (iii) of section 2.4, there exists an isometry of the lattice exchanging any two velocity vectors $\mathbf{c}_i$ and $\mathbf{c}_j$; the vector $\mathbf{u} = 0$ being also trivially invariant, the mean population N_i is independent of i. Thus $f_{\text{FD}}\left(h(\rho,0)\right) = d$ and $\mathbf{q}(\rho,0) = 0$.

Furthermore, it follows from parity-invariance ($\mathbf{u} \to -\mathbf{u},\ \mathbf{c}_i \to -\mathbf{c}_i$) that

$$h(\rho,-\mathbf{u}) = h(\rho,\mathbf{u}), \quad \mathbf{q}(\rho,-\mathbf{u}) = -\mathbf{q}(\rho,\mathbf{u}). \tag{4.11}$$

We now expand h and $\mathbf{q}$ in powers of $\mathbf{u}$

$$\begin{aligned} h(\rho,\mathbf{u}) &= h_0 + h_2 u^2 + O(u^4) \\ q_\alpha(\rho,\mathbf{u}) &= q_1 u_\alpha + O(u^3), \end{aligned} \tag{4.12}$$

where h_0, h_2, and q_1 depend on ρ. The fact that h_2 and q_1 are scalars rather than second order tensors is a consequence of the isotropy of second order tensors (property P4 of section 2.4). We substitute (4.12) into (4.8) and expand the mean populations in powers of $\mathbf{u}$

$$N_i = f_{FD} + q_1 f'_{FD}\,\mathbf{u}\cdot\mathbf{c}_i + h_2 f'_{FD} u^2 + \frac{1}{2} q_1^2 f''_{FD} (\mathbf{u}\cdot\mathbf{c}_i)^2 + O(u^3). \tag{4.13}$$

Here, f_{FD}, f'_{FD}, and f''_{FD} are the values at h_0 of the Fermi-Dirac function and its first and second derivatives. From (4.13) we calculate the density $\rho = \sum_i N_i$ and the mass current $\rho\mathbf{u} = \sum_i \mathbf{c}_i N_i$, using the velocity moment relations (P6 of section 2.4). Identification gives h_0, h_2, and q_1 in terms of ρ. This is then used to calculate the equilibrium mean population up to second order in $\mathbf{u}$; we obtain

$$N_i^{eq}(\rho, \mathbf{u}) = \frac{\rho}{b} + \frac{\rho D}{c^2 b} c_{i\alpha} u_\alpha + \rho G(\rho) Q_{i\alpha\beta} u_\alpha u_\beta + O(u^3), \tag{4.14}$$

where

$$G(\rho) = \frac{D^2}{2c^4 b}\,\frac{b - 2\rho}{b - \rho} \quad \text{and} \quad Q_{i\alpha\beta} = c_{i\alpha} c_{i\beta} - \frac{c^2}{D}\delta_{\alpha\beta}. \tag{4.15}$$

In (4.14) the superscript "eq" stresses that the mean population are evaluated at equilibrium.

Note that the coefficient $G(\rho)$ of the quadratic term vanishes for $\rho = b/2$, that is, when the density of particles and holes are the same. This result, which holds more generally for the coefficients of any even power of $\mathbf{u}$, follows by duality: N_i^{eq} goes into $1 - N_i^{eq}$ and $\mathbf{u}$ into $-\mathbf{u}$ at $\rho = b/2$. It does not matter whether or not the collision rules are duality-invariant, as long as they satisfy semi-detailed balance, since the equilibrium is then universal.

5. Macrodynamical equations

In the "real world", fluid dynamics may be viewed as the glueing of *local* thermodynamic equilibria with slowly varying parameters.[62,63] Lattice gases also admit equilibrium solutions.[4] These have continuously adjustable parameters, the mean values of the conserved quantities, namely mass and momentum. On a very large lattice, we can set up local equilibria with density and mass current slowly changing in space and time. From the conservation relations we shall derive by a multi-scale technique *macrodynamical* equations, that is PDE's for the large scale and long time behaviour of density and mass current.

We consider a lattice gas satisfying all the assumptions of section 2.4. We denote by $\rho(\mathbf{r}_*)$ and $\mathbf{u}(\mathbf{r}_*)$ the density and (mean) velocity[5] at lattice node $\mathbf{r}_*$. We assume that these quantities are changing on a spatial scale ϵ^{-1} (in units of lattice constant). This requires that the lattice size L be itself at least $O(\epsilon^{-1})$. Eventually, we let $\epsilon \rightarrow 0$. The spatial change is assumed to be sufficiently regular to allow interpolations for the purpose of calculating derivatives.[6] When time and space are treated as continuous, they are denoted t and $\mathbf{r}$. We further assume that the density is $O(1)$ and that the velocity is small compared to the particle speed c.[7] We expect the following phenomena:

1) relaxation to local equilibrium on time scale ϵ^0,

2) density perturbations propagating as sound waves on time scale ϵ^{-1},

3) diffusive (and possibly advective) effects on time scale ϵ^{-2}.

We thus use a three time formalism: t_* (discrete), $t_1 = \epsilon t_*$, and $t_2 = \epsilon^2 t_*$, the latter two being treated as continuous variables. We use two space variables: $\mathbf{r}_*$ (discrete) and $\mathbf{r}_1 = \epsilon \mathbf{r}_*$ (continuous).

Let us denote by $N_i^{(0)}(\mathbf{r}_*)$ the mean equilibrium populations based on the local value

---

[4] The qualification "thermodynamic" is not so appropriate since there is no relevant energy variable

[5] Henceforth we shall just write "velocity", since this *mean velocity* changes in space.

[6] The interpolations can be done via the Fourier representation if the lattice is periodic.

[7] Eventually, we shall assume the velocity to be $O(\epsilon)$, but at this point it is more convenient to keep ϵ and u as independent expansion parameters.

of ρ and $\mathbf{u}$. They are given by (4.14). The actual mean populations $N_i(t,\mathbf{r})$ will be close to the equilibrium values and may be expanded in powers of ϵ:

$$N_i = N_i^{(0)}(t,\mathbf{r}) + \epsilon N_i^{(1)}(t,\mathbf{r}) + O(\epsilon^2). \tag{5.1}$$

The corrections should not contribute to the local values of density and mean momentum; thus

$$\sum_i N_i^{(1)}(t,\mathbf{r}) = 0 \quad \text{and} \quad \sum_i \mathbf{c}_i N_i^{(1)}(t,\mathbf{r}) = 0. \tag{5.2}$$

We now start from the exact conservation relations (3.26) and (3.27) and expand both the N_i's and the finite differences in powers of ϵ. Note that all finite differences must be expanded to *second* order; otherwise, the viscous terms are not correctly captured. Time and space derivatives will be denoted ∂_t and $\partial_\mathbf{r} = \{\partial_\alpha,\ \alpha = 1, ..., D\}$. For the multi-scale formalism, we make the substitutions

$$\partial_t \rightarrow \epsilon\partial_{t_1} + \epsilon^2\partial_{t_2} \quad \text{and} \quad \partial_\mathbf{r} \rightarrow \epsilon\partial_{\mathbf{r}_1}. \tag{5.3}$$

The components of $\partial_{\mathbf{r}_1}$ will be denoted $\partial_{1\alpha}$.

To leading order, $O(\epsilon)$, we obtain

$$\partial_{t_1}\sum_i N_i^{(0)} + \partial_{1\beta}\sum_i c_{i\beta}N_i^{(0)} = 0, \tag{5.4}$$

and

$$\partial_{t_1}\sum_i c_{i\alpha}N_i^{(0)} + \partial_{1\beta}\sum_i c_{i\alpha}c_{i\beta}N_i^{(0)} = 0. \tag{5.5}$$

We now substitute the equilibrium values (4.14) for the $N_i^{(0)}$'s and use the velocity moment relations P6 of section 2.4. We obtain the "macrodynamical Euler equations"

$$\partial_{t_1}\rho + \partial_{1\beta}(\rho u_\beta) = 0, \tag{5.6}$$

and

$$\partial_{t_1}(\rho u_\alpha) + \partial_{1\beta}P_{\alpha\beta} = 0. \tag{5.7}$$

$P_{\alpha\beta}$ is the momentum-flux tensor,[8]

$$\begin{aligned} P_{\alpha\beta} &\equiv \sum_i c_{i\alpha} c_{i\beta} N_i^{eq} \\ &= \frac{c^2}{D}\rho\delta_{\alpha\beta} + \rho G(\rho) T_{\alpha\beta\gamma\delta}\, u_\gamma u_\delta + O(u^4), \end{aligned} \tag{5.8}$$

with

$$T_{\alpha\beta\gamma\delta} = \sum_i c_{i\alpha} c_{i\beta} Q_{i\gamma\delta}, \tag{5.9}$$

and $G(\rho)$ and $Q_{i\gamma\delta}$ given by (4.15) of section 4. Note that the correction term in the r.h.s. of (5.8) is $O(u^4)$ rather than $O(u^3)$; indeed, it follows from the parity-invariance of the lattice gas that first order spatial derivative terms do not contain odd powers of u.

We now proceed to the next order, $O(\epsilon^2)$. We expand (3.26) and (3.27) to second order, collecting all $O(\epsilon^2)$ terms, we obtain

$$\begin{aligned} &\partial_{t_2} \sum_i N_i^{(0)} + \frac{1}{2}\partial_{t_1}\partial_{t_1} \sum_i N_i^{(0)} + \partial_{t_1}\partial_{1\beta} \sum_i c_{i\beta} N_i^{(0)} + \\ &\frac{1}{2}\partial_{1\beta}\partial_{1\gamma} \sum_i c_{i\beta} c_{i\gamma} N_i^{(0)} + \partial_{t_1} \sum_i N_i^{(1)} + \partial_{1\beta} \sum_i c_{i\beta} N_i^{(1)} = 0, \end{aligned} \tag{5.10}$$

and

$$\begin{aligned} &\partial_{t_2} \sum_i c_{i\alpha} N_i^{(0)} + \frac{1}{2}\partial_{t_1}\partial_{t_1} \sum_i c_{i\alpha} N_i^{(0)} + \partial_{t_1}\partial_{1\beta} \sum_i c_{i\alpha} c_{i\beta} N_i^{(0)} + \\ &\frac{1}{2}\partial_{1\beta}\partial_{1\gamma} \sum_i c_{i\alpha} c_{i\beta} c_{i\gamma} N_i^{(0)} + \partial_{t_1} \sum_i c_{i\alpha} N_i^{(1)} + \partial_{1\beta} \sum_i c_{i\alpha} c_{i\beta} N_i^{(1)} = 0. \end{aligned} \tag{5.11}$$

By (5.2) $\sum_i N_i^{(1)} = 0$ and $\sum_i c_{i\alpha} N_i^{(1)} = 0$. For the $N_i^{(0)}$'s, we substitute their low-speed equilibrium form (4.14), leaving out $O(u^2)$ terms. Reexpressing derivatives of ρ and ρu with respect to t_1 in terms of space derivatives, using (5.6)-(5.7), we obtain

$$\partial_{t_2}\rho = 0, \tag{5.12}$$

and

$$\partial_{t_2}(\rho u_\alpha) + \partial_{1\beta}\left(\sum_i c_{i\alpha} c_{i\beta} N_i^{(1)} + \frac{D}{2c^2 b} T_{\alpha\beta\gamma\delta}\partial_{1\gamma}(\rho u_\delta)\right) = O(u^2). \tag{5.13}$$

---

[8] Actually, this is only the *leading order* approximation to the momentum-flux.

Eq. (5.12) tells us that there is no mass diffusion (there is a single species of particles). Eq. (5.13) describes the momentum diffusion over long ($O(\epsilon^{-2})$) time-scales. It has two contributions. The term involving $T_{\alpha\beta\gamma\delta}$ comes from particle propagation and we shall comment on it later.

The other term in (5.13) involves the deviations $N_i^{(1)}$ from the equilibrium mean populations. $N_i^{(1)}$ vanishes when the equilibrium is uniform. It must therefore be a linear combination of gradients (with respect to $\mathbf{r}_1$) of ρ and $\rho\mathbf{u}$. Linear response theory is needed to calculate the coefficients. At this point, we shall only make use of symmetry arguments to reduce the number of coefficients. We assume that $\mathbf{u}$ is small, so that, to leading order equilibria are invariant under the isometry group $\mathcal{G}$ of the lattice (see section 2.4). Since the gradient of ρ is a vector and the gradient of $\rho\mathbf{u}$ is a second order tensor, properties P2 and P3 of section 2.4 allow us to write

$$N_i^{(1)} = \sigma \mathbf{c}_{i\alpha} \partial_{1\alpha}\rho + (\psi \mathbf{c}_{i\alpha}\mathbf{c}_{i\beta} + \chi\delta_{\alpha\beta})\,\partial_{1\alpha}(\rho u_\beta). \tag{5.14}$$

By eq. (5.2), we have $\sigma = 0$ and $c^2\psi + D\chi = 0$. Note that ψ should depend on ρ, but not on $\mathbf{u}$, since it is evaluated at $\mathbf{u} = 0$. Substituting the expression for $N_i^{(1)}$ into (5.13), we obtain

$$\partial_{t_2}(\rho u_\alpha) + \partial_{1\beta}\left[\left(\psi(\rho) + \frac{D}{2c^2 b}\right) T_{\alpha\beta\gamma\delta}\partial_{1\gamma}(\rho u_\delta)\right] = O(u^2). \tag{5.15}$$

In the sequel, it will be more convenient to collapse the set of four equations, governing the evolution of ρ and $\rho\mathbf{u}$ on $O(\epsilon^{-1})$ and $O(\epsilon^{-2})$ time-scales, into a pair of equations, written in terms of the original variables t and $\mathbf{r}$ (in their continuous version). We thus obtain the *macrodynamical equations*

$$\partial_t\rho + \partial_\beta(\rho u_\beta) = 0, \tag{5.16}$$

$$\partial_t(\rho u_\alpha) + \partial_\beta\left(\rho G(\rho) T_{\alpha\beta\gamma\delta} u_\gamma u_\delta + \frac{c^2}{D}\rho\delta_{\alpha\beta}\right) + \partial_\beta\left[\left(\psi(\rho) + \frac{D}{2c^2 b}\right) T_{\alpha\beta\gamma\delta}\partial_\gamma(\rho u_\delta)\right] \tag{5.17}$$
$$= O(\epsilon u^3) + O(\epsilon^2 u^2) + O(\epsilon^3 u).$$

The equivalence of (5.16) and (5.17) to (5.6), (5.7), (5.12), and (5.15) follows by (5.3). Note that (5.16) is the standard density equation of fluid mechanics and that (5.17) already has a strong resemblance to the Navier-Stokes equations.

6. Recovering isotropy

The macrodynamical equations (5.16)-(5.17) are not fully isotropic. The presence of a lattice with discrete rotational symmetries is still felt through the tensor

$$T_{\alpha\beta\gamma\delta} = \sum_i c_{i\alpha} c_{i\beta} Q_{i\gamma\delta} = \sum_i c_{i\alpha} c_{i\beta} \left(c_{i\gamma} c_{i\delta} - \frac{c^2}{D} \delta_{\gamma\delta} \right), \tag{6.1}$$

appearing in both the nonlinear and diffusive terms of (5.17). Furthermore, the higher order terms in the r.h.s. of (5.17) have no reason to be isotropic. This should not worry us since they will eventually turn out to be irrelevant. Contrary to translational discreteness, rotational discreteness cannot go away under the macroscopic limit; the latter involves large scales but not in any way "large angles" since the group of rotations is compact.

We have seen in section 2.4 that tensors up to third order, having the same invariance group $\mathcal{G}$ as the discrete velocity set are isotropic. Not so for tensors of fourth order such as $T_{\alpha\beta\gamma\delta}$. Indeed, for the HPP model (section 2.1) explicit calculation of the momentum-flux tensor, given by (5.8), is quite straightforward. The result is

$$P_{11} = \rho G(\rho)(u_1^2 - u_2^2) + \frac{\rho}{2} + O(u^4), \qquad P_{22} = \rho G(\rho)(u_2^2 - u_1^2) + \frac{\rho}{2} + O(u^4), \tag{6.2}$$

$$P_{12} = P_{21} = 0, \tag{6.3}$$

with

$$G(\rho) = \frac{2-\rho}{4-\rho}. \tag{6.4}$$

The only second order tensors quadratic in the velocity being $u_\alpha u_\beta$ and $\mathbf{u}\cdot\mathbf{u}\,\delta_{\alpha\beta}$, the tensor $P_{\alpha\beta}$ is not isotropic.

In order to eventually obtain the Navier-Stokes equations, the tensor $T_{\alpha\beta\gamma\delta}$ given by (6.1) must be isotropic that is, *invariant under the full orthogonal group*. This tensor is pairwise symmetrical in (α, β) and (γ, δ); from (6.1), it follows that it satisfies

$$\sum_\gamma T_{\alpha\beta\gamma\gamma} = 0, \qquad \sum_{\alpha\beta} T_{\alpha\beta\alpha\beta} = bc^4 \left(1 - \frac{1}{D}\right). \tag{6.5}$$

When the tensor $T_{\alpha\beta\gamma\delta}$ is isotropic, these properties uniquely constrain it to be of the following form:

$$T_{\alpha\beta\gamma\delta} = \frac{bc^4}{D(D+2)}\left(\delta_{\alpha\gamma}\delta_{\beta\delta} + \delta_{\alpha\delta}\delta_{\beta\gamma} - \frac{2}{D}\delta_{\alpha\beta}\delta_{\gamma\delta}\right). \tag{6.6}$$

For general group-theoretical material concerning the isotropy of tensors with discrete symmetries in the context of lattice gases, we refer the reader to Ref. 39. Crucial observations for obtaining the two and three-dimensional Navier-Stokes equations are the isotropy of pairwise symmetrical tensors for the triangular FHP lattice in two dimensions and the face-centered-hypercubic (FCHC) lattice in four dimensions, and thus also for the pseudo-4-D three-dimensional model. We give now elementary proofs of these results.

In two dimensions, it is convenient to consider $T_{\alpha\beta\gamma\delta}$ as a linear map from the space E of two-by-two real symmetrical matrices into itself:

$$T: \quad A_{\alpha\beta} \longmapsto T_{\alpha\beta\gamma\delta}A_{\gamma\delta}. \tag{6.7}$$

A basis of the space E is formed by the matrices P_1, P_2, and P_3, associated with the orthogonal projections onto the x_1-axis and onto two other directions at $2\pi/3$ and $4\pi/3$. In this representation, an arbitrary E-matrix may be written as

$$A = \chi_1 P_1 + \chi_2 P_2 + \chi_3 P_3, \tag{6.8}$$

and T becomes a three-by-three matrix T_{ab}, $(a, b = 1, 2, 3)$. The key observation is that the hexagonal group (rotations by multiples of $\pi/3$) becomes the permutation group of P_1, P_2, and P_3. Thus T_{ab} is invariant under arbitrary permutations of the coordinates, i.e. is of the form

$$T_{ab} = \phi\, \mathrm{diag}_{ab}(1,1,1) + \chi 1_{ab}, \tag{6.9}$$

where $\mathrm{diag}_{ab}(1,1,1)$ is the diagonal matrix with entries one and 1_{ab} is the matrix with all entries equal to one, and ϕ and χ are arbitrary scalars. From (6.8) we have

$$\mathrm{tr}(A) = \chi_1 + \chi_2 + \chi_3, \tag{6.10}$$

where tr denotes the trace. We also note that

$$P_1 + P_2 + P_3 = (3/2)I, \tag{6.11}$$

where I is the identity (check it for the unit vectors of the x_1 and x_2 axis). Using (6.10) and (6.11), we can rewrite (6.9) as

$$T: \quad A \mapsto \phi A + \frac{3}{2}\chi\,\mathrm{tr}(A)I. \tag{6.12}$$

Reverting to tensor notations, this becomes

$$T_{\alpha\beta\gamma\delta} = \frac{\phi}{2}\left(\delta_{\alpha\gamma}\delta_{\beta\delta} + \delta_{\alpha\delta}\delta_{\beta\gamma}\right) + \frac{3\chi}{2}\delta_{\alpha\beta}\delta_{\gamma\delta}, \tag{6.13}$$

which is obviously isotropic.

We turn to the four-dimensional case, using the FCHC model of section 2.3. Invariance under permutations of coordinates and reversal of any coordinate implies that the most general possible form for $T_{\alpha\beta\gamma\delta}$ is

$$T_{\alpha\beta\gamma\delta} = \phi\delta_{\alpha\beta}\delta_{\beta\gamma}\delta_{\gamma\delta} + \chi\left(\delta_{\alpha\gamma}\delta_{\beta\delta} + \delta_{\alpha\delta}\delta_{\beta\gamma}\right) + \psi\delta_{\alpha\beta}\delta_{\gamma\delta}. \tag{6.14}$$

The χ and ψ terms are already isotropic. The vanishing of ϕ is a consequence of the invariance of the velocity set under the symmetry Σ with respect to the hyperplane $x_1 + x_2 + x_3 + x_4 = 0$, that is

$$x_\alpha \mapsto x_\alpha - \sigma, \quad \sigma = \frac{1}{2}\sum_\alpha x_\alpha. \tag{6.15}$$

Indeed, consider the vector $v_\alpha = (2, 0, 0, 0)$. Contracting the ϕ term four times with v_α, we obtain 16ϕ; the image of v_α under Σ is $w_\alpha = (1, -1, -1, -1)$, which contracted four times with the ϕ term gives 4ϕ. Thus invariance requires $\phi = 0$, which proves isotropy.

We return to the general D-dimensional case, assuming isotropy. Substituting (6.6) into the macrodynamical momentum equation (5.17), we obtain

$$\partial_t(\rho u_\alpha) + \partial_\beta\Big(\rho g(\rho)u_\alpha u_\beta\Big) + \partial_\alpha\left(c_s^2\rho\left(1 - g(\rho)\frac{u^2}{c^2}\right)\right)$$
$$= \partial_\beta\left[\left(\nu_c(\rho) + \nu_p\right)\left(\partial_\alpha(\rho u_\beta) + \partial_\beta(\rho u_\alpha) - \frac{2}{D}\delta_{\alpha\beta}\partial_\gamma(\rho u_\gamma)\right)\right] + O(\epsilon u^3) + O(\epsilon^2 u^2) + O(\epsilon^3 u), \tag{6.16}$$

with

$$g(\rho) = \frac{D}{D+2}\frac{b-2\rho}{b-\rho}, \quad c_s^2 = \frac{c^2}{D}, \quad \nu_c(\rho) = -\frac{bc^4}{D(D+2)}\psi(\rho), \quad \nu_p = -\frac{c^2}{2(D+2)}. \quad (6.17)$$

Note that $g(\rho)$ appearing in (6.17) is not the same as $G(\rho)$ introduced in (4.15). Note also that $\psi(\rho)$, which was introduced in section 5, is still to be determined (cf. section 8).

We have now recovered macroscopic isotropy; equation (6.16) is very closely related to the fluid dynamical momentum (Navier-Stokes) equations. We postpone all further remarks to the next section.

7. Fluid dynamical régimes

Let us rewrite the macrodynamical equations for mass and momentum, derived in the previous sections in a compact form which brings out their similarities with the equations of fluid dynamics:

$$\partial_t\rho + \partial_\beta\,(\rho u_\beta) = 0, \quad (7.1)$$

$$\partial_t\,(\rho u_\alpha) + \partial_\beta P_{\alpha\beta} = \partial_\beta S_{\alpha\beta} + O(\epsilon u^3) + O(\epsilon^2 u^2) + O(\epsilon^3 u). \quad (7.2)$$

The momentum-flux tensor $P_{\alpha\beta}$ and the viscous stress tensor $S_{\alpha\beta}$ are given by

$$P_{\alpha\beta} = c_s^2\rho\left(1 - g(\rho)\frac{u^2}{c^2}\right)\delta_{\alpha\beta} + \rho g(\rho) u_\alpha u_\beta, \quad (7.3)$$

and

$$\begin{aligned} S_{\alpha\beta} &= \nu(\rho)\left(\partial_\alpha(\rho u_\beta) + \partial_\beta(\rho u_\alpha) - \frac{2}{D}\delta_{\alpha\beta}\partial_\gamma(\rho u_\gamma)\right) \\ \nu(\rho) &= \nu_c(\rho) + \nu_p, \end{aligned} \quad (7.4)$$

where $g(\rho)$, c_s^2, ν_c, and ν_p are defined in (6.17). Their values for the FHP-I and FCHC models are given below

$$\begin{aligned} g(\rho) &= \frac{3-\rho}{6-\rho}, \quad c_s^2 = \frac{1}{2}, \quad \nu_c(\rho) = -\frac{3}{4}\psi(\rho), \quad \nu_p = -\frac{1}{8}, \quad \text{for FHP-I} \\ g(\rho) &= \frac{4}{3}\frac{12-\rho}{24-\rho}, \quad c_s^2 = \frac{1}{2}, \quad \nu_c(\rho) = -4\psi(\rho), \quad \nu_p = -\frac{1}{6}, \quad \text{for FCHC.} \end{aligned} \quad (7.5)$$

Various remarks are now in order. When the velocity $\mathbf{u}$ is very small, the momentum-flux tensor reduces to a diagonal pressure term $p\delta_{\alpha\beta}$ with the pressure given by the "isothermal" relation

$$p = c_s^2 \rho. \tag{7.6}$$

From this, we infer that the *speed of sound* should be c_s, namely $1/\sqrt{2}$ for FHP-I and FCHC.

The momentum-flux tensor in the "real world" is $P_{\alpha\beta} = p\delta_{\alpha\beta} + \rho u_\alpha u_\beta$. This form is a consequence of Galilean invariance, which allows one to relate thermodynamic equilibria with vanishing and nonvanishing mean velocities. The lattice gas momentum-flux tensor (7.3) with nonvanishing velocity differs by an additive term in the pressure and a multiplicative density-dependent factor $g(\rho)$ in the advection term. We shall see later in this section how Galilean invariance can nevertheless be recovered.

Eq. (7.4) is the stress-strain relation for a Newtonian fluid having kinematic viscosity $\nu_c + \nu_p$ and *vanishing bulk viscosity.*[64] The traceless character of $S_{\alpha\beta}$ (which implies this vanishing of the bulk viscosity) comes from the traceless character of $Q_{i\alpha\beta}$, defined by (4.15); this result would be upset by the presence of rest-particles such as exist in the models FHP-II and III (cf. Appendix E). The kinematic viscosity has two contributions. One is the "collision viscosity" ν_c, not yet determined, which depends on the details of the collisions and is positive (cf. section 8). The other one is the "propagation viscosity" ν_p, which is *negative* and does not involve the collisions. The presence of such a negative propagation viscosity is an effect of the lattice discreteness (cf. Ref. 42).

The general strategy by which standard fluid dynamical equations are derived from (7.1)-(7.2) is to rescale the space, time and velocity variables in such a way as to make undesirable terms irrelevant as $\epsilon \to 0$. Three different régimes will be considered in the following subsections. They correspond respectively to sound propagation, to sound propagation with slow damping, and to incompressible (Navier-Stokes) fluid dynamics.

7.1 Sound propagation

Consider a weak perturbation of the equilibrium solution with density ρ_0 and velocity zero. We write

$$\rho = \rho_0 + \rho'. \tag{7.7}$$

In a suitable limit we expect that the only relevant terms in (7.1)-(7.2) will be[9]

$$\begin{aligned} &\partial_t \rho' + \rho_0 \nabla \cdot \mathbf{u} = 0 \\ &\rho_0 \partial_t \mathbf{u} + c_s^2 \nabla \rho' = 0. \end{aligned} \tag{7.8}$$

Formally, this régime is obtained by setting

$$\mathbf{r} = \epsilon^{-1}\mathbf{r}_1, \quad t = \epsilon^{-1}t_1, \quad \rho' = \epsilon^a \rho_1', \quad \mathbf{u} = \epsilon^a \mathbf{U}, \quad a > 0. \tag{7.9}$$

It is then straightforward to check that the leading order terms take the form of eqs. (7.8) (in the rescaled variables). Eliminating **u** in (7.8), we obtain the scalar wave equation

$$\frac{\partial^2}{\partial t^2}\rho' - c_s^2 \nabla^2 \rho' = 0. \tag{7.10}$$

In other words, density and velocity perturbations with amplitudes $o(1)$ on temporal and spatial scales $O(\epsilon)$ propagate as sound waves with speed c_s.[10] Since the present régime of undamped sound waves involves only tensors of second order, it also applies to the HPP model.

7.2 Damped sound

Another régime includes the viscous damping term, so that instead of (7.8) we should have

$$\begin{aligned} &\partial_t \rho' + \rho_0 \nabla \cdot \mathbf{u} = 0 \\ &\rho_0 \partial_t \mathbf{u} + c_s^2 \nabla \rho' = \rho_0 \nu(\rho_0) \left(\nabla^2 \mathbf{u} + \frac{D-2}{D} \nabla\nabla \cdot \mathbf{u} \right). \end{aligned} \tag{7.11}$$

To obtain this régime we proceed as in section 7.1 and include an additional time $t_2 = \epsilon^2 t$. Furthermore , in the scaling relation (7.9) we now require $a > 1$, that is, **u** and ρ' should

---

[9] From here on we use vector notation whenever possible.

[10] We have used here the Landau $O()$ and $o()$ notation.

be $o(\epsilon)$; otherwise the nonlinear term becomes also relevant. Note that the damping is now on a time scale $O(\epsilon^{-2})$. Since propagation and damping are on time-scales involving different powers of ϵ, it is not possible to describe them in a single equation without mixing orders.

7.3 *Incompressible fluid dynamics: the Navier-Stokes equations*

It is known that many features of low Mach number[11] flows in an ordinary gas can be described by the incompressible Navier-Stokes equation

$$\begin{aligned} &\partial_t \mathbf{u} + \mathbf{u}\cdot\nabla\mathbf{u} = -\nabla p + \nu\nabla^2\mathbf{u} \\ &\nabla\cdot\mathbf{u} = 0. \end{aligned} \tag{7.12}$$

In the "real world", the incompressible Navier-Stokes equation can be derived from the full compressible equations, using a Mach number expansion. There are some fine points in this expansion for which we refer the interested reader to Ref. 65. Ignoring these, the essential observation is that, to leading order, density variations become irrelevant everywhere, except in the pressure term; the latter becomes slaved to the nonlinear term by the incompressibility constraint.

Just the same kind of expansion (with the same difficulties) can be applied to lattice gas dynamics. We start from (7.1)-(7.2) and freeze the density by setting it equal to the constant and uniform value ρ_0 everywhere except in the pressure term where we keep the density fluctuations. We also ignore all higher order terms $O(\epsilon^3 u)$, etc. This produces the following set of equations

$$\begin{aligned} &\rho_0\partial_t\mathbf{u} + \rho_0 g(\rho_0)\mathbf{u}\cdot\nabla\mathbf{u} = -c_s^2\nabla\left(\rho' - \rho_0 g(\rho_0)\frac{u^2}{c^2}\right) + \rho_0\nu(\rho_0)\nabla^2\mathbf{u} \\ &\nabla\cdot\mathbf{u} = 0. \end{aligned} \tag{7.13}$$

The resulting equations (7.13) differ from (7.12) only by the presence of the factor $g(\rho_0)$ in front of the advection term $\mathbf{u}\cdot\nabla\mathbf{u}$. As it stands (7.13) is not Galilean invariant. This of course reflects the lack of Galilean invariance at the lattice level. Similarly, the vanishing

---

[11] The Mach number is the ratio of a characteristic flow velocity to the speed of sound.

of $g(\rho_0)$ when the density per cell $d = \rho_0/b$ is equal to $1/2$, i.e. for equal mean numbers of particles and holes, reflects a duality-invariance of the lattice gas without counterpart in the "real world" (cf. end of section 4.2). However, as soon as $d < 1/2$, it is straightforward to reduce (7.13) to the true Navier-Stokes equations (7.12); it suffices to rescale time and viscosity:

$$t \to \frac{t}{g(\rho_0)}, \qquad \nu \to g(\rho_0)\nu. \tag{7.14}$$

Now we show that there is actually a rescaling of variables which reduces the macro-dynamical equations to the incompressible Navier-Stokes equations. We set

$$\mathbf{r} = \epsilon^{-1}\mathbf{r}_1, \quad t = \frac{\epsilon^{-2}T}{g(\rho_0)}, \quad \mathbf{u} = \epsilon\mathbf{U}, \quad \rho' - \rho_0 g(\rho_0)\frac{u^2}{c^2} = \frac{\rho_0 g(\rho_0)}{c_s^2}\epsilon^2 P', \quad \nu = g(\rho_0)\nu'. \tag{7.15}$$

Thus, all the relevant terms are $O(\epsilon^2)$ in (7.1) and $O(\epsilon^3)$ in (7.2). The higher order terms in the r.h.s. of (7.2) are $O(\epsilon^4)$ or smaller. In this way we obtain, to leading order (∇_1 denotes the gradient with respect to $\mathbf{r}_1$)

$$\begin{aligned} &\partial_T \mathbf{U} + \mathbf{U}\cdot\nabla_1\mathbf{U} = -\nabla_1 P' + \nu'\nabla_1^2\mathbf{U} \\ &\nabla_1\cdot\mathbf{U} = 0, \end{aligned} \tag{7.16}$$

which are exactly the incompressible Navier-Stokes equations.

Various comments are now made. The expansion leading to (7.16) is a large-scale and low Mach number expansion (the former is here inversely proportional to the latter). It also follows from the scaling relations (7.15) that the Reynolds number is kept fixed. It is not possible within our framework to have an asymptotic régime leading to nonlinear compressible equations at finite Mach number. Indeed, the speed of sound is here a *finite* fraction of the particle speed and it is essential that the macroscopic velocity be small compared to particle speed, so as not to be contaminated by higher order nonlinearities. It is noteworthy that models can be constructed having many rest-particles (zero-velocity) with arbitrarily low speed of sound.

In a pure Navier-Stokes context, the non-Galilean invariance at the microscopic level is not a serious difficulty; as we have seen, Galilean invariance is recovered macroscopically, just by rescaling the time variable. However, when the models discussed here are generalized to include for example multi-phase flow or buoyancy effects, a more serious problem may arise because the advection term of scalar quantities such as chemical concentrations or temperature involves usually a factor $g(\rho)$ different from that of the nonlinear advection term in the Navier-Stokes equations. Various solutions to this problem have been proposed.[48,66]

There is a variant of our formalism, leading also to the incompressible Navier-Stokes equations, but in terms of the *mass current* $\mathbf{j} = \rho\mathbf{u}$ rather than the velocity $\mathbf{u}$. The analog of (7.13) (without rescaling) is then

$$\begin{aligned} &\partial_t \mathbf{j} + \frac{g(\rho_0)}{\rho_0}\mathbf{j}\cdot\nabla\mathbf{j} = -c_s^2\nabla\rho' + \nu(\rho_0)\nabla^2\mathbf{j} \\ &\nabla\cdot\mathbf{j} = 0. \end{aligned} \tag{7.17}$$

Since $\mathbf{j}$ and $g(\rho_0)/\rho_0$ change sign under duality, (7.17) brings out duality-invariance.[12] A more decisive advantage of the $\mathbf{j}$-representation is that it gives a better approximation to the *steady state* Navier-Stokes equations when the Mach number is only moderately small. This is because in the steady state the continuity equation implies exactly $\nabla\cdot\mathbf{j} = 0$.

In three dimensions, when we use the pseudo-4-D FCHC model, there are three independent space variables $\mathbf{r} = (x_1, x_2, x_3)$ but four velocity components

$$\mathbf{U}_f = (\mathbf{U}, U_4) = (U_1, U_2, U_3, U_4). \tag{7.18}$$

The four-velocity $\mathbf{U}_f$ satisfies the four-dimensional Navier-Stokes equations with no x_4-dependence. Thus, the three-velocity $\mathbf{U}$ satisfies the three-dimensional Navier-Stokes equa-

[12] In the u-representation duality-invariance is broken because we have decided to work with the velocity of *particles* rather than with that of *holes*.

tions (7.16),[13] while U_4 satisfies (note that the pressure term drops out)

$$\partial_T U_4 + \mathbf{U} \cdot \nabla_1 U_4 = \nu' \nabla_1^2 U_4. \tag{7.19}$$

This is the equation for a *passive scalar* with unit Schmidt number (ratio of viscosity to diffusivity).

Finally, we refer the reader to Appendix D for the inclusion of body forces in the Navier-Stokes equations.

8. The viscosity

All the macroscopic equations derived in section 7 have a universal form, which does not depend on the details of collisions. The kinematic shear viscosity ν, which we shall henceforth call the viscosity, does not possess this universality. Transport coefficients such as the viscosity characterize the linear response of equilibrium solutions to small externally imposed perturbations. It is known in Statistical Mechanics that the relaxation (or dissipation) of external perturbations is connected to the fluctuations at equilibrium via *fluctuation-dissipation* relations. Such relations have a counterpart for lattice gases. Two quite different approaches are known. In section 8.1, following a suggestion already made in Ref. 23, we present the "noisy" hydrodynamics view-point, in the spirit of Landau and Lifschitz.(67,68) Another approach, in the spirit of Kubo(69) and Green,(70) using a Liouville equation formalism, may be found in Ref. 43. In section 8.2 we introduce the lattice analog of the Boltzmann approximation, which allows an explicit calculation of the viscosity. In section 8.3 we discuss some implications for the Reynolds numbers of incompressible flows simulated on lattice gases.

---

[13] Since the velocity set of the pseudo-4-D model is the same as in four dimensions, isotropy is ensured for all fourth order tensors depending only on the velocity set. Thus the nonlinear term has the correct isotropic form. The viscous term is isotropic within the Boltzmann approximation (cf. section 8.2); otherwise, deviations from isotropy are expected to be small.(2)

8.1 Fluctuation-dissipation relation and ''noisy'' hydrodynamics

We first explain the basic ideas in words. Spontaneous fluctuations at equilibrium involve modes of all possible scales. The fluctuations of very large scales should have their dynamics governed by the macroscopic equations derived in sections 5-7. Such fluctuations are also expected to be very weak, so that *linear* hydrodynamics should apply. Large-scale spontaneous fluctuations are constantly regenerated, and in a random manner; this regeneration is provided by a random force (noise) term which can be identified and expressed in terms of the fluctuating microscopic variables. If this random force has a short correlation-time (i.e. small compared to the life-time of the large-scale fluctuations under investigation), then each large-scale mode v has its dynamics governed by a Langevin equation.[14] It follows that the variance $\langle v^2 \rangle$ can be expressed in terms of the damping coefficient γ (related to the viscosity) and of the time-correlation function of the random force. Alternatively, the variance $\langle v^2 \rangle$ can be calculated from the known one-time equilibrium properties. Identification gives the viscosity in terms of equilibrium time-correlation functions. This is the general programme that we now carry out for the special case of lattice gases. We restrict ourselves to equilibrium solutions with zero mean velocity.

We shall use in this section the following notation. The density ρ and the mass current $\mathbf{j}$ are no longer given by their expressions (3.24) in terms of the mean populations; instead, they are defined in terms of the fluctuating Boolean field

$$\rho(t_*, \mathbf{r}_*) = \sum_i n_i(t_*, \mathbf{r}_*), \qquad \mathbf{j}(t_*, \mathbf{r}_*) = \sum_i \mathbf{c}_i n_i(t_*, \mathbf{r}_*). \tag{8.1}$$

We denote by $\tilde{n}_i$ the fluctuating part of the Boolean field, defined by

$$n_i(t_*, \mathbf{r}_*) = d + \tilde{n}_i(t_*, \mathbf{r}_*), \tag{8.2}$$

where d is the density per cell.

We introduce *meso-averaged* fields by taking *spatial* averages over a a distance ϵ^{-1}.[15] These will be denoted by angular brackets with the subscript *ma*. The meso-averages of

[14] For the case of lattice gases, we shall actually obtain a finite difference equation.
[15] More precisely, by dropping spatial Fourier components with wavenumber $k > \epsilon$.

n_i, ρ, and $\mathbf{j}$ are denoted $\bar{n}_i$, $\bar{\rho}$, and $\bar{\mathbf{j}}$ respectively. Locally, the equilibrium relation (4.14) should hold approximately for the meso-averaged populations. We thus write

$$\bar{n}_i = \frac{\bar{\rho}}{b} + \frac{D}{c^2 b}\bar{\mathbf{j}} \cdot \mathbf{c}_i + \delta_i + \bar{n}_i^{(1)}(t_\star, \mathbf{r}_\star). \tag{8.3}$$

δ_i represents the (still unknown) input from non-hydrodynamic fluctuations; $\bar{n}_i^{(1)}$ is the contribution analogous to $\epsilon N_i^{(1)}$ in (5.1), arising from the gradients of meso-averages. Note that in (8.3) we dropped contributions nonlinear in the mass current; indeed, we should be able to determine the viscosity from just linear hydrodynamics.[16]

We now derive the equations for noisy hydrodynamics. As usual, we start from the microscopic conservation relations (3.16) and (3.17) and we take their meso-averages:

$$\sum_i [\bar{n}_i(t_\star + 1, \mathbf{r}_\star + \mathbf{c}_i) - \bar{n}_i(t_\star, \mathbf{r}_\star)] = 0, \tag{8.4}$$

$$\sum_i \mathbf{c}_i [\bar{n}_i(t_\star + 1, \mathbf{r}_\star + \mathbf{c}_i) - \bar{n}_i(t_\star, \mathbf{r}_\star)] = 0. \tag{8.5}$$

Substituting (8.3) into (8.5), we obtain

$$\begin{aligned} &\frac{1}{b}\sum_i \mathbf{c}_i[\bar{\rho}(t_\star + 1, \mathbf{r}_\star + \mathbf{c}_i) - \bar{\rho}(t_\star, \mathbf{r}_\star)] + \frac{D}{c^2 b}\sum_i \mathbf{c}_i\,\mathbf{c}_i \cdot [\bar{\mathbf{j}}(t_\star + 1, \mathbf{r}_\star + \mathbf{c}_i) - \bar{\mathbf{j}}(t_\star, \mathbf{r}_\star)] \\ &+ \sum_i \mathbf{c}_i[\bar{n}_i^{(1)}(t_\star + 1, \mathbf{r}_\star + \mathbf{c}_i) - \bar{n}_i^{(1)}(t_\star, \mathbf{r}_\star)] = \mathbf{f}(t_\star, \mathbf{r}_\star), \end{aligned} \tag{8.6}$$

where

$$\mathbf{f}(t_\star, \mathbf{r}_\star) = -\sum_i \mathbf{c}_i[\delta_i(t_\star + 1, \mathbf{r}_\star + \mathbf{c}_i) - \delta_i(t_\star, \mathbf{r}_\star)] \tag{8.7}$$

is the random force. Using (8.1), (8.2), (8.3), (8.4), and (8.5) , we can also write (to leading order in gradients)

$$\mathbf{f}(t_\star, \mathbf{r}_\star) = \left\langle \frac{1}{c^2 b}\sum_{ij} \left(c^2\mathbf{c}_i + D\mathbf{c}_i \cdot \mathbf{c}_j\,\mathbf{c}_i\right) [\bar{n}_j(t_\star + 1, \mathbf{r}_\star + \mathbf{c}_i) - \bar{n}_j(t_\star + 1, \mathbf{r}_\star + \mathbf{c}_j)] \right\rangle_{ma}. \tag{8.8}$$

The l.h.s. of (8.6) is expanded in powers of gradients (i.e. of ϵ), as we have done in section 5. However, we keep finite differences rather than derivatives in time because

[16] This is not exactly true in two dimensions as we shall see below.

of the presence of the rapidly varying random force. Since we only want to identify the shear viscosity (the bulk viscosity is zero), it suffices to extract the solenoidal part of the hydrodynamical equation. For this and other reasons it is better to work in Fourier space. We define the (spatial) Fourier transform of the fluctuating Boolean field by

$$\bar{n}_i(t_*, \mathbf{r}_*) = \sum_{\mathbf{k}} e^{i\mathbf{k}\cdot\mathbf{r}_*}\hat{n}(t_*, \mathbf{k}), \tag{8.9}$$

where the components of $\mathbf{k}$ are multiples of 2π divided by the lattice periodicities in the various directions. We similarly define $\hat{\mathbf{j}}$ and $\hat{\mathbf{f}}$, the Fourier transforms of the mass current and the random force. Their solenoidal parts, projection on the hyperplane perpendicular to $\mathbf{k}$, are denoted $\hat{\mathbf{j}}_\perp$ and $\hat{\mathbf{f}}_\perp$.

To leading order in k, we obtain from (8.8), using (2.5)

$$\hat{\mathbf{f}}_\perp(t_*, \mathbf{k}) = -\sum_j i\mathbf{k}\cdot\mathbf{c}_j\left(\mathbf{c}_j - \frac{\mathbf{c}_j\cdot\mathbf{k}\mathbf{k}}{k^2}\right)\hat{n}_j(t_* + 1, \mathbf{k}). \tag{8.10}$$

The meso-averaging is just the restriction that $k < \epsilon$. Fourier transforming (8.6) and taking the solenoidal part, we obtain for small k

$$\hat{\mathbf{j}}_\perp(t_* + 1, \mathbf{k}) - \hat{\mathbf{j}}_\perp(t_*, \mathbf{k}) + \nu k^2\hat{\mathbf{j}}_\perp(t_*, \mathbf{k}) = \hat{\mathbf{f}}_\perp(t_*, \mathbf{k}). \tag{8.11}$$

This is our discrete Langevin equation. Note that ν is the (total) viscosity $\nu = \nu_c + \nu_p$. In principle we must expand to second order in k to obtain the viscous terms, but we could as well have written the l.h.s of (8.11) a priori, since we want to use (8.11) to *determine* the viscosity. It is straightforward to solve the linear finite-difference equation (8.11). From the solution, we calculate the variance of $\hat{\mathbf{j}}_\perp$ and obtain, when the viscous damping time $1/(\nu k^2)$ is large compared to the correlation time of the random force

$$\left\langle\left|\hat{\mathbf{j}}_\perp(t_*, \mathbf{k})\right|^2\right\rangle = \frac{1}{2\nu k^2}\sum_{t_*=-\infty}^{t_*=+\infty}\langle\mathbf{f}_\perp(t_*, \mathbf{k})\cdot\mathbf{f}_\perp^*(t_*, \mathbf{k})\rangle, \tag{8.12}$$

where the asterisk denotes complex conjugation. The variance of $\hat{\mathbf{j}}_\perp$ can also be calculated directly using (8.1) and

$$\begin{aligned}\langle\bar{n}_i(t_*, \rho_*)\bar{n}_j(t_*, 0)\rangle &= \langle\bar{n}_i^2\rangle\delta_{ij}\delta_{\rho_*}\\ \langle\bar{n}_i^2\rangle &= \langle n_i^2\rangle - \langle n_i\rangle^2 = d - d^2,\end{aligned} \tag{8.13}$$

where δ_{ρ_*} denotes a Kronecker delta in the spatial separation ρ_*. We obtain

$$\left\langle \left| \hat{\mathrm{j}}_\perp(t_*, \mathbf{k}) \right|^2 \right\rangle = \frac{1}{V} bc^2 d(1-d) \frac{D-1}{D}, \tag{8.14}$$

where V denotes the total number of lattice points in the periodicity volume. Thus, the l.h.s. of (8.12) is $\mathbf{k}$-independent. We evaluate the r.h.s of (8.12) in the limit $k \to 0$, using (8.10). We skip some intermediate steps in which we (i) use the stationariness of the fluctuations at equilibrium, (ii) use the isotropy of second and fourth order symmetrical tensors, (iii) interchange the $\mathbf{k} \to 0$ limit and the infinite summation over t_*.[17] Identifying the two expressions (8.12) and (8.14), we obtain for the viscosity

$$\begin{aligned} \nu &= \frac{D}{2(D-1)(D+2)} \frac{1}{bc^2} \frac{1}{d(1-d)} \frac{1}{V} \sum_{t_*=-\infty}^{t_*=+\infty} \sum_{ij\alpha\beta} Q_{i\alpha\beta} Q_{j\alpha\beta} \left\langle \hat{n}_i(t_*, 0) \hat{n}_j^*(0,0) \right\rangle \\ &= \frac{D}{2(D-1)(D+2)} \frac{1}{bc^2} \frac{1}{d(1-d)} \sum_{t_*=-\infty}^{t_*=+\infty} \sum_{\rho_* \in \mathcal{L}} \sum_{ij\alpha\beta} Q_{i\alpha\beta} Q_{j\alpha\beta} \left\langle \bar{n}_i(t_*, \rho_*) \bar{n}_j(0,0) \right\rangle, \end{aligned} \tag{8.15}$$

with

$$Q_{i\alpha\beta} = c_{i\alpha} c_{i\beta} - \frac{c^2}{D} \delta_{\alpha\beta}. \tag{8.16}$$

This completes the fluctuation-dissipation calculation of the viscosity. A consequence of the Fourier-space representation (the upper half of (8.15)) is the positivity of the viscosity; indeed, the viscosity is, within a positive factor, the time-summation of the auto-correlation of $\sum_i Q_{i\alpha\beta} \hat{n}_i(t_*, 0)$.

Several comments are now in order. It is easily checked that the $t_* = 0$ contribution to the viscosity (lower part of (8.15)) is $c^2/(2(D+2))$, that is, just the opposite of the "propagation viscosity" ν_p introduced in section 7. The viscosity is the sum of the collision viscosity ν_c and of ν_p. Using the identity

$$\sum_{t_*=-\infty}^{t_*=+\infty} Z(t_*) = 2 \sum_{t_*=0}^{t_*=+\infty} Z(t_*) \; - \; Z(0), \tag{8.17}$$

(for an even function $Z(t_*)$), we find that ν_c has a representation similar to (8.15) (lower part), with an additional factor of 2 and the summation over t_* extending only from 0 to

[17] This is equivalent to assuming that the viscosity is finite, cf. below.

∞. We thereby recover an expression derived in Ref. 43, using a discrete variant of the Green-Kubo formalism. It is reassuring to have two completely different derivations of the viscosity, since we consider our fluctuation-dissipation derivation somewhat delicate.

It is of interest that the fluctuation-dissipation derivation gives directly the (total) viscosity. This suggests that the splitting into collision and propagation viscosities is an artefact of our multi-scale formalism.

There is no closed form representation of the correlation function $\langle \bar{n}_i(t_*, \rho_*)\bar{n}_j(0,0)\rangle$, except for short times. However, (8.15) is a good starting point for a Monte-Carlo calculation of the viscosity (cf. Ref. 43).

In our derivation we have dropped all contributions from *nonlinear* terms in the mass current j. Is this justified? If we reinstate the nonlinear terms, we obtain, for the solenoidal part of the meso-averaged mass current, the Navier-Stokes equations (7.17) of section 8 with the additional random force, given in the Fourier representation by (8.10). On macroscopic scales this force may be considered as δ-correlated in time. Its spectrum follows, for small k, a k^{D+1} power-law.[18] The Navier-Stokes equations with this kind of power-law forcing is one of the few problems in nonlinear Statistical Fluid Mechanics which can be systematically analyzed by renormalization group methods.[71,72] For $D > 2$, the nonlinear term is irrelevant for small k so that our calculation of the viscosity is legitimate. At the "crossover" dimension $D = 2$, the nonlinear term becomes "marginal"; it produces a renormalization of the viscosity which is then logarithmically scale-dependent. Thus, in the limit of infinite scale-separation, the viscosity becomes infinite in two-dimensions. This is an instance of the known divergence of transport coefficients in two-dimensional Statistical Mechanics.[68,73] Alternatively, the divergence of the viscosity in two-dimensions can be viewed as due to the presence of a "long-time-tail", proportional to $t_*^{-D/2}$, in the correlation function appearing in (8.15). Attempts have been made to observe long-time-tails and scale-dependence of the viscosity in Monte-Carlo simulations of lattice gas

[18] A factor k^2 comes from the average squared Fourier amplitude and another factor k^{D-1} from the D-dimensional volume element.

models.[8,23,43,44] This is not easy because (i) the effects show up only at very long times (or large scales) and may then be hidden by Monte-Carlo noise (insufficient averaging); (ii) the effects should get weaker as the number b of cells per node increases (cf. end of section 8.2).

Finally, the noisy hydrodynamics formalism can be used to estimate to what extent the microscopic noise contaminates the hydrodynamic macroscopic signal. Estimates, assuming the signal to be meso-averaged in space and time, have been made in the context of fully developed incompressible two- and three-dimensional turbulence.[19] It has been found that in two dimensions noise is relevant only at scales less than the dissipation scale, while in three dimensions this happens only far out in the dissipation range.[74]

8.2 The lattice Boltzmann approximation

Explicit calculation of transport coefficients can be done for lattice gases, using the Boltzmann approximation. In this approximation one assumes that *particles entering a collision process have no prior correlations*. The microdynamical formalism of section 3.1 is particularly well suited for deriving what we shall call the *lattice Boltzmann equation*. We take the ensemble average of eq. (3.9). The Boolean variables n_i become the mean populations N_i. The average of the collision function Δ_i can be completely factorized, thanks to the Boltzmann approximation. We obtain

$$N_i(t_* + 1, \mathbf{r}_* + \mathbf{c}_i) = N_i(t_*, \mathbf{r}_*) + \Delta_i^{\mathrm{Boltz}}$$

$$\Delta_i^{\mathrm{Boltz}} = \sum_{s,s'} (s'_i - s_i) A(s \to s') \prod_j N_j^{s_j} (1 - N_j)^{(1-s_j)}. \tag{8.18}$$

Here, all the N_j's are evaluated at t_* and $\mathbf{r}_*$. The $A(s \to s')$'s, the transition probabilities introduced in section 2.4, are the averages of the Boolean transition variables $\xi_{ss'}$. Note that the (Boltzmann) *collision function* $\Delta_i^{\mathrm{Boltz}}$ vanishes at equilibrium.

The Boltzmann approximation in ordinary gases is associated with low density situations, when the mean-free path is so large that particles entering a collision come mostly from distant uncorrelated regions. The Boltzmann approximation for a lattice gas appears

---

[19] Note that in the incompressible case, only solenoidal noise is relevant.

to have a very broad validity, not particularly restricted to low densities.[20] We shall come back to the matter at the end of this section.

Our lattice Boltzmann equation (8.18) is a *finite difference* equation. There is a differential version of it, obtained by Taylor-expanding the finite differences to first order, namely

$$\partial_t N_i + \mathbf{c}_i \cdot \nabla N_i = \Delta_i^{\mathrm{Boltz}}, \tag{8.19}$$

where $\Delta_i^{\mathrm{Boltz}}$ is defined as in (8.18). Boltzmann equations of the form (8.19) have been extensively studied as discrete velocity approximations to the ordinary Boltzmann equation.[15–1] The (differential) Boltzmann formalism has been applied to various lattice gas models.[35,39] This formalism correctly captures all hydrodynamic phenomena involving only first order derivatives. Indeed, for these, we have seen that only the equilibrium solutions matter, and the latter are completely factorized. Diffusive phenomena involve second order derivatives. Hence, the *propagation viscosities* (cf. section 7), which are an effect of lattice-discreteness, are not captured by the (differential) Boltzmann equation. At low densities, where collision viscosities dominate over propagation viscosities, the discrepancy is irrelevant.

We do not intend to engage in extended discussions of the consequences of the lattice Boltzmann equation because most of the derivation of the hydrodynamical equations is independent of this approximation. There are however two important results which follow from the lattice Boltzmann equation. The first one concerns the irreversible approach to equilibrium. It is derived by adapting an H-theorem formalism to the fully discrete context (see Appendix F by Hénon).

The second result is an explicit derivation of the viscosity. From the Boltzmann equation this is usually done by a Chapman-Enskog formalism[75,76] (see also Gatignol's monography, Ref. 17). This formalism is easily adapted to the lattice Boltzmann equation.[77]

---

[20] Even at low densities, the Boltzmann approximation may not be valid. Indeed, without effectively changing the dynamics, we can reduce the density by arbitrary large factors by having the particles initially located on a sub-lattice with some large periodicity; these are however pathologically unstable configurations.

With the general multi-scale formalism of sections 5-7, we have already covered a substantial fraction of the ground. Furthermore an alternative derivation, which stays completely at the microscopic level is presented in this volume by Hénon who also discusses consequences of his explicit viscosity-formula.[42] We shall therefore be brief.

The problem of the viscosity amounts to finding the coefficient ψ relating the gradient of the mass current $\rho\mathbf{u}$ to the first order perturbation $N_i^{(1)}$ of the mean population, through (cf. eq. (5.14) of section 5)

$$\begin{aligned} N_i^{(1)} &= \psi Q_{i\alpha\beta}\partial_{1\alpha}(\rho u_\beta) \\ Q_{i\alpha\beta} &= \mathrm{c}_{i\alpha}\mathrm{c}_{i\beta} - \frac{c^2}{D}\delta_{\alpha\beta}. \end{aligned} \tag{8.20}$$

We start from (5.1) with $N_i^{(0)}$ given by (4.14). We substitute into the lattice Boltzmann equation (8.18) and identify the terms $O(\epsilon)$. For this we Taylor-expand finite differences to first order, use (5.6) and (5.7) to express time-derivatives in terms of space-derivatives, and ignore all terms beyond the linear ones in the velocity. We obtain

$$\frac{D}{bc^2}Q_{i\alpha\beta}\partial_{1\alpha}(\rho u_\beta) = \sum_j \mathcal{A}_{ij}N_j^{(1)}. \tag{8.21}$$

Here,

$$\mathcal{A}_{ij} = \left[\frac{\partial \Delta_i^{\mathrm{Boltz}}}{\partial N_j}\right]_{N_i=\rho/b}, \tag{8.22}$$

is the *linearized collision matrix*, evaluated at the zero-velocity equilibrium, which can be expressed in compact form as[42]

$$\mathcal{A}_{ij} = -\frac{1}{2}\sum_{ss'}(s_i - s_i')A(s \to s')d^{p-1}(1-d)^{b-p-1}(s_j - s_j'), \quad p = \sum_i s_i. \tag{8.23}$$

We eliminate $N_i^{(1)}$ between (8.20) and (8.21), to obtain

$$\left[\frac{D}{bc^2}Q_{i\alpha\beta} - \psi\sum_j \mathcal{A}_{ij}Q_{j\alpha\beta}\right]\partial_{1\alpha}(\rho u_\beta) = 0. \tag{8.24}$$

This should hold for arbitrary gradients of the mass current. Thus, the quantity between square brackets vanishes. This means that, for any (α,β), $Q_{i\alpha\beta}$, considered as a vector with components labelled by i, is an eigenvector of the linearized collision matrix with eigenvalue $D/(bc^2\psi)$; a direct proof of this may be derived from the $\mathcal{G}$-invariance. From (8.24) we can easily caculate ψ; the simplest method is to multiply the vanishing square bracket by $Q_{i\alpha\beta}$ and sum over i, α, and β. If, in addition, we assume the isotropy of fourth order tensors, we can use (6.17) to obtain a closed-form expression for the collision viscosity

$$\nu_c = -\frac{c^2}{D+2} \frac{\sum_{i\alpha\beta} Q_{i\alpha\beta}^2}{\sum_{ij\alpha\beta} Q_{i\alpha\beta} A_{ij} Q_{j\alpha\beta}}. \tag{8.25}$$

In Appendix E we give explicit formulae calculated from (8.25) for the viscosities of the FHP models (including those with rest-particles which require minor amendements of our formalism).

We finally address the question of the validity of the lattice Boltzmann equation. Comparisons of the viscosities obtained from simulations(25,29,31,33) or Monte-Carlo calculations(77) with the predictions of the lattice Boltzmann approximation suggest that the validity of the latter is not limited to low densities. We know that equilibrium solutions are factorized and that transport coefficients can be calculated with arbitrarily weak macroscopic gradients. However, this cannot be the basis for the validity of the Boltzmann approximation: a weak *macroscopic* gradient implies that the probability of changing the state of a given node from its equilibrium value is small; but when such a change takes place, it produces a strong *microscopic* perturbation in its environment. Otherwise there would be no (weak) divergence of the viscosity in two dimensions; indeed, the Boltzmann approximation does not capture noise-induced renormalization effects (cf. end of section 8.1). A more likely explanation of the success of the lattice Boltzmann approximation may be that it is the leading order in some kind of $1/b$ expansion, where b is the number of velocity cells at each node. At the moment, we can only support this by the following heuristic argument. Deviations from Boltzmann require correlations between particles entering a collision. The latter arise from previous collisions;[21] when b is large the weight pertaining to such events ought to be small.

---

[21] Collisions produce correlations whenever the particles are not exactly at equilibrium.

8.3 The Reynolds number

Knowing the kinematic shear viscosity in terms of the density and the collision rules, we can calculate the Reynolds number associated to a large-scale flow.

A natural unit of length is the lattice constant (distance of adjacent nodes), which has been taken equal to one for the two-dimensional HPP and FHP models. The four-dimensional FCHC model has a lattice constant of $\sqrt{2}$, but its three-dimensional projected version, the pseudo-4-D FCHC model, resides on a cubic lattice which has also unit lattice constant. The time necessary for microscopic information to propagate from one node to its connecting neighbours defines a natural unit of time. We then have a natural unit of velocity: the speed necessary to travel the lattice constant (or the projected lattice constant for the pseudo-4-D model) in a unit time. In these units, the characteristic scale and velocity of the flow will be denoted by ℓ_0 and u_0.

The standard definition of the Reynolds number is

$$R = \frac{\text{characteristic scale} \times \text{characteristic velocity}}{\text{kinematic shear viscosity}}. \tag{8.26}$$

In deriving the Navier-Stokes equations in section 7.3, we rescaled space, time, velocity, pressure and viscosity (cf. eq. (7.15)). The rescaling of space (by ϵ) and of velocity (by ϵ^{-1}) cancel in the numerator of (8.26). The rescaled viscosity is $\nu'(\rho_0) = \nu(\rho_0)/g(\rho_0)$. Hence, the Reynolds number is

$$R = \ell_0 u_0 \frac{g(\rho_0)}{\nu(\rho_0)}. \tag{8.27}$$

In order to operate in an incompressible régime, the velocity u_0 should be small compared to the speed of sound c_s. The latter is model-dependent: $c_s = 1/\sqrt{2}$ for FHP-I and FCHC, $c_s = \sqrt{3/7}$ for FHP-II and FHP-III (cf. section 7 and Appendix E). Let us therefore reexpress the Reynolds number in terms of the Mach number

$$M = \frac{u_0}{c_s}. \tag{8.28}$$

We obtain

$$R = M \ell_0 R_*(\rho_0), \tag{8.29}$$

where

$$R_*(\rho_0) = \frac{c_s g(\rho_0)}{\nu(\rho_0)} \qquad (8.30)$$

contains all the *local* information.

In flow simulations using lattice gases, it is of interest to operate at the density which maximizes R_*. Let us work this out for the simplest case of FHP-I. For the viscosity, we use the lattice Boltzmann value given in Appendix E. We have

$$g(\rho_0) = \frac{1}{2}\frac{1-2d}{1-d}, \quad \nu(\rho_0) = \frac{1}{12d(1-d)^3} - \frac{1}{8}, \quad d = \frac{\rho_0}{6}. \qquad (8.31)$$

Here, d is the mean density per cell. Substituting in (8.30), we find that

$$R_*^{\max} = \max R_* = 0.387, \quad \text{for} \quad d = d_{\max} = 0.187. \qquad (8.32)$$

Results for FHP-II and FHP-III are given in Appendix E. Note that a gain of about a factor 6 is achieved in going from FHP-I to FHP-III, because the latter includes many more collisions. For the pseudo-4-D FCHC model there is work in progress on the optimization of collisions. It is already known that $R_*^{\max}$ is at least 6.4.(78)

High Reynolds number incompressible turbulent flows have a whole range of scales. The smallest effectively excited scale is called the *dissipation scale* and denoted ℓ_d. It is then of interest to find how many lattice constants are contained in ℓ_d, since this will determine how effective lattice gases are in simulating high Reynolds number flows.(1,36) For this, let ℓ_0 denote the integral scale of the flow. Between ℓ_0, ℓ_d and the Reynolds number R, there is the following relation

$$\frac{\ell_d}{\ell_0} = CR^{-m}. \qquad (8.33)$$

$m = 1/2$ in two dimensions and $m = 3/4$ in three dimensions; C is a dimensionless constant not given by theory. In two dimensions, (8.33) is a consequence of the Batchelor-Kraichnan(79,80) phenomenological theory of the enstrophy cascade, which is well supported by numerical simulations.(81) In three dimensions, (8.33) follows from the Kolmogorov(82) phenomenological theory of the energy cascade, which is well supported[22] by experimental data.[(83)] Using (8.29) and (8.33) and assuming that R_* has its maximum value $R_*^{\max}$, we obtain

$$\ell_d = C\,(MR_*^{\max})^{-\frac{1}{2}}\,\ell_0^{\frac{1}{2}} = C\,(MR_*^{\max})^{-1}\,R^{\frac{1}{2}} \quad \text{in 2-D,} \tag{8.34}$$

and

$$\ell_d = C\,(MR_*^{\max})^{-\frac{3}{4}}\,\ell_0^{\frac{1}{4}} = C\,(MR_*^{\max})^{-1}\,R^{\frac{1}{4}} \quad \text{in 3-D.} \tag{8.35}$$

In all cases, we see that $\ell_d \rightarrow \infty$ as $R \rightarrow \infty$, but more slowly in three than in two dimensions. We are thus assured that at high Reynolds numbers the separation of scale between the lattice constant and ℓ_d, necessary for hydrodynamic behaviour is satisfied. Having it too well satisfied may however be a mixed blessing, as stressed in Ref. 36. Indeed, in hydrodynamic simulations using lattice gases it is not desirable to have too much irrelevant microscopic information. We note that ℓ_0 appears in eqs. (8.34)-(8.35) with a larger exponent in the 2-D case; thus the above mentioned problem is most severe for large lattices in two dimensions.

The highest Reynolds number which can be simulated by lattice gas methods in three dimensions can be estimated as follows. We take $M = 0.3$, a Mach number at which compressibility effects can be safely ignored;[(84)] we take the maximum known value $R_*^{\max} = 6.4$ for the FCHC model, and we take $\ell_0 = 10^3$, a fairly large value which implies a memory requirement of at least 24 gigabits; from (8.29) we find that the maximum Reynods number is about two thousands. It is of interest both in two and three dimensions to try to decrease the viscosity, thereby increasing $R_*^{\max}$. Note that it is not correct to infer from dimensional analysis that necessarily $R_*^{\max}$ must be $O(1)$. $R_*^{\max}$ is very much a function of the *complexity* of collisions. For example, by going from FHP-I to FCHC (which can also be projected down to two dimensions), $R_*^{\max}$ increases more than sixteen times.

---

[22] Small intermittency corrections which would slightly increase the exponent m cannot be ruled out.

9. Conclusion

In Statistical Mechanics there are many instances where two models, microscopically quite different, have the same large-scale properties. For example, the Ising model and a real Ferromagnet have presumably the same large-scale critical behaviour. Similarly, the lattice gases studied in this paper, such as FHP and FCHC, are macroscopically indistiguishable from real fluids. This provides us with an attractive alternative to the traditional simulations of Fluid Mechanics. In lattice gas simulations, we just manipulate bits representing occupation of microscopic cells. The physical interpretation need not be in terms of particles moving and colliding. The idea can clearly be extended to include processes such as chemical reactions or multi-phase flow.[53–57] An open question is whether there are cellular automata implementations of processes which in the real world do not have a discrete microscopic origin, such as propagation of e.m. waves. More generally, what are the P.D.E.'s which can be *efficiently* implemented on cellular automata? We emphasize *efficiently*, because there are always brute force implementations: replace derivatives by finite differences on a regular grid and use finite floating point truncations of the continuous fields. The result may be viewed as a cellular automaton, but one in which there is no "bit democracy", insofar as there is a rigid hierarchical order between the bits.

Our derivation of hydrodynamics from the microdynamics leaves room for improvement. A key assumption made in section 4.1 may be formulated as follows. Among the invariant measures of the microdynamical equations, only the completely factorized ones (which play the role, here, of the microcanonical ensemble) are relevant in the limit of large lattices. On a finite lattice with deterministic and invertible updating rules, we expect that there are many other invariant measures. Indeed, phase space is a finite set and updating is a permutation of this set; it is thus unlikely that there should be a closed orbit going through all points. So, we do not expect the discrete equivalent of an ergodic theorem. Anyway, ergodic results should be irrelevant. On the one hand, on an $L \times L$ lattice with b bits per node its takes 2^{bL^2} updates to visit all configurations (if they are accessible). On the other hand, we know (from simulations) that local equilibrium is achieved in a few updates and global equilibrium is achieved on a diffusive time scale (approximately L^2). We believe that, on large lattices, the factorized equilibrium distributions constitute some kind of "fixed point" to which there is rapid convergence of the iterated Boolean map defined by the microdynamical equations of section 3.1. Understanding this process should clarify the mechanism of irreversibility in lattice gases and, eventually, in real gases.

Acknowledgements

Many colleagues have contributed to this work by their suggestions as well as by their questions. They include V. Arnold, H. Cabannes, R. Caflish, P. Clavin, A. Deudon, G. Doolen, R. Gatignol, F. Hayot, M. Hénon, S. Kaniel, R. Kraichnan, J. Lebowitz, D. Levermore, N. Margolus, J.-L. Oneto, S. Orszag, H. Rose, J. Searby, Z. She, Ya. Sinai, T. Toffoli, G. Vichniac, S. Wolfram, V. Yakhot, and S. Zaleski. This work was supported by European Community grant ST-2J-029-1-F, CNRS-Los Alamos grant PICS "Cellular Automata Hydrodynamics", and DOE grant KC-04-01-030.

Appendix A. Basic symmetries of HPP, FHP, and FCHC models

We show that the models HPP, FHP, and FCHC, introduced in section 2, satisfy the symmetry assumptions (i) through (iv) of section 2.4. Assumptions (i) and (ii) are obvious for all three models. Let us consider (iii) and (iv) successively for the three models.

HPP

Let us take the x_1 axis in the direction of the vector $\mathbf{c}_1$. The isometry group $\mathcal{G}$ of the velocity set is generated by permutations of the x_1 and x_2 coordinates and reversals of any of them. Clearly, any two vectors $\mathbf{c}_i$ and $\mathbf{c}_j$ can be exchanged by some isometry, so that assumptions (iii) holds. Consider a particular vector, say, $\mathbf{c}_1$. The subgroup $\mathcal{G}_1$, leaving $\mathbf{c}_1$ invariant reduces to the identity and reversal of x_2; this implies parts (a) and (b) of assumption (iv).

FHP

Let us take the x_1 axis in the direction of $\mathbf{c}_1$. The isometry group $\mathcal{G}$ is now generated by rotations of $\pi/3$ and reversal of the x_2 coordinate. Assumption (iii) is obvious. The subgroup $\mathcal{G}_1$ reduces again to the identity and the reversal of x_2, so that (iv) follows.

FCHC

The FCHC lattice was defined in section 2.3 with explicit reference to coordinates x_1, x_2, x_3, and x_4. In this coordinate system, the velocity set is formed of

$$\begin{aligned} &(\pm1,\pm1,0,0), \quad (\pm1,0,\pm1,0), \quad (\pm1,0,0,\pm1) \\ &(0,\pm1,\pm1,0), \quad (0,\pm1,0,\pm1), \quad (0,0,\pm1,\pm1). \end{aligned} \tag{A.1}$$

By the orthonormal change of variables

$$\begin{pmatrix} y_1 \\ y_2 \\ y_3 \\ y_4 \end{pmatrix} = \frac{1}{\sqrt{2}} \begin{pmatrix} 1 & 1 & 0 & 0 \\ -1 & 1 & 0 & 0 \\ 0 & 0 & 1 & 1 \\ 0 & 0 & -1 & 1 \end{pmatrix} \begin{pmatrix} x_1 \\ x_2 \\ x_3 \\ x_4 \end{pmatrix}, \tag{A.2}$$

the velocity set becomes

$$(\pm\sqrt{2},0,0,0), \quad (0,\pm\sqrt{2},0,0), \quad (0,0,\pm\sqrt{2},0), \quad (0,0,0,\pm\sqrt{2}),$$

$$(\pm\frac{1}{\sqrt{2}}, \pm\frac{1}{\sqrt{2}}, \pm\frac{1}{\sqrt{2}}, \pm\frac{1}{\sqrt{2}}). \tag{A.3}$$

The isometry group $\mathcal{G}$ is generated by permutations and reversals of the x_α coordinates and by the symmetry with respect to the hyperplane $x_1 + x_2 + x_3 + x_4 = 0$, which is conveniently written in terms of y_α coordinates as

$$\Sigma: \quad (y_1, y_2, y_3, y_4) \mapsto (-y_3, y_2, -y_1, y_4). \tag{A.4}$$

Assumption (iii) is obvious in any of the coordinate systems. As for assumption (iv), let us consider the subgroup $\mathcal{G}_1$ leaving invariant, say, the vector with y_α coordinates $(0, 0, 0, 1/\sqrt{2})$. The restriction of $\mathcal{G}_1$ to the hyperplane $y_4 = 0$ is generated by the identity, permutations, and reversals of y_1, y_2, and y_3. Assumptions (a) and (b) follow readily.

Appendix B. Symmetry-related properties

Using assumptions (i) through (iv) of section 2.4, we prove properties P1-P6

P1 *Parity-invariance.* The set of velocity vectors is invariant under space-reversal.

Indeed, on a Bravais lattice, vectors connecting neighbouring nodes come in opposite pairs.

P2 Any set of i-dependent vectors $v_{i\alpha}$, which is $\mathcal{G}$-invariant, is of the form $\lambda c_{i\alpha}$.

We write $\mathbf{v}_i$ as the sum of its projection on $\mathbf{c}_i$ and of a vector perpendicular to $\mathbf{c}_i$. This decomposition being $\mathcal{G}$-invariant, the latter vector vanishes by (iv - a).

P3 Any set of i-dependent tensors $t_{i\alpha\beta}$, which is $\mathcal{G}$-invariant, is of the form $\lambda c_{i\alpha} c_{i\beta} - \mu\delta_{\alpha\beta}$.

To the tensors $t_{i\alpha\beta}$, we associate the linear operators $T_i : x_\alpha \mapsto t_{i\alpha\beta} x_\beta$. $\mathcal{G}$-invariance means that the T_i's commute with any lattice isometry leaving $\mathbf{c}_i$ invariant. We now write the $\mathcal{G}$-invariant decomposition

$$T_i = P_i T_i P_i + (I - P_i) T_i P_i + P_i T_i (I - P_i) + (I - P_i) T_i (I - P_i), \tag{B.1}$$

where I is the identity in $\mathbf{R}^D$ and P_i is the orthogonal projection on $\mathbf{c}_i$. The second operator in (B.1), applied to an arbitrary vector $\mathbf{w}$, gives

$$(I - P_i) T_i P_i \mathbf{w} = \frac{\mathbf{w} \cdot \mathbf{c}_i}{c^2} (I - P_i) T_i \mathbf{c}_i. \tag{B.2}$$

The vectors $(I - P_i)T_i\mathbf{c}_i$ are $\mathcal{G}$-invariant and orthogonal to $\mathbf{c}_i$, and thus vanish by (iv-a). The third operator in (B.1) vanishes for similar reasons (use the $\mathcal{G}$-invariance of the transposed of the T_i's). The fourth operator in (B.1) is, by (iv-b) proportional to I_i, the identity in the subspace orthogonal to $\mathbf{c}_i$. Since $I = I_i + P_i$, the proof is completed.

We mention that we obtained P3 by trying to formalize a result used by Hénon[(42)] in deriving a closed-form viscosity formula.

P4 *Isotropy of second order tensors.* Any $\mathcal{G}$-invariant tensor $t_{\alpha\beta}$ is of the form $\mu\delta_{\alpha\beta}$.

This is a special case of P3, when there is no i-dependence.

P5 Any $\mathcal{G}$-invariant third order tensor vanishes.

This follows from P1 (parity invariance).

P6 *Velocity moments.* Odd order velocity moments vanish. The second order velocity moment is given by

$$\sum_i c_{i\alpha} c_{i\beta} = \frac{bc^2}{D}\delta_{\alpha\beta}. \qquad (B.3)$$

The vanishing of odd order moments is a consequence of P1. (B.3) follows from P4 and the identity

$$\sum_i c_{i\alpha} c_{i\alpha} = bc^2. \qquad (B.4)$$

Appendix C. Equilibrium solutions

We prove the

Lemma. The following statements are equivalent:

(a) The N_i's are a solution of

$$\prod_j N_j^{s'_j}(1-N_j)^{(1-s'_j)} = \sum_s A(s \to s') \prod_j N_j^{s_j}(1-N_j)^{(1-s_j)}, \quad \forall s'. \tag{C.1}$$

(b) The N_i's are a solution of the set of b equations

$$\Delta_i(N) \equiv \sum_{ss'}(s'_i - s_i)A(s \to s') \prod_j N_j^{s_j}(1-N_j)^{(1-s_j)} = 0, \quad \forall i. \tag{C.2}$$

(c) The N_i's are given by the Fermi-Dirac distribution

$$N_i = \frac{1}{1+\exp(h + \mathbf{q}\cdot\mathbf{c}_i)}, \tag{C.3}$$

where h is an arbitrary real number and $\mathbf{q}$ is an arbitrary D-dimensional vector.

Proof that (a) implies (b).

We multiply (C.1) by s'_i and sum over all states s' to obtain

$$\sum_{s'} s'_i \prod_j N_j^{s'_j}(1-N_j)^{(1-s'_j)} = \sum_{ss'} s'_i A(s \to s') \prod_j N_j^{s_j}(1-N_j)^{(1-s_j)}. \tag{C.4}$$

In the l.h.s. of (C.4) we change the dummy variable s' into s and decorate it with a factor $A(s \to s')$, summed over s', which is one by normalization of probability. Transferring everything into the r.h.s., we obtain (C.2). Note that the l.h.s of (C.2) resembles the "collision function" Δ_i of section 3.1 (eq. (3.9)), but is evaluated with the mean populations instead of the Boolean populations n_i. The relation $\Delta_i = 0$ expresses that there is no change in the mean populations under collisions.

Proof that (b) implies (c).

We define

$$\check{N}_i \equiv \frac{N_i}{1-N_i}, \tag{C.5}$$

$$\Pi \equiv \prod_j (1-N_j). \tag{C.6}$$

Eq. (C.2) may be written

$$\Delta_i/\Pi = \sum_{ss'}(s'_i - s_i)A(s \to s')\prod_j \tilde{N}_j^{s_j} = 0. \qquad (C.7)$$

We now make use of a trick employed in proving H-Theorems in discrete velocity models (see Ref. 17, p.29). We multiply (C.7) by $\log \tilde{N}_i$, sum over i, and use

$$\sum_i (s'_i - s_i)\log \tilde{N}_i = \log \frac{\prod_j \tilde{N}_j^{s'_j}}{\prod_j \tilde{N}_j^{s_j}}, \qquad (C.8)$$

to obtain

$$\sum_{ss'} A(s \to s')\log\left(\frac{\prod_j \tilde{N}_j^{s'_j}}{\prod_j \tilde{N}_j^{s_j}}\right)\prod_j \tilde{N}_j^{s_j} = 0. \qquad (C.9)$$

Semi-detailed balance ($\sum_s A(s \to s') = \sum_{s'} A(s \to s') = 1$) implies that

$$\sum_{ss'} A(s \to s')\left(\prod_j \tilde{N}_j^{s_j} - \prod_j \tilde{N}_j^{s'_j}\right) = 0. \qquad (C.10)$$

Combining (C.9) and (C.10), we obtain

$$\sum_{ss'} A(s \to s')\left[\log\left(\frac{\prod_j \tilde{N}_j^{s'_j}}{\prod_j \tilde{N}_j^{s_j}}\right)\prod_j \tilde{N}_j^{s_j} + \prod_j \tilde{N}_j^{s_j} - \prod_j \tilde{N}_j^{s'_j}\right] = 0. \qquad (C.11)$$

We make use of the relation ($x > 0$, $y > 0$)

$$y\log\frac{x}{y} + y - x = -\int_x^y \log\frac{t}{x}\,dt \le 0, \qquad (C.12)$$

equality being achieved only when $x = y$. The l.h.s. of (C.11) is a linear combination of expressions of the form (C.12) with nonnegative weights $A(s \to s')$. For it to vanish, we must have

$$\prod_j \tilde{N}_j^{s_j} = \prod_j \tilde{N}_j^{s'_j}, \quad \text{whenever } A(s \to s') \neq 0. \qquad (C.13)$$

This is equivalent to

$$\sum_i \log(\tilde{N}_i)(s'_i - s_i)A(s \to s') = 0 \quad \forall s,\, s'. \qquad (C.14)$$

(C.13) means that $\log \check{N}_i$ is a collision invariant. We now use assumption (v) of section 2.4, concerning the absence of spurious invariants, to conclude that

$$\log \check{N}_i = -(h + \mathbf{q} \cdot \mathbf{c}_i), \tag{C.15}$$

which is the most general collision invariant (a linear combination of the mass invariant and of the D momentum invariants). Reverting to the mean populations $N_i = \check{N}_i/(1+\check{N}_i)$, we obtain (C.3).

Proof that (c) implies (a).

(C.3) implies

$$\sum_j \log(\check{N}_j)(s_j - s'_j) = 0, \quad \text{whenever } A(s \to s') \neq 0. \tag{C.16}$$

This implies

$$\sum_s A(s \to s') \left(\prod_j \check{N}_j^{(s_j - s'_j)} - 1 \right) = 0. \tag{C.17}$$

Using semi-detailed balance, this may be written as

$$1 = \sum_s A(s \to s') \frac{\prod_j \check{N}_j^{s_j}}{\prod_j \check{N}_j^{s'_j}}. \tag{C.18}$$

Reverting to the N_j's, we obtain (C.1). This completes the proof of the equivalence lemma.

Appendix D. Inclusion of body-forces

Using the same notation as in section 7.3, we wish to obtain a Navier-Stokes equation with a body-force $\mathbf{f}$, that is

$$\begin{aligned} &\partial_T \mathbf{U} + \mathbf{U} \cdot \nabla_1 \mathbf{U} = -\nabla_1 P' + \nu' \nabla_1^2 \mathbf{U} + \mathbf{f} \\ &\nabla_1 \cdot \mathbf{U} = 0. \end{aligned} \tag{D.1}$$

The force $\mathbf{f}$ may depend on space and time and can be velocity-independent (case I; e.g. gravity) or linear in the velocity $\mathbf{U}$ (case II; e.g. Coriolis force). The idea is to introduce a bias in the transition rules so as to give a net momentum input. Since all the terms in the Navier-Stokes momentum equation are $O(\epsilon^3)$ and the hydrodynamic velocity is $O(\epsilon)$ (before rescaling), the bias should be $O(\epsilon^3)$ for case I and $O(\epsilon^2)$ for case II.

We give now the modified form of the microdynamical equation (3.9) appropriate for body-forces. We introduce, in addition to the Boolean (transition) variables $\xi_{ss'}$ of section 3.1, the Boolean variables $\xi'_{ss'}$ such that

$$\langle \xi'_{ss'} \rangle = B(s \to s'). \tag{D.2}$$

The $B(s \to s')$'s are a set of transition probabilities associated to the body- force; they satisfy normalization

$$\sum_{s'} B(s \to s') = 1, \tag{D.3}$$

and mass conservation

$$\sum_i (s'_i - s_i) B(s \to s') = 0, \quad \forall s, s'. \tag{D.4}$$

They do not satisfy momentum conservation, semi-detailed balance and $\mathcal{G}$-invariance. The $\xi'_{ss'}$'s are chosen independently at each discrete time and node and the $B(s \to s')$'s may depend on space and time; further constraints will be given below. We also need a Boolean variable ζ which acts as a *switch*: when $\zeta = 0$ the force is off and the usual transition rules apply. The mean of ζ is given by

$$\begin{aligned} &\langle \zeta \rangle = \rho_0 g(\rho_0) \epsilon^n \\ &n = 3 \quad \text{case I}, \qquad n = 2 \quad \text{case II}. \end{aligned} \tag{D.5}$$

This will take care of the scaling factors arising from the change of variables (7.15). The modified microdynamical equation is now

$$n_i(t_\star + 1, \mathbf{r}_\star + \mathbf{c}_i) = n_i + \Delta_i(n)$$
$$\Delta_i(n) = \sum_{s,s'} (s'_i - s_i)\left((1-\zeta)\xi_{ss'} + \zeta\xi'_{ss'}\right) \prod_j n_j^{s_j}(1-n_j)^{(1-s_j)}. \qquad (D.6)$$

Let us evaluate the body-force resulting from the insuficientadditional ξ' term. For this we multiply by $\mathbf{c}_i$ and average over the equilibrium distribution; deviations from equilibrium arising from hydrodynamic gradients are irrelevant. We ignore the ζ-factor since it just provides the scaling factor.

We begin with case I. The average is then evaluated over the zero-velocity equilibrium distribution with density per cell d; we obtain

$$\mathbf{f} = \sum_{s,s',i} \mathbf{c}_i(s'_i - s_i)B(s \to s')\left(\frac{d}{1-d}\right)^p (1-d)^b, \quad p = \sum_j s_j, \qquad (D.7)$$

where b is the number of cells per node. Equation (D.7) is the additional constraint on the $B(s \to s')$'s for case I. If $\mathbf{f}$ is space and/or time-dependent, so are the $B(s \to s')$'s. It is easy to check that for any given vector $\mathbf{f}$ there exist Boolean transition variables $\xi'_{ss'}$ of mean $B(s \to s')$ satisfying (D.7). When $\mathbf{f}$ is in the direction of a particular velocity vector, say $\mathbf{c}_{i_0}$, we can flip particles with velocity $-\mathbf{c}_{i_0}$ into particles with velocity $\mathbf{c}_{i_0}$ whenever this is possible, while leaving all other particles unchanged. This is done with a probability dependent on the amplitude of the force. Other directions of the force are handled by superposition.

We turn to case II. We wish to obtain a force of the form

$$f_\alpha = C_{\alpha\beta}U_\beta, \qquad (D.8)$$

where $C_{\alpha\beta}$ is a D-dimensional matrix. When the velocity $\mathbf{U}$ vanishes, the body-force should also vanish; this requires

$$\sum_{s,s',i} \mathbf{c}_i(s'_i - s_i)B(s \to s')\left(\frac{d}{1-d}\right)^p (1-d)^b = 0, \quad p = \sum_j s_j. \qquad (D.9)$$

With nonvanishing velocity we must use the corresponding equilibrium populations given to relevant order by (cf. (4.14))

$$N_i = d + \frac{dD}{c^2} c_{i\alpha} u_\alpha. \tag{D.10}$$

Here we have used the unscaled velocity $\mathbf{u}$. Below, we shall however use $\mathbf{U}$ since the scaling factor is taken care of by the Boolean switch ζ. Using (D.10) in (D.6), we find that the average momentum imparted by $\xi'_{ss'}$ transitions is to leading order linear in $\mathbf{U}$. Identifying with (D.8), we find that the $B(s \to s')$'s must satisfy the following constraints

$$C_{\alpha\beta} = \frac{D}{c^2}(1-d)^{b-1} \sum_{s,s',i} c_{i\alpha}(s'_i - s_i) B(s \to s') \left(\frac{d}{1-d}\right)^p \sum_j s_j c_{j\beta}, \quad p = \sum_j s_j. \tag{D.11}$$

Equations (D.9)and (D.11) are the additional constraints on the $B(s \to s')$'s for case II.

As an illustration, consider the case of the pseudo-4-D FCHC model with a Coriolis force $2\mathbf{\Omega} \wedge \mathbf{U}$, where $\mathbf{\Omega}$ is in the x_3-direction. A possible implementation for the $\xi'_{ss'}$ transitions is through rotation by $\pi/2$ around the x_3-axis of those particles having their velocity perpendicular to this axis (with a probability dependent on Ω).

Appendix E. Catalog of results for FHP models

The purpose of this appendix is to summarize all known analytic results for the FHP models, including the models II and III which have rest-particles. Adapting the theory to cases with at most one rest-particle is quite straightforward if one includes the rest-particle velocity, namely vector zero. Our derivations made extensive use of properties P1 to P6 of section 2.4. With rest-particles, P1, P2, P4, and P5 are unchanged. In P3, λ and μ have usually different values for *moving* and rest-particles. P6 becomes

$$\sum_i \mathbf{c}_{i\alpha}\mathbf{c}_{i\beta} = \frac{(b-1)c^2}{D}\delta_{\alpha\beta}, \tag{E.1}$$

where b is still the number of bits, so that $b-1$ is the number of particles *moving* with speed c.

In Table 1 below, we give results in terms of the mean density per cell d for the following quantities: the mean density ρ_0, the coefficient $g(\rho_0)$ rescaling the nonlinear term in the Navier-Stokes equation (cf. for example (7.13)), the kinematic shear viscosity ν, the kinematic bulk viscosity ζ, the maximum value $R_*^{\max}$ of the coefficient R_* appearing in the Reynolds number (cf. (8.29)), and $d_{\max}$, the density at which the Reynolds number is maximum. The viscosities ν and ζ are calculated within the lattice Boltzmann approximation (cf. section 8.2). $\rho_0\zeta$ is the dynamic bulk viscosity; when it does not vanish, as is the case with rest-particles, eq. (7.11) becomes

$$\begin{aligned} &\partial_t\rho' + \rho_0\nabla\cdot\mathbf{u} = 0 \\ &\rho_0\partial_t\mathbf{u} + c_s^2\nabla\rho' = \rho_0\nu\left(\nabla^2\mathbf{u} + \frac{D-2}{D}\nabla\nabla\cdot\mathbf{u}\right) + \rho_0\zeta\nabla\nabla\cdot\mathbf{u}. \end{aligned} \tag{E.2}$$

	FHP-I	FHP-II	FHP-III
ρ_0	$6d$	$7d$	$7d$
c_s	$\frac{1}{\sqrt{2}}$	$\sqrt{\frac{3}{7}}$	$\sqrt{\frac{3}{7}}$

g	$\frac{1}{2}\frac{1-2d}{1-d}$	$\frac{7}{12}\frac{1-2d}{1-d}$	$\frac{7}{12}\frac{1-2d}{1-d}$
ν	$\frac{1}{12}\frac{1}{d(1-d)^3}-\frac{1}{8}$	$\frac{1}{28}\frac{1}{d(1-d)^3}\frac{1}{1-4d/7}-\frac{1}{8}$	$\frac{1}{28}\frac{1}{d(1-d)}\frac{1}{1-8d(1-d)/7}-\frac{1}{8}$
ζ	0	$\frac{1}{98}\frac{1}{d(1-d)^4}-\frac{1}{28}$	$\frac{1}{98}\frac{1}{d(1-d)}\frac{1}{1-2d(1-d)}-\frac{1}{28}$
$R_*^{\max}$	0.387	1.08	2.22
$d_{\max}$	0.187	0.179	0.285

Appendix F. An H Theorem for lattice gases

by M. Hénon, Observatoire de Nice.

1. *Notation and basic equations*

We number from 1 to b the cells at a given node (b is the number of different velocity vectors). It is not necessary that the velocity moduli are equal. Also it will not be necessary to specify any symmetry for the lattice or for the collision rules. Finally, we will not make use of the conservation of the number of particles or of the momentum, so that the proof is applicable to lattices where these conservation laws are violated.

We write $s_i = 1$ if particle i is present in the *input state*, 0 if it is absent. An input state is thus defined by $s = (s_1, \ldots, s_b)$. The number of distinct input states is 2^b.

We call $P(s)$ the probability of an input state s. We have

$$\sum_s P(s) = 1. \tag{F.1}$$

We call N_i the probability that particle i is present. We have

$$N_i = \sum_s s_i P(s), \qquad 1 - N_i = \sum_s (1 - s_i) P(s). \tag{F.2}$$

We define in the same way s'_i, $s' = (s'_1, \ldots, s'_b)$, $P'(s')$, N'_i for the *output state*.

We call $A(s \to s')$ the probability that an input state s is changed into an output state s' by the collision. We have

$$P'(s') = \sum_s P(s) A(s \to s'). \tag{F.3}$$

We have of course

$$\sum_{s'} A(s \to s') = 1, \tag{F.4}$$

where the sum is over all output states. We will assume that the collision rules obey *semi-detailed balancing*, i.e. that we have also

$$\sum_s A(s \to s') = 1. \tag{F.5}$$

2. *Local theorem*

Lemma 1. *If $f(x)$ is a convex function ($d^2f/dx^2 > 0$), then*

$$\sum_{s'} f[P'(s')] \leq \sum_{s} f[P(s)]. \tag{F.6}$$

Proof: from general properties of convex functions we have

$$f\left[\frac{\sum_s q(s)P(s)}{\sum_s q(s)}\right] \leq \frac{\sum_s q(s)f[P(s)]}{\sum_s q(s)}, \tag{F.7}$$

where the $q(s)$ are arbitrary positive or zero coefficients. Taking $q(s) = A(s \to s')$, with s' given, and using (F.3) and (F.5), we obtain

$$f[P'(s')] \leq \sum_{s} A(s, s')f[P(s)]. \tag{F.8}$$

Summing over s' and using (F.4), we obtain (F.6).

Lemma 2. *The following inequality holds:*

$$\sum_{s'} P'(s') \ln P'(s') \leq \sum_{s} P(s) \ln P(s). \tag{F.9}$$

Proof: we apply Lemma 1 with $f(x) = x \ln x$.

Lemma 3. *The following inequality holds:*

$$\sum_{s} P(s) \ln P(s) \geq \sum_{i=1}^{b} \left[N_i \ln N_i + (1 - N_i) \ln(1 - N_i)\right]. \tag{F.10}$$

The equality holds if and only if

$$P(s_1, \ldots, s_b) = \prod_{i=1}^{b} N_i^{s_i}(1 - N_i)^{1-s_i}. \tag{F.11}$$

Proof (inspired by Ref. 85): The right-hand side of (F.10) can be written, using (F.2):

$$\sum_{i=1}^{b} \sum_{s} \left[s_i P(s) \ln N_i + (1 - s_i)P(s) \ln(1 - N_i)\right], \tag{F.12}$$

or

$$\sum_{s} P(s) \ln \left[\prod_{i=1}^{b} N_i^{s_i}(1 - N_i)^{1-s_i}\right]. \tag{F.13}$$

Therefore (F.10) can also be written

$$\sum_s P(s)\ln\left[\frac{\prod_{i=1}^b N_i^{s_i}(1-N_i)^{1-s_i}}{P(s)}\right] \le 0. \tag{F.14}$$

We have, for any x:

$$\ln x \le x - 1, \tag{F.15}$$

where the equality holds only if $x = 1$. Therefore

$$\ln\left[\frac{\prod_{i=1}^b N_i^{s_i}(1-N_i)^{1-s_i}}{P(s)}\right] \le \frac{\prod_{i=1}^b N_i^{s_i}(1-N_i)^{1-s_i}}{P(s)} - 1. \tag{F.16}$$

Multiplying this by $P(s)$ and summing over s, we obtain the desired result.

The relation (F.11) corresponds to the Boltzmann approximation (independence of input particles).

Local H theorem. *If the collision rules satisfy semi-detailed balancing, and in the Boltzmann approximation, the following inequality holds:*

$$\sum_{i=1}^b [N_i' \ln N_i' + (1-N_i')\ln(1-N_i')] \le \sum_{i=1}^b [N_i \ln N_i + (1-N_i)\ln(1-N_i)]. \tag{F.17)}$$

Proof: from Lemma 3 we have

$$\sum_s P(s)\ln P(s) = \sum_{i=1}^b [N_i \ln N_i + (1-N_i)\ln(1-N_i)]. \tag{F.18}$$

Combining with Lemma 2:

$$\sum_{s'} P'(s')\ln P'(s') \le \sum_{i=1}^b [N_i \ln N_i + (1-N_i)\ln(1-N_i)]. \tag{F.19}$$

Finally, applying Lemma 3 to the N_i''s and the P''s, we obtain (F.17).

We remark that both conditions of the theorem are necessary; one can easily find counterexamples if one or the other is not satisfied. Consider for instance a node of the HPP lattice with probabilities before collision: $P(1,0,1,0) = 1/2$, $P(0,1,0,0) = 1/2$. We have: $N_1 = 1/2$, $N_2 = 1/2$, $N_3 = 1/2$, $N_4 = 0$; The Boltzmann approximation is not satisfied. We take the usual HPP collision rules. The probabilities after collision are then $P'(0,1,0,1) = 1/2$, $P'(0,1,0,0) = 1/2$. From this we deduce $N_1' = 0$, $N_2' = 1$, $N_3' = 0$, $N_4' = 1/2$, and it can be immediately verified that the left-hand member of (F.17) is larger than the right-hand member.

Similarly, let us modify the collision rules and keep only one kind of collision: $(1,0,1,0)$ gives $(0,1,0,1)$, but not conversely. Semi-detailed balancing is not satisfied. Take for instance $N_1 = N_2 = N_3 = N_4 = 1/2$. We assume that the Boltzmann approximation holds, therefore $P(s) = 1/16$ for all s. We deduce: $P'(1,0,1,0) = 0$; $P'(0,1,0,1) = 2/16$; $P'(s') = 1/16$ for the other s'; $N_1' = N_3' = 7/16$, $N_2' = N_4' = 9/16$; and here again the inequality (F.17) is violated.

3. *Global theorem*

First we sum (F.17) over all lattice nodes. We obtain a sum over all cells at all lattice nodes; their total number will be denoted by r:

$$\sum_{j=1}^{r}\left[N'^{(j)}\ln N'^{(j)} + (1-N'^{(j)})\ln(1-N'^{(j)})\right] \leq \sum_{j=1}^{r}\left[N^{(j)}\ln N^{(j)} + (1-N^{(j)})\ln(1-N^{(j)})\right]. \tag{F.20}$$

Next we remark that this sum is invariant under propagation. We can therefore extend the theorem to an arbitrary number of time steps, and we obtain (with the same hypotheses as for the local theorem):

Global H theorem. *The function*

$$\sum_{j=1}^{r}\left[N^{(j)}\ln N^{(j)} + (1-N^{(j)})\ln(1-N^{(j)})\right] \tag{F.21}$$

is non-increasing as the lattice gas evolves.

4. *Interpretation in terms of information theory*

Consider a probability distribution over ν possible cases: $p_1, \ldots, p_\nu$. The associated information is

$$\log_2 \nu + \sum_{i=1}^{\nu} p_i \log_2 p_i. \tag{F.22}$$

This information has a minimal value 0 if all cases have the same probability: $p_1 = \cdots = p_\nu = 1/\nu$. It has a maximal value $\log_2 \nu$ if one of the p_i is 1 while the others are 0, i.e. for a deterministic choice between the ν cases.

We come back to lattices. $P(s)$ represents a probability distribution on 2^b cases, and therefore an information

$$b + \sum_s P(s) \log_2 P(s). \tag{F.23}$$

Thus, Lemma 2 expresses the following property: if semi-detailed balancing is satisfied, then the information contained in the P can only remain constant or decrease in a collision.

From the P's, we can compute the N_i's by the formulas (F.2), but the converse is not generally true; in other words, the P's contain more information than the N_i's. Lemma 3 expresses this fact.

In the particular case of the Boltzmann approximation, the particles are considered as independent, and therefore the P's contain no more information than the N_i's. We have then the equality in (F.10).

The proof of the local H theorem can therefore be interpreted as follows: (i) initially the N_i's are given; this represents a given information; (ii) we compute the corresponding P's in the Boltzmann approximation; the information does not change; (iii) we compute the collision and obtain the P''s; the information decreases or stays constant; (iv) we compute the N''s from the P''s: here again the information decreases or stays constant.

REFERENCES

1. U. Frisch, B. Hasslacher, and Y. Pomeau, *Phys. Rev. Lett.* **56**:1505 (1986).
2. D. d'Humières, P. Lallemand, and U. Frisch, *Europhys. Lett.* **2**:291 (1986).
3. J. von Neumann, *Theory of Self-Reproducting Automata*, (Univ. of Illinois press, 1966).
4. E.R. Berlekamp, J.H. Conway, and R.K. Guy, *Winning Ways for Your Mathematical Plays*, (Academic Press, 1984), vol. 2.
5. S. Wolfram, *Review of Modern Physics* **55**:601, (1983).
6. S. Wolfram, *Theory and Applications of Cellular Automata*, (World Scientific, 1986).
7. Y. Pomeau, *J. Phys.* **A17**:L415 (1984).
8. N. Margolus, T. Toffoli, and G. Vichniac, *Phys. Rev. Lett.* **56**:1694 (1986).
9. N. Margolus and T. Toffoli, Cellular automata machines, *these Proceedings* (1987).
10. W.D. Hillis, *The Connection Machine*, (MIT Press, 1985).
11. J.P. Boon and S. Yip, *Molecular Hydrodynamics*, (McGraw-Hill, 1980).
12. D.C. Rapaport and F. Clementi, *Phys. Rev. Lett.* **57**:695 (1986).
13. E. Meiburg, *Phys. Fluids* **29**:3107 (1986).
14. L. Hannon, private communications 1986.
15. J.E. Broadwell, *Phys. Fluids* **7**:1243 (1964).
16. S. Harris, *Phys. Fluids* **9**:1328 (1966).
17. R. Gatignol, *Théorie Cinétique des Gaz à Répartition Discrète des Vitesses*, Lecture Notes in Physics vol. 36, (Springer, Berlin, 1975).
18. R. Gatignol, *Complex Systems* **1**:708, (1987).
19. H. Cabannes, *Mech. Res. Comm.* **12**:289 (1985).
20. R. Caflisch, Asymptotics of the Boltzmann equation and fluid dynamics, lecture notes from CISM summer school on *Kinetic Theory and Gas Dynamics*, (Udine, Italy, June 1986).
21. L.P. Kadanoff and J. Swift, *Phys. Rev.* **165**:310 (1968).
22. J. Hardy, Y. Pomeau, and O. de Pazzis, *J. Math. Phys.* **14**:1746 (1973).

23. J. Hardy, O. de Pazzis, and Y. Pomeau, *Phys. Rev.* **A13**:1949 (1976).

24. J. Hardy and Y. Pomeau, *J. Math. Phys.* **13**:1042 (1972).

25. D. d'Humières, P. Lallemand, and T. Shimomura, An experimental study of lattice gas hydrodynamics, Los Alamos Report LA-UR-85-4051 (1985).

26. D. d'Humières, Y. Pomeau, and P. Lallemand, *C. R. Acad. Sci. Paris II* **301**:1391 (1985).

27. J. Salem and S. Wolfram, in *Theory and Applications of Cellular Automata*, S. Wolfram. ed. (World Scientific, 1986), p. 362.

28. D. d'Humières and P. Lallemand, *C. R. Acad. Sci. Paris II* **302**:983 (1986).

29. D. d'Humières and P. Lallemand, *Physica* **140A**:337 (1986).

30. D. d'Humières and P. Lallemand, *Helvetica Physica Acta* **59**:1231 (1986).

31. D. d'Humières and P. Lallemand, *Complex Systems* **1**:598, (1987).

32. M. Hénon, *Complex Systems* **1**:475, (1987).

33. J.-P. Rivet, *C. R. Acad. Sci. Paris II*, **305**:751, (1987).

34. A. Clouqueur and D. d'Humières, *Complex Systems* **1**:584, (1987).

35. J.-P. Rivet and U. Frisch, *C. R. Acad. Sci. Paris II* **302**:267 (1986).

36. V. Yakhot and S. Orszag, *Phys. Rev. Lett.* **56**:169 (1986).

37. V. Yakhot, B. Bayly, and S. Orszag, *Phys. Fluids* **29**:2025 (1986).

38. U. Frisch and J.-P. Rivet, *C. R. Acad. Sci. Paris II* **303**:1065 (1986).

39. S. Wolfram, *J. Stat. Phys.* **45**:471 (1986).

40. F. Hayot, *Physica D* **26**:210, (1987).

41. F. Hayot, *Complex Systems* **1**:752, (1987).

42. M. Hénon, *Complex Systems* **1**:762, (1987).

43. J.-P. Rivet, *Complex Systems* **1**:838, (1987).

44. L. Kadanoff, G. McNamara, and G. Zanetti, Size-dependence of the shear viscosity for a two-dimensional lattice gas, preprint Univ. Chicago (1987).

45. D. Tarnowski, *La Recherche* **174**:272 (1986).

46. D. d'Humières, P. Lallemand, and Y. Pomeau, *bull. Soc. Franç. Phys.* **60**:14 (1986).

47. L. Kadanoff, *Physics Today*, September 1986, p.7.

48. C. Burges and S. Zaleski, *Complex Systems* **1**:31 (1987).

49. D.H. Rothmann, Modeling seismic P-waves with cellular automata, preprint, MIT Earth Resources Lab. (1986).

50. D. Montgomery and G. Doolen, *Complex Systems* **1**:830, (1987).

51. H. Chen and W. Matthaeus, Cellular automaton formulation of passive scalar dynamics, preprint, Bartol Res. Found. Univ of Delaware (1987).

52. H. Chen and W. Matthaeus, A new cellular automaton model for magnetohydrodynamics, preprint, Bartol Res. Found. Univ of Delaware (1987).

53. A. de Masi, P.A. Ferrari, and J.L. Lebowitz, *Phys. Rev. Lett.* **55**:1947 (1985).

54. J.L. Lebowitz, *Physica* **140A**:232 (1986).

55. A. de Masi, P.A. Ferrari, and J.L. Lebowitz, Reaction-Diffusion equations for interacting particle systems, *Physica* (1986), in press.

56. P. Clavin, D. d'Humières, P. Lallemand, and Y. Pomeau, *C. R. Acad. Sci. Paris II* **303**:1169 (1986).

57. P. Clavin, P. Lallemand, Y. Pomeau, ans J. Searby, Simulation of free boundaries in flow systems: a new proposal based on latttice-gas models, *J. Fluid Mech.* (1987), in press.

58. B.M. Boghosian and C.D. Levermore, *Complex Systems* **1**:17 (1987).

59. D. Levermore, private communication (1986).

60. N. Ashcroft and D. Mermin, *Solid State Physics*, (Holt-Saunders International Editions, 1976).

61. G. Doolen, private communication (1986).

62. S. Chapman and T.G. Cowling, *The Mathematical Theory of Non-Uniform Gases* (Cambridge Univ. Press, 1939).

63. G. Uhlenbeck and G. Ford *Lectures in Statistical Mechanics* (American Math. Soc. 1963).

64. L. Landau and E. Lifschitz, *Fluid Mechanics* (Pergamon 1981).

65. A. Majda, *Compressible Fluid Flow and Systems of Conservation Laws in Several Space Variables* (Springer Verlag, 1984).

66. D. d'Humières, P. Lallemand, and J. Searby, *Complex Systems* **1**:632, (1987).

67. L. Landau and E. Lifschitz, *Zh. Eksp. Teor. Fiz.* **32**:918 (1975)[Sov. Phys.-JETP **5**:512 (1957)].

68. Y. Pomeau and P. Résibois, *Phys. Rep.* **19C**:64 (1975).

69. R. Kubo, *J. Phys. Soc. Jpn.* **7**:439 (1957).

70. H.S. Green, *J. Math. Phys* **2**:344 (1961).

71. D. Forster, D.R. Nelson, and M.J. Stephen, *Phys. Rev.* **A16**:732 (1977).

72. J.D. Fournier and U. Frisch, *Phys. Rev.* **A28**:1000 (1983).

73. B.J. Alder and T.E. Wainwright, *Phys. Rev.* **A1**:18 (1970).

74. V. Yakhot, S. Orszag, and U. Frisch, unpublished (1986).

75. S. Chapman, *Philos. Trans. Roy. Soc. London* Ser. **A216**:279 (1916); **217**:115 (1917).

76. D. Enskog, *Kinetische Theorie der Vorgänge in Mässig Verdünnten Gasen*, Dissertation, Uppsala (1917).

77. J.-P. Rivet, *Gas sur Réseaux*, internal report, Observatoire de Nice (1986).

78. M. Hénon, private communication (1986).

79. G.K. Batchelor, *Phys. Fluids (Suppl. 2)* **12**:233 (1969).

80. R.H. Kraichnan, *Phys. Fluids* **10**:1417 (1967).

81. M.-E. Brachet, M. Meneguzzi, and P.-L. Sulem, *Phys. Rev. Lett.* **56**:353 (1986).

82. A.W. Kolmogorov, *C. R. Acad. Sci. USSR* **30**:301,538 (1941).

83. A.S. Monin and A.M. Yaglom, *Statistical Fluid Mechanics* J.L. Lumley. ed. (MIT Press, 1975) vol. 2; revised and augmented edition of the russian original *Statisticheskaya Gidromekhanika* (Nauka, Moscow 1965).

84. T. Passot and A. Pouquet, Numerical simulations of compressible homogeneous fluids in the turbulent régime, *J. Fluid Mech.* (1987), in press.

85. R. M. Fano, 1961, *Transmission of Information*, (Wiley, 1961), p.43.